Introduction to Megavoltage X-Ray Dose Computation Algorithms

Series in Medical Physics and Biomedical Engineering

Series Editors: John G. Webster, E. Russell Ritenour, Slavik Tabakov, and Kwan Hoong Ng

Introduction to Megavoltage X-Ray Dose Computation Algorithms

Edited by

Jerry J. Battista

Emeritus Professor
Departments of Oncology and Medical Biophysics
University of Western Ontario
London, Ontario, Canada

CRC Press
Taylor & Francis Group
Boca Raton London New York

CRC Press is an imprint of the
Taylor & Francis Group, an **informa** business

CRC Press
Taylor & Francis Group
6000 Broken Sound Parkway NW, Suite 300
Boca Raton, FL 33487-2742

First issued in paperback 2021

© 2019 by Taylor & Francis Group, LLC
CRC Press is an imprint of Taylor & Francis Group, an Informa business

No claim to original U.S. Government works

Version Date: 20181203

ISBN 13: 978-0-367-78051-7 (pbk)
ISBN 13: 978-1-138-05684-8 (hbk)

Visit the Taylor & Francis Web site at
http://www.taylorandfrancis.com

and the CRC Press Web site at
http://www.crcpress.com

Michael Bryan Sharpe 1965-2016

This book is dedicated to the memory of
Michael as a friend and colleague,
for his lasting contributions to our lives and to
the field of medical physics.

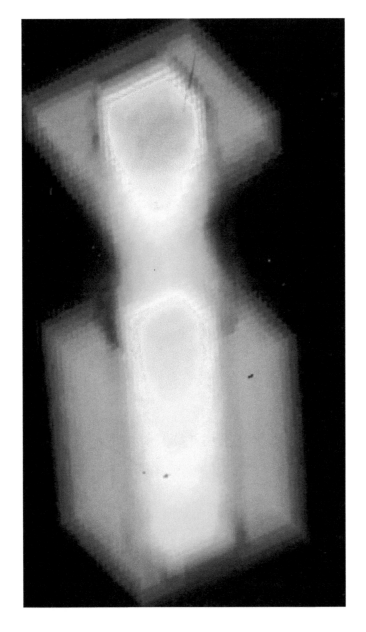

Megavoltage x-ray dose distribution in a 3D lung phantom, from the Ph.D. research of Michael Sharpe (*circa* 1995).
In the low-density section, the central dose is reduced and the beam penumbra is degraded.

Contents

Preface

WORLDWIDE projections indicate that annual cancer incidence will rise to nearly 23 million new cases by the year 2030, with approximately 50% of patients requiring access to radiation therapy. It is estimated that 2.5 million patients per year would benefit from improved loco-regional tumour control and 950,000 would achieve long term survival attributable to the radiation.

This book deals with a critical element of radiation treatment planning that can impact the lives of cancer patients. Accurate prediction and placement of a high-dose region onto the target volume, while minimizing collateral damage to healthy tissue, is very important. Three-dimensional (3D) dose computation algorithms simulate the passage of radiation through heterogeneous tissue, accounting for the attenuation and scattering of the x-rays. Historically, computerized treatment planning systems focused on producing a static dose distribution based on a pre-treatment 3D snapshot of the anatomy of the patient. The emerging trend is to forecast the dose distribution that will actually be delivered *in vivo* upon completion of treatment. This will be accomplished through imaging with CT scanners or MRI that are now integrated with modern treatment machines. This imaging enables reconstruction of the dose distribution *du jour*, and accumulation of dose throughout a treatment course. Targeting has also greatly improved with four-dimensional (4D) imaging and treatment planning that accounts for tissue movement and deformation caused by respiratory motion. Looking ahead, functional and molecular imaging techniques will help identify the critical subset of cancer cells and micro-environmental conditions that currently impair treatment efficacy. With unabated exponential gains in cost-effective computing, dose calculation algorithms will ultimately fuse with human radiobiological models to deliver long-awaited biologically-adaptive radiation therapy (BART).

The title of this book conveys its primary educational purpose as an introductory source of information. The targeted audience is future medical physicists, including graduate students, as well as clinical physics residents. The experienced medical physicist might also benefit from having this book on hand as a supplementary teaching resource and handbook. The book may also be of value for educating technical staff involved in supporting users of treatment planning software.

In the experience of the authors, there is a translational gap between completion of a graduate course in radiological physics and the clinical application of dose computation algorithms. This is evident when trainees participate in clinically-oriented activity or prepare for professional board certification exams. They review the published literature but often only manage to develop a "grey box" understanding of algorithms. Furthermore, the underlying physics may be camouflaged or manipulated in commercial software products. The situation is only partially resolved in the vendor's software manuals – documents that tend to restate the literature almost verbatim without adding fresh insight or stating assumptions made in the computer implementation. **The purpose of this book is to fill in some of the missing details in the physics and mathematical infrastructure that limit the understanding of modern dose calculation algorithms.**

The book is divided into seven chapters. The first two chapters may be bypassed by experienced clinical physicists working in oncology. Chapter 1 establishes the rationale for radiation treatment planning and explains the spatio-temporal connection between dose gradients and fractionated radiobiology. This perspective is often confused when each topic is presented separately in unrelated textbooks. Chapter 2 provides "what you need to know" in radiological physics to understand the strengths and limitations of dose computation algorithms. It is a recap of what most graduate students will have learned previously in university and college courses. Chapter 3 provides a historical overview of semi-empirical algorithms that evolved to three algorithms in clinical use today: convolution-superposition methods, stochastic Monte Carlo simulation, and deterministic solution of the linearized Boltzmann radiation transport equation. A reader might detect a geographic bias in Chapter 3 for which the editor makes no apology; Canadian medical physicists played a significant scientific role in this evolution. Chapters 4 to 6 constitute the core of this book, with considerable mathematical detail and data on algorithm performance. Chapter 7 presents a future outlook with emphasis on 4D dose reconstruction enabled by faster computers.

This book adheres to a student-friendly presentation inspired by the famous physicist and communicator, Richard P. Feynman. We also adopted an intuitive didactic philosophy inspired by H.M. Schey, author of the book *Div Grad Curl and All That*. He emphasized the notion of introducing concepts before paralysing the student with a barrage of equations. Overarching ideas are therefore described initially, often using analogies and metaphors from everyday experience. This is then followed by mathematical formulae with ample physical interpretation and continuity across chapters. Tacit assumptions are exposed. Original colour figures have been created with descriptive captions to accelerate learning. In some cases, humour and story-telling were interlaced with difficult concepts to lighten the cognitive load. It is hoped that this mode of learning will also improve the recall of knowledge during clinical applications of dose computation algorithms.

Note that MATLAB® is referenced in Chapter 5. It is a registered trademark of The MathWorks, Inc. For product information please contact: The MathWorks Inc., 3 Apple Hill Drive Natick, MA, 01760-2098 USA. Tel: 508-647-7000 Fax: 508-647-7001 E-mail: info@mathworks.com Web: www.mathworks.com

Acknowledgements

The idea of writing this book originated with students and residents that we have encountered in lecture rooms, labs, clinical sessions, and at conferences. Our goal was to accelerate their learning of clinically-applied radiation dosimetry. We dedicate this book to Michael Bryan Sharpe, a former Ph.D. student in our program, who left this world far too early because of cancer. He left an indelible positive influence on the field of clinical physics. We thank the following recent students at the University of Western Ontario who beta-tested early versions of this book or provided updated information from their own research portfolio: Michael MacFarlane, Jason Vickress, Kurtis Dekker, and Katie Marissa Parkins.

The editor is indebted to his dedicated team of co-authors. They are busy clinical physicists who agreed to extend their workday and sacrifice personal time to meet an aggressive timetable of 18 months. George Hajdok deserves special recognition and extra credit. He not only authored a major chapter, co-authored another, but he also provided much needed local technical support on LaTeX coding. He produced many of the original figures and tables required by other authors. He consolidated the chapters into a final book format for submission to the publisher, while coping with the unexpected early arrival of baby Harper!

The authors are grateful to their families for continuous steady support during the busy writing phases. We especially thank our spouses, Leigh Battista, Wanping Chen, Linda Rempel, and Nicole Hajdok for their patience while our minds wandered off into computational hyperspace. An additional Chen family member, Lynna, also provided some editing improvements to an early version of Chapter 4. We could not have completed this task without their patience, support and encouragement. They merit virtual co-authorship.

We thank all members of the Physics and Engineering Department of the London Regional Cancer Program for providing backup clinical services to our authors during book-writing absences from the clinic. Rob Barnett, chief physicist of the London Regional Cancer Program permitted flexible work schedules and coverage during intense writing periods. Glenn Bauman, Chief of Oncology, provided clinical insights and perfected the content of Chapter 1. Barbara Barons provided general assistance in locating and coordinating busy clinical staff. Carol Johnson and Jeff Kempe prepared some original data and graphics for the book. Jonatan Snir, Linada Kaci, and John Patrick provided photographs for Chapter 1. Doug Hoover reviewed some chapters with a keen eye for detecting faulty equations. Donna Murrell provided helpful feedback on sections of Chapter 4. Matt Mulligan provided technical expertise and discussion on Monte Carlo methods. Gabriel Boldt of the London Health Sciences Library assisted with tracking down old articles that pre-dated the age of digital journals. Andrea McNiven of the Princess Margaret Cancer Centre in Toronto provided valuable feedback on Chapter 7.

The editor is particularly grateful to Dr. "Jack" Cunningham who was very influential as a mentor and leader in the development of dose calculation algorithms. We also appreciated the assistance of other pioneers in this field. They kindly critiqued and improved our work: Rock Mackie, Anders Ahnesjö, Alex Bielajew and Todd Wareing. Anders provided a detailed, timely, and critical review of Chapter 4, just before embarking on his summer vacation! Alex offered important suggestions for improving Chapter 5 during a busy semester of teaching at the University of Michigan. Todd responded to last-minute requests with encouraging positive feedback on Chapter 6 while travelling on business to Helsinki. We also generally benefited from fruitful e-mail or phone exchanges with Dave Rogers, Pedro Andreo, Simon Thomas, Lorenzo Brualla, Jan Seuntjens, Tony Popescu, Syed Ahmed, Moti Pautel, Boyd McCurdy, Ben Mijnheer, Hans Moravec, Karl Rupp, and Ross Mitchell.

Jake Van Dyk provided up-to-date concepts and data on dose uncertainty from his recent work at the International Atomic Energy Agency (IAEA); this set the stage for Chapter 1. He has been a source of inspiration as a colleague, researcher, and prolific writer. Ben Mijnheer's recent experience with his book *Clinical 3D Dosimetry in Modern Radiation Therapy* was openly shared; his advice was very helpful to our project. Patrick McDermott's book *Tutorials in Radiotherapy Physics* provided an exceptional primer on dose calculation algorithms and served as a model. Both of these books were produced by the same publisher as for this book.

We acknowledge the staff of the Taylor & Francis Group in England and the CRC Press in the United States. We express gratitude to Francesca McGowan who initially convinced the editor to produce this book with the help of collaborators. Rebecca Davies, Georgia Harrison, Kirsten Barr, and Karen Simon navigated the authors through the book preparation and extended final production phases. We thank Alice Mulhern who did a superb job in meticulous proof-reading of the initial version of the book. Shashi Kumar assisted with last-minute technical support on LaTeX formatting issues. Scott Shamblin and Jonathan Pennell finalized the front cover design with steady suggestions of graphical elements from the authors. Collectively, these efforts greatly improved the content and appearance of our book.

About the Series

The *Series in Medical Physics and Biomedical Engineering* describes the applications of physical sciences, engineering, and mathematics in medicine and clinical research.

The series seeks (but is not restricted to) publications in the following topics:

- Artificial organs
- Assistive technology
- Bioinformatics
- Bioinstrumentation
- Biomaterials
- Biomechanics
- Biomedical engineering
- Clinical engineering
- Imaging
- Implants
- Medical computing and mathematics
- Medical/surgical devices

- Patient monitoring
- Physiological measurement
- Prosthetics
- Radiation protection, health physics, and dosimetry
- Regulatory issues
- Rehabilitation engineering
- Sports medicine
- Systems physiology
- Telemedicine
- Tissue engineering
- Treatment

THE INTERNATIONAL ORGANIZATION FOR MEDICAL PHYSICS

The International Organization for Medical Physics (IOMP) represents over 18,000 medical physicists worldwide and has a membership of 80 national and 6 regional organisations, together with a number of corporate members. Individual medical physicists of all national member organisations are also automatically members.

The mission of IOMP is to advance medical physics practice worldwide by disseminating scientific and technical information, fostering the educational and professional development of medical physics and promoting the highest quality medical physics services for patients.

A World Congress on Medical Physics and Biomedical Engineering is held every three years in cooperation with International Federation for Medical and Biological Engineering (IFMBE) and International Union for Physics and Engineering Sciences in Medicine (IUPESM). A regionally based international conference, the International Congress of Medical Physics (ICMP) is held between world congresses. IOMP also sponsors international conferences, workshops and courses.

The IOMP has several programmes to assist medical physicists in developing countries. The joint IOMP Library Programme supports 75 active libraries in 43 developing countries, and the Used Equipment Programme coordinates equipment donations. The Travel Assistance Programme provides a limited number of grants to enable physicists to attend the world congresses.

IOMP co-sponsors the *Journal of Applied Clinical Medical Physics*. The IOMP publishes, twice a year, an electronic bulletin, *Medical Physics World*. IOMP also publishes e-Zine, an electronic news letter about six times a year. IOMP has an agreement with Taylor & Francis for the publication of the *Medical Physics and Biomedical Engineering* series of textbooks. IOMP members receive a discount.

IOMP collaborates with international organisations, such as the World Health Organisations (WHO), the International Atomic Energy Agency (IAEA) and other international professional bodies such as the International Radiation Protection Association (IRPA) and the International Commission on Radiological Protection (ICRP), to promote the development of medical physics and the safe use of radiation and medical devices.

Guidance on education, training and professional development of medical physicists is issued by IOMP, which is collaborating with other professional organisations in development of a professional certification system for medical physicists that can be implemented on a global basis.

The IOMP website (www.iomp.org) contains information on all the activities of the IOMP, policy statements 1 and 2 and the 'IOMP: Review and Way Forward' which outlines all the activities of IOMP and plans for the future.

Editor

Dr. Jerry J. Battista earned his Ph.D. in Medical Biophysics at the University of Toronto in 1977 under the supervision of Dr. Michael Bronskill. His thesis project involved Compton-scatter tomography at the same time as the advent of x-ray computed tomography (CT). As a post-doctoral fellow at Princess Margaret Cancer Centre, he gained clinical physics experience alongside Dr. John R. Cunningham. Jerry relocated to the Cross Cancer Institute and University of Alberta in 1979. His team developed one of the first 3D treatment planning systems and introduced the concept of convolution-based algorithms. He has published over 125 peer-reviewed journal articles and co-authored major collaborative grants in partnership with industry. With colleagues, he holds patents related to fast inverse dose optimization (FIDO).

Dr. Battista directed Medical Physics Research at the London (Canada) Regional Cancer Program, and served as Chair of Medical Biophysics at the University of Western Ontario. His clear presentations, enthusiastic style, and use of vivid analogies brought physics concepts to a wide range of audiences. Hopefully readers will get a glimpse of this unique skill in this book. He has received the university's top honour for excellence in teaching and was nominated by students for national recognition. He has developed novel laboratory instruments and teaching modules related to CT scanning (i.e. DeskCAT^{TM} scanner). He has mentored 30 graduate students who have impacted the field of contemporary radiotherapy.

Jerry is a Fellow of the Canadian College of Physicists in Medicine, Canadian Organization of Medical Physics, and American Association of Physicists in Medicine. He is Emeritus Professor at Western and provincial coordinator of medical physics education programs for Cancer Care Ontario. Jerry received the Kirkby Award from the Canadian Association of Physicists (CAP) and Gold Medal from the Canadian Organization of Medical Physicists (COMP) for lifetime contributions to society.

Contributors

Jerry J. Battista, PhD, FCCPM, FCOMP, FAAPM
Emeritus Professor
Departments of Oncology and of Medical Biophysics
University of Western Ontario
Former Chair, Department of Medical Biophysics
University of Western Ontario
Former Director, Physics Research and Education
London Regional Cancer Program, London Health Sciences Centre
London, Ontario, Canada

Jeff Z. Chen, PhD, FCCPM
Associate Professor
Departments of Oncology and of Medical Biophysics
University of Western Ontario
Senior Clinical Physicist
London Regional Cancer Program, London Health Sciences Centre
London, Ontario, Canada

Stephen Sawchuk, PhD, FCCPM, ABR
Assistant Professor
Departments of Oncology and of Medical Biophysics
University of Western Ontario
Senior Clinical Physicist
London Regional Cancer Program, London Health Sciences Centre
London, Ontario, Canada

George Hajdok, PhD, MCCPM
Assistant Professor
Departments of Oncology and of Medical Biophysics
University of Western Ontario
Clinical Physicist
London Regional Cancer Program, London Health Sciences Centre
London, Ontario, Canada

Foreword

I owe much of the success in my career to Dr. Jerry Battista. My foreword to this book, in large part written and edited by Jerry, is very personal. I first met Jerry in 1980. He was a junior professor at the University of Alberta and medical physicist at the Cross Cancer Institute, and not that much older than I was – a first year physics graduate student. Jerry then led one of the first systematic programs to develop a 3D radiation oncology planning system *de novo* based on CT information, at a time when CT scanners were largely used only for diagnostic purposes. It was not generally known at the time whether CT would add much value to radiation treatment planning; in fact, the cost effectiveness of this strategy was strongly debated. Jerry expanded on the hypothesis of Nobel Prize winner Allan Cormack that the location of tumour and soft tissue revealed in a CT scan would enable much more precise dose targeting of the tumour while avoiding critical normal tissues. He confirmed that CT density information (covered in Chapter 3 of this book) would allow much more accurate dose calculations, and furthermore that follow-up scans could assess treatment response to radiation therapy. Jerry was already a superlative science communicator. He gave an institute-wide lecture in the Fall of 1980 called "CAT and Mouse". The title attracted many curious attendees to hear about how fibrotic changes in the lungs of mice could be foreshadowed by density changes in CT, well before clinical radiation pneumonitis symptoms appeared. He said that the CT scanner would someday be used for radiation therapy planning of all patients. This was quite a leap of faith since at the time since practically no patients were scanned for explicit purposes of treatment planning and precise radiation dosimetry.

Much later in my career, I co-invented and developed a treatment method called *tomotherapy*, which combined CT scanning and radiotherapy. I credit Jerry's enthusiasm for recognizing CT's importance to the clinical adoption of tomotherapy. For full disclosure, I note that Jerry and his colleague Jake Van Dyk were such strong vocal proponents for tomotherapy that the London Regional Cancer Centre acquired one of the first three tomotherapy prototypes. It must also be noted that a rival approach, linear accelerator equipped with a cone-beam CT scanner, was also co-invented by one of Jerry's Ph.D. students, David Jaffray. I estimate that vendors of linear accelerators equipped with therapy room CT systems have sold about 15 billion dollars worth of product worldwide. The fact that Jerry was our common supervisor is no coincidence.

"Rock" Mackie, Jerry Battista, and David Jaffray

Jerry taught me the most interesting course that I took in the Physics Department at the University of Alberta, i.e. *Radiological Science and Radiology*. His notes made it easy to understand difficult concepts like Fourier analysis. He found a particularly simple way of deriving the CT central slice theorem that I in turn used throughout my career. When I drafted my lecture notes for my class on Radiological Physics and Introduction to Dosimetry at the University of Wisconsin (Madison), I drew from notes of Jerry's lectures. His path through radiological physics was more logical than the standard textbook that I used for my class. I taught that class to more than 800 graduate students during my career. If students are like offspring, then Jerry is grandfather to thousands of medical physicists! This book is making his impact on treatment planning education exponentially even greater. Chapter 2, written jointly with George Hajdok, is a concise summary of core radiation physics that is necessary to understand the underlying principles of megavoltage x-ray dose calculations. This book has inherited the characteristic smooth flow within and across chapters.

For my first few years in Alberta, Jerry was not my Ph.D. supervisor. I began working on a simple model of radiation transport through matter to describe radiation dose distributions and it showed promise for accurately modelling the build-up and penumbra of megavoltage photon beams. I remember standing at a blackboard in our lab writing down some primitive algebraic expressions and not being very confident in my approach. Dr. Battista walked by, stopped and asked me what I was doing. My explanation to him was rather heuristic. I used algorithmic notation rather than formal mathematics. I understood the physics intuitively but could not express it clearly in an equation. He quickly spotted that for the case of a uniform density absorber, my model could be described as a convolution integral, which he understood well because it was a standard approach used for predicting spatial blurring in digital images. With varying density tissue, the more general formulation we called a superposition integral. At the end of what turned into an hour at the blackboard, he convinced me that the convolution-superposition concept was an important insight and should be pursued. The name stuck and it has become the most used formulation for dose calculation in commercial CT-based treatment planning systems up to this time. Chapter 4 does a superb job of cross-linking the various 2D and 3D implementations of the convolution-superposition model. This contemporary method is used in millions of treatment plans per year. If Jerry had not come by when he did and stopped to talk to a rookie student who he did not then supervise, my life would have been substantially different. Chapter 4, written by Jeff Chen, might not exist and millions of patients may not have benefited from more accurate dose calculations over the past few decades.

Chapter 3 presents a unique historical bridge between the convolution method and semi-empirical precursor methods developed by Dr. J.R Cunningham. He is given overdue credit for his formulation of pencil beam models based on 2D convolution. The chapter uses interesting analogies to explain dose concepts and fully prepares the reader for subsequent more complex chapters (4 through to 6).

The Monte Carlo simulation of multiple x-ray interactions is detailed in Chapter 5, written by Stephen Sawchuk. The topic of Monte Carlo simulation reminds me of a story from the beginning of my final year of graduate studies at the University of Alberta. My original Ph.D. supervisor, Dr. John Scrimger, undertook a sabbatical in England and Jerry took over supervising me. Jerry accompanied me and several other students to the AAPM meeting in New York city in the summer of 1983. After the meeting, both of us with a day to spare before flying back to Edmonton, went out to a Belmont Park racetrack on Long Island (Jerry's love of Monte Carlo simulation is based in part on his youthful enthusiasm for handicapping horse races and playing poker). While waiting for the evening races, we decided to go to Jones Beach, which faces New York City. We started talking about algorithmic approaches for fast Monte Carlo calculations for electron beams. One idea was based on pre-canning results from detailed Monte Carlo simulation into a fast voxel-based lookup database and then predicting the group behaviour of particles traversing voxels. We sketched out the details in our bathing suits on the sand with sticks like prehistoric men might have done in animal hides twenty millennia ago. Remembering the experience still makes me chuckle. This sand-inscribed algorithm that we called the Macro-Monte Carlo (MMC) method has served as the foundation for two of my Ph.D. students' dissertations as well others at the University of Berne and at Lawrence Livermore National Laboratory.

Chapter 6, written by George Hajdok, details the deterministic solution to the radiation transport equations that describe and constrain the spread of radiant energy across the patient space. It is a mathematically challenging chapter that requires the readers to shift their thinking from real space to a hyper-dimensional phase space. Dr. Hajdok brings a rigorous comprehensive treatment with a smooth flowing perspective.

Chapter 7 reflects Jerry's vision of the future in 4D dose calculations and deformable dose reconstruction, using transit dosimetry (his first peer-reviewed publication) and on-board CT scans. If the past is any predictor, the reader should pay close attention to this chapter and prepare for forecasted developments. The most exciting promise is the fusion of accurate *in vivo* 3D/4D dosimetry with measured clinical outcomes data to refine models of human response to radiation – a fulfilment of a *CAT and Man* prophecy made in his *CAT and Mouse* lecture!

Jerry is an emeritus professor at the University of Western Ontario after a distinguished career as educator and administrator. This book on radiation treatment planning is a capstone of his medical physics career. It is destined to become a modern classic in the field of medical physics and a necessary read for anyone who wants to understand details of dose computation algorithms used in radiation oncology.

Thomas Rockwell Mackie
Emeritus Professor
University of Wisconsin (Madison)

CHAPTER 1

Introduction

Jerry J. Battista

London Health Sciences Centre and University of Western Ontario

CONTENTS

1.1 OVERVIEW

ALL forms of cancer therapy are aimed at a common goal – to eradicate tumours while minimizing collateral injury to surrounding normal organs. During the early days of radiotherapy, this was certainly a most challenging proposition often described as *aiming invisible radiation at an invisible target*. Fortunately, radiation oncology was one of the first medical specialities to embrace computerization and it remains poised to take full advantage of emerging technology. One of the largest impacts was the development of digital three-dimensional (3D) imaging. The rapid adoption of x-ray computed tomography (CT) not only improved the targeting of disease, but it also provided quantitative maps of *in vivo* tissue densities for accurate computation of dose distributions (Van Dyk and Battista 2014). Ultrasound, magnetic resonance imaging (MRI) and radioisotope imaging all contribute to detecting internal disease. Further advances will continue to impact the clinical practice of radiation oncology (Jaffray et al. 2018). As an example, CT or MRI systems have been integrated with treatment machines for dose delivery of unprecedented precision (Lagendijk et al. 2014).

The Merriam-Webster dictionary defines an algorithm as "a procedure for solving a mathematical problem in a finite number of steps that frequently involves repetition of an operation; broadly, a step-by-step procedure for solving a problem or accomplishing some end, especially by means of a computer". Dose computation algorithms constitute critical elements of a radiation treatment planning system. While ergonomic software tools for displaying, contouring, or aligning images are important, a clinically-viable treatment plan must rely on a trustworthy dose algorithm. Inaccurate dosimetry can mislead clinical decision-making and taint the interpretation of clinical trials, potentially affecting the lives of many cancer patients. While *la raison d'être* for dose algorithms has been the optimization of dose distributions during treatment planning, the range of applications is now rapidly expanding. Algorithms are used to re-compute dose distributions *du jour* to adapt to changing anatomy seen by on-board imaging. When coupled with deformable image registration algorithms, the accumulation of dose in mobile tissue voxels can be tracked and dose can be re-optimized throughout a course of treatment. In the long term, highly controlled dose delivery will facilitate development and testing of new human radiobiology models.

1.1.1 A Brief History of Radiation Treatment Planning

Computerization of the radiotherapy planning process has a rich historical track (Rinks 2012; Cunningham 1989). External beam treatment planning in the 1950s involved manual overlay of semi-transparent isodose charts onto an external contour of the patient measured mechanically in a single plane and automatically re-plotted onto paper using a pantograph device. The treatment planner would then join the dots of equal dose sums to map the composite isodose pattern from several beams. One of the earliest forms of automated treatment planning was developed by Tsien in 1955 (Tsien 1955). Isodose data were encoded on a stack of punch cards for data entry. The external contour of the patient was drawn in polar coordinates and beam directions were decided. The beam energy was most often selected as cobalt-60 (Smith et al. 1964). The punch cards were sorted automatically by an adding machine that summed up the dose at a grid of points within the patient's external contour. The isodose distribution was then plotted and a simple treatment plan could be produced in 15 minutes rather than a few hours!

A panel of consultants met at the International Atomic Energy Agency (IAEA) in Vienna in 1965 and 1967 to exchange knowledge on dose calculation methods and general applications of computers in radiotherapy (IAEA 1966, 1968). The front cover of their final report is shown in Figure 1.1. The report described potential roles of computers in radiotherapy departments, including clinical treatment planning for teletherapy and brachytherapy. A list of thirty computer programs was compiled. One article described a Monte Carlo simulation of multiple photon scattering within a water absorber – a precursor to future modelling of scatter kernels. The pivotal idea of splitting radiation dose into its primary and scattered components was dated back to the kilovoltage era. It was re-asserted as an important strategy in advancing dose computation capabilities forward. The compendium also listed programs for radiobiological optimization of treatment plans. The panel chairman concluded his lead article with the following remarkable foresight and challenge:

> *"It is likely that, at present, the weakest link...is now the informa-*
> *tion about the patient: localization, extent of tumour and body in-*
> *homogeneities, as well as the means of accurate and reproducible*
> *beam direction. It is very likely that the computer can also aid*
> *this."*
>
> Dr. J.R Cunningham

A virtual humanoid phantom was also described in the proceedings and it could be scaled to account for the external size of a particular patient (Busch 1968). This phantom is seen on the book cover (Figure 1.1) and this pre-dated 3D displays and printing. It foreshadowed the use of anatomical transverse slices for individual patients developed later in CT scanning during the 1970-80 period of accelerated growth. A sample 3D dose distribution is shown in Figure 1.1. The author noted that "the program was too slow and would be re-written in FORTRAN to run on an IBM 7040 computer. The new program will be able to calculate any therapy field by computing elementary fields of 1×1 cm^2." This beam decomposition technique is essentially the pencil beam model of later decades.

Atlases of radiation dose distributions were later published by the IAEA (Mac-Donald 1965), including those used for rotational therapy which is a revived mode of contemporary treatment delivery (e.g. VMAT and tomotherapy). Early developments were aimed at improving international access to treatment planning resources by sharing cobalt-60 data sets. This included rapid retrieval of measured data for each beam size, with secondary corrections for the patient's external shape and major internal tissue inhomogeneities such as in lungs. In the mid-1960s, powerful computer workstations were still not available within the constraints of hospital space and budgets. An interim solution was provided through the Ontario Cancer Treatment and Research Foundation in Canada. Remote terminals were installed across a provincial network of several cancer centres served by a time-shared central IBM computer. Treatment planning software was developed at Princess Margaret

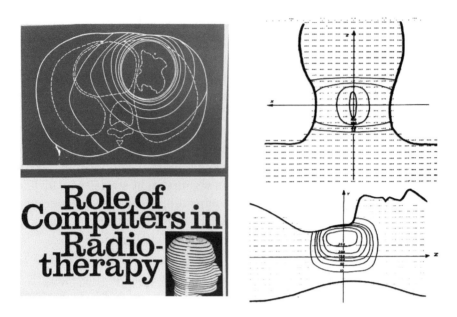

Figure 1.1: IAEA report that forecast the role of computers in the field of radiation oncology (IAEA 1968). A virtual humanoid phantom, shown on the cover, was used for early 3D dose calculations (Busch 1968).

Hospital in Toronto and shared across the cancer centres. Isodose distributions were printed with alpha-numeric characters on noisy teletype terminals. A dedicated program console (PC) eventually emerged for stand-alone use with interactive input devices and monochrome graphics display. By the early 1970s, this type of workstation evolved into commercial products for 2-dimensional (2D) planning, including the RAD-8, PC-12 (Artronix), and TP-11 (AECL) systems. In time, video displays and printers were introduced with grey-tone and colour capabilities.

The next major thrust towards 3D treatment planning came with the introduction of CT head scans in 1972. The applications to cancer diagnosis and treatment planning became obvious (Battista et al. 1980). Whole body systems were then designed by numerous vendors such as General Electric, Picker, Siemens, Philips, Toshiba, and Artronix. Some were installed in cancer treatment centres, with specialized "beam's-eye-view" software, gradually replacing traditional radiographic treatment simulators. The slice-by-slice images were ideal for contouring of targets and organs-at-risk, and meshed well within existing 2D treatment planning software architecture. The assumption of a homogeneous patient composed solely of water-like tissue, however, became untenable. The limitations of 2D approaches led to investments by companies and national agencies, such as the National Cancer Institute in the USA, the CART group initiative in Scandinavia (i.e. Helax system), and the OMEGA group in North America. Major upgrades to dose computation capabil-

Figure 1.2: Annual publication rate for three types of dose calculation algorithms. Data for published journal articles, including abstracts and conference proceedings, were extracted from the SCOPUS database (www.scopus.com).

ities followed with consideration of heterogeneous tissue described in 3-dimensional data structures (Sontag and Cunningham 1978; Wong et al. 1984). The similarity of concepts used in medical image reconstruction and dose optimization through beam intensity modulation was noted (Bortfeld et al. 1990). A new generation of *convolution* algorithms soon developed (Mackie et al. 1984). For higher energy x-ray beams, the spread of energy away from x-ray interaction sites required consideration of the range of more energetic electrons (Mackie et al. 1988; Ahnesjö et al. 1987; Yu et al. 1995). Cost-effective workstations with enhanced computing power caused a resurgence of interest in Monte Carlo simulations (Andreo 1991). This technique was implemented for modelling radiation particles emerging from linear accelerators (Rogers et al. 1995, 2009), providing the data necessary to bootstrap the new breed of algorithms. Figure 1.2 shows the annual publication rate for the current generation of dose algorithms. Between 1980 and 2017, the total publications were 293, 677, and 97 for convolution-superposition, Monte Carlo, and Boltzmann transport algorithms, respectively. The dominance of Monte Carlo publications is not surprising because this method has been used extensively as a quality assurance and benchmarking tool to validate other algorithms. The resurgence of activity in Boltzmann methods is seen in the open literature and in textbooks (Andreo et al. 2017; MacDermott 2016). These revitalized algorithms will be the main subjects in Chapters 5 and 6.

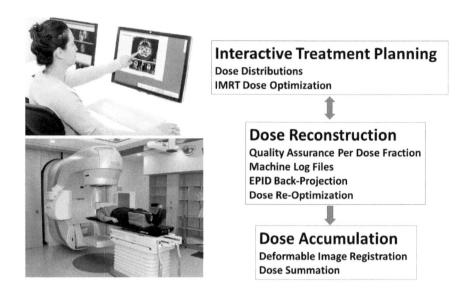

Figure 1.3: Applications of dose computation algorithms. Photos courtesy of Jonatan Snir, Linada Kaci, and John Patrick of the London Regional Cancer Program.

1.1.2 Expanding Role of Dose Computation Algorithms

Principle #1: "Dose computation algorithms play an important role in treatment planning, delivery, and assessment of clinical outcomes"

Figure 1.3 illustrates the expanded applications of dose computation algorithms. The arrival of intensity-modulated radiation therapy (IMRT) placed much greater demands on treatment planning. The decomposition of beams into beamlets with different intensities escalated the number of degrees of freedom for optimization, beyond the reach of mental arithmetic. Multiple iterative cycles were needed to make the dose distribution converge to prescribed clinical dose-volume constraints. In this book, the term IMRT is used in the broadest sense to include fixed-field and arc therapy modes of delivery including VMAT and tomotherapy.

With the introduction of on-line image guidance systems on treatment machines, the possibility now exists for recomputing the dose distribution per treatment session based on current anatomy. Using log files that record the momentary status of a treatment machine during irradiation, the entrance beam fluence incident on the patient can be determined to compute the dose distribution from CT scans obtained at treatment time (Katsuta et al. 2017). Alternatively, images from an electronic portal imaging device (EPID) can be used to reconstruct the delivered dose distribution (Mijnheer et al. 2015). By tracking the movements of tissue voxels with a deformable image registration procedure, the accumulating dose distribution can be tested for convergence to the treatment plan objectives (Wong et al. 2018). If clinically-significant deviations are detected, an adaptive correction can be applied in remaining fractions to suppress hotter dose regions in normal tissues, or enhance cooler dose regions in the target volume. In summary, the role of dose calculation algorithms will expand with enhanced presence at the treatment console.

1.1.3 Where Physics Intersects Biology

1.1.3.1 Cell Survival Curves

Radiation oncology rests upon the foundations of fundamental radiological physics and radiation biology. The transfer of energy from x-rays to tissue cells results in atomic ionization events with potential damage of *all* exposed cells. Double-strand breaks of DNA cause chromosome breaks that form critical lesions that prevent cell replication or lead to cell transformation and cancer, if left unrepaired. At low doses, the damage is caused mainly by *single tracks* of charged particles passing through chromosomal space. In this mode of action, exponential cell killing occurs as a function of dose (i.e. αD where D is the dose). At higher doses, additional damage can be caused by *pairs of tracks* that are spatially and temporally correlated (i.e. βD^2). The double-track events gives rise to curvature in logarithmic cell survival curves. Figure 1.4A shows hypothetical curves for tumour and normal cells exposed to the *same* acute levels of radiation, plotted on a semi-logarithmic scale. Normal cells exhibit improved survival at lower acute doses which can be interpreted as better repair capability for sublethal damage. The expected fraction of surviving cells, $S(D)$, is well predicted by Poisson statistics and the Linear-Quadratic (LQ) model of cell killing

$$S(D) = \exp[-(\alpha D + \beta D^2)] \tag{1.1}$$

Selective dose targeting of tumour cells can be achieved by application of *dose gradients* that reduce the dose and damage to adjacent normal tissues. The dose gradient between target and normal tissue regions can be accentuated, for example, by using IMRT. Furthermore, dose can be delivered in small dose fractions spaced over time. This exploits the differential rates of recovery from genomic damage in normal and tumour cells. Figure 1.4B exhibits survival curves for the same cell characteristics as in Figure 1.4A, but the total dose (D) has been split into n smaller doses of size d (Fowler et al. 2015). There are two significant differences compared with single-shot exposure: (1) d is set at 2 Gy for the tumour cells, while d is reduced to 1 Gy for normal tissue cells, and (2) a sufficient time gap between dose fractions allows for recovery of normal cells and re-oxygenation of hypoxic tumour cells, restoring their radiosensitivity to x-rays. The resulting survival of normal cells is greatly enhanced compared with tumour cells after the total dose is delivered. The simplest equation for cell survival, $S(d, n)$, for a *multi-fraction* exposure is predicted by compounding the single-fraction effect

$$S(d, n) = \exp[-n(\alpha d + \beta d^2)] \tag{1.2}$$

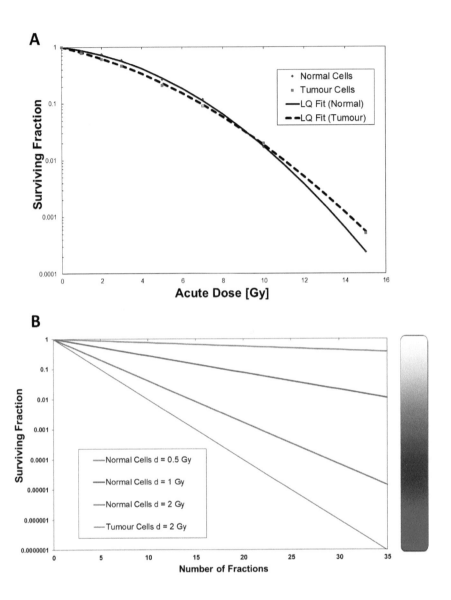

Figure 1.4: A. Cell survival curves for tumour and normal cells exposed to the same acute doses of x-rays. B. Survival curves for cells exposed to fractionated doses (d) with an applied dose gradient illustrated by the colour strip on the right. For tumour cells, $\alpha = 0.2$ and $\beta = 0.02$. For normal cells, $\alpha = 0.1$ and $\beta = 0.03$.

In summary, the cellular reaction to a dose of radiation is characterized by the α/β ratio which describes the extent of bouncing back from injury. A wide range of α/β values has been observed across tumour types and normal tissues (Bentzen and Joiner 2009). For early-reacting tissue and most tumours except prostate (Vogelius and Bentzen 2013), α/β is high (i.e > 10) and cell killing is essentially logarithmic with dose. There is minimal recovery from radiation damage so that fractionation size is less important. Conversely, for late-reacting tissue and prostate tumours, α/β is depressed (i.e. < 3), implying sensitivity to the size of dose fractions.

1.1.3.2 Biologically Effective Dose

Mathematical models based on the LQ concept are used for comparing dose prescriptions with different fractionation schemes (Fowler 2010). The Biologically Effective Dose (BED) is a surrogate for the cell-killing effect of radiation. It converts a total dose (D) delivered as an n-fractionated course of therapy to a single effective dose value that would produce the same level of cell killing, *as if* it were delivered by an infinite series of smaller dose fractions, or continuously at very low dose rate (i.e. $d \ll \alpha/\beta$). For a given cell type, treatments to the same BED result in the same level of cell survival. The simplest form of the BED is given by the following expression, assuming that the total dose, D, is split evenly into n dose fractions of size, d, as follows:

$$\text{BED}(n, D, \alpha/\beta) = D\left(1 + \frac{D}{n(\alpha/\beta)}\right) = D\left(1 + \frac{d}{(\alpha/\beta)}\right) \qquad (1.3)$$

Figure 1.5 shows the effect of changing α/β values and the dose fraction size ($d = D/n$) for two types of hypothetical treatments, delivered in 35 standard fractions (2 Gy) or 3 hypo-fractions (16 Gy) of ablative radiosurgery. The plots show the impact of deviations in dose fraction sizes (d) and radiobiological parameters (α/β). The influence of fraction size is especially pronounced for low values of α/β.

Equation 1.3 is especially useful for converting various dose prescriptions to a common frame of reference. It can also be used to assess the composite cell-killing effects from split courses of radiotherapy, including brachytherapy. More comprehensive forms of the equation are required to consider repair, cell age effects, re-oxygenation, and re-population that modulate cellular response to radiation (Hall 2006; Fowler et al. 2015; Lindblom et al. 2015; Joiner and Van der Kogel 2016).

1.1.3.3 Tumour Control and Normal Tissue Complication Probability

Figure 1.6A shows an application of dose gradients to a hypothetical lung tumour, with avoidance of contra-lateral normal lung damage. A pair of dose-response curves is shown in Figure 1.6B. The Normal Tissue Complication Probability (NTCP) is intentionally positioned unfavourably to the left side of the Tumour Control Probability (TCP) curve, as is often the case clinically. Various mathematical models

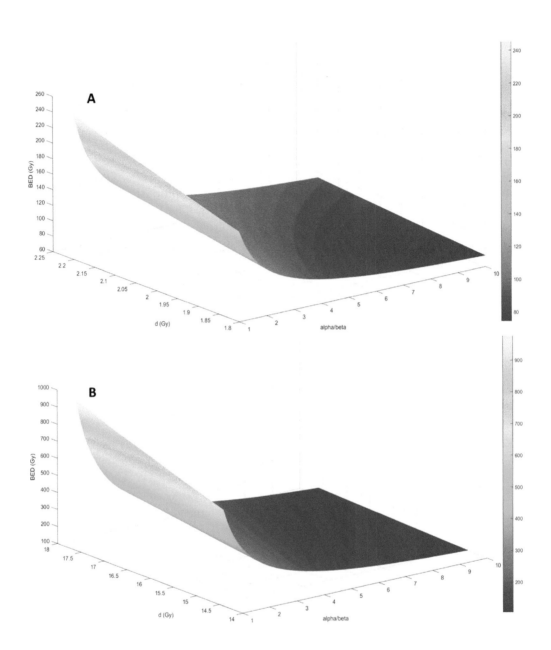

Figure 1.5: BED landscape for A. Conventional treatment with 35 fractions with a nominal fraction size (d) of 2 Gy. B. Radiosurgery treatment with 3 fractions and nominal fraction size (d) of 16 Gy. Graphics courtesy of Jeff Kempe, London Regional Cancer Program.

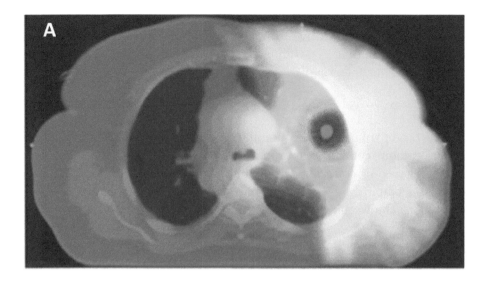

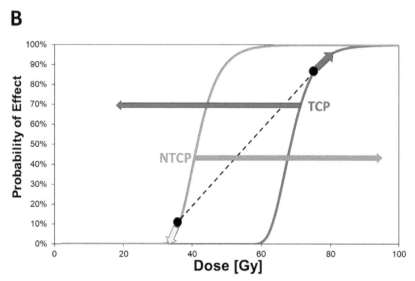

Figure 1.6: A. Dose distribution for a lung tumour. Graphics courtesy of Brandon Disher. B. Typical NTCP and TCP curves. The effect of a dose gradient (dashed line) is shown for a hypothetical tumour (short red arrow) and organ-at-risk (short green arrow). Alternatively, an exaggerated lateral shifting of the curves is also shown, assuming tumour radiosensitization (long red arrow) and normal tissue radioprotection (long green arrow). Either strategy can expand the therapeutic dose window. In clinical practice, the two strategies are often used in combination.

for calculating TCP and NTCP curves have evolved to explore different clinical scenarios (Fowler et al. 2015; Moiseenko et al. 2005).

We start with the most intuitive form of the TCP curve that follows from the Poisson probability of finding N_S tumour cells alive after a radiation exposure of N_0 cells to a dose, D, as follows:

$$\text{TCP}(D, V_T) = \exp[-N_S] = \exp[-N_0\,S] = \exp[-V_T\,\rho_T\,S] \approx 1 - V_T\,\rho_T\,S \quad (1.4)$$

where V_T is the target volume with a tumour cell density of ρ_T. If $N_S < 1$, the probability of local tumour control is approximately $1 - N_S$. Intuitively, if $N_S = 0.1$, there is a 90% chance of controlling the tumour. The following dual-purpose expression that uses empirical D_{50} and γ_{50} parameters is suitable for either TCP or NTCP calculations (Moiseenko et al. 2005; IAEA 2016):

$$\text{TCP or NTCP}\,(D, D_{50}, \gamma_{50}) = \frac{1}{1 + \exp[4\gamma_{50}(1 - \frac{D}{D_{50}})]} \approx \frac{1}{[1 + (\frac{D_{50}}{D})^{4\gamma_{50}}]} \quad (1.5)$$

where the lateral position of each of the response curves is specified by a radiosensitivity parameter D_{50}, i.e. the dose that produces a 50% bio-response. At this dose level, the chance of a favourable output is 50/50 – as in flipping a coin. The slope of each curve is specified by the normalized parameter γ_{50} – the relative change (%) in response to a 1% relative change in dose near D_{50}. The γ_{50} value characterizes the sharpness of the switching point and sensitivity to uncertainty in dose. The values of these two parameters depend on the tissue and tumour type, as well as their organizational structure (series or parallel) and the possible use of radiosensitizer or radioprotective drugs. Other NTCP models exist for incorporating the tissue features and are summarized by Moiseenko et al. (Moiseenko et al. 2005). The dose gap between the D_{50} values for tumour and normal tissues represents the *therapeutic dose window* for treatment planning. In Figure 1.6B, the D_{50} value for normal lung tolerance is approximately 40 Gy, while control of lung tumours requires a dose of > 70 Gy. The therapeutic dose window is unfortunately negative (i.e. $40-70$ Gy $= -30$ Gy). Two strategies can deal with this unfavourable situation:

(a) Displace the two *doses*. This is a physical strategy. A clinical colleague expressed it most succinctly – "Hit the tumour, and miss the patient!". Where the dose gradient is steepest, the separation of the doses is greatest. The target dose can be maintained while reducing toxicity levels, or the target dose can be escalated without increasing toxicity (see Figure 1.6B). This requires a *trustworthy* dose computation algorithm that accurately models dose gradients in tissue.

Principle #2: "Treatment planning aims to optimize *dose gradients* across the target and normal tissue zones"

(b) Displace the two *curves*, i.e. moving the NTCP curve to the right and the TCP curve to the left (see Figure 1.6B). This is a radiobiological strategy. The combination of radiotherapy with chemotherapy, for example, can enhance the radiosensitivity of tumour cells. Radioprotective drugs, on the other hand, can shift the normal tissue curve to the right for greater radiotolerance; for a parallel-structured organ, partial irradiation of an organ also shifts the normal curve to the right (Bentzen et al. 2010).

The present state of knowledge in predicting tumour control and normal tissue reactions has been reviewed recently (Tommasino et al. 2017). Prediction of TCP and NTCP is weakened by inadequate knowledge of *in-vivo* patient-specific parameters (D_{50}, γ_{50}). For example, in the case of non-small-cell lung cancer, the values of D_{50} can double from 45 Gy to 90 Gy, depending on dose fractionation, re-oxygenation rate, and intra-fraction repair rate (Lindblom et al. 2015). The predicted values are also affected indirectly by the delineation of anatomical volumes influenced by the type of imaging modalities used (e.g. CT, MRI, PET, including hybrid imaging). In radiosurgery of lung lesions (Guckenberger et al. 2016), the variations in computed dose distributions due to the dose algorithm selection also affect TCP predictions (Chetty et al. 2013; Chen 2014; Liang et al. 2016; Zhou et al. 2017). This is not surprising because when small fields of high energy x-rays traverse low-density tissue, there is a *reduction* in target dose and spillage of unwanted dose to nearby normal tissue. This anomalous effect is opposite to the increase in dose expected intuitively from greater beam penetration through a low-density medium. The effect is caused by charged particle disequilibrium that will be explained in Chapter 2. In Figure 1.7, predicted TCP values, averaged over 133 cases of non-small-cell lung cancer, are shown for different dose algorithms and target sizes. The *clinically-observed* rates of local tumour control are in reasonable agreement with TCP values predicted by 3D algorithms capable of modelling charged particle transport.

A recent clinical study of 205 patients, however, did *not* show an improvement in local tumour control and survival for cases planned with more accurate dose computation algorithms (Bibault et al. 2015). This demonstrates that accuracy in dose is necessary but not always sufficient for improvement of clinical outcomes; other uncertainties can overshadow the gain in accuracy, as we shall explore in Section 1.2. In the specific case of small isolated tumours, disequilibrium conditions confounded the definition of the target volume (Section 1.2.4) and masked the gains from improved dosimetry.

Another study focused on the effect of dose algorithm selection on NTCP values for 80 clinical cases at 4 tumour sites (breast, lung, head and neck, prostate) (Bufacchi et al. 2013). Two algorithms, such as the Anisotropic Analytical Algorithm and Pencil Beam Convolution (Varian Medical Systems, Palo Alto, CA), were contrasted and significant differences were found in predicted NTCP values for organs-at-risk.

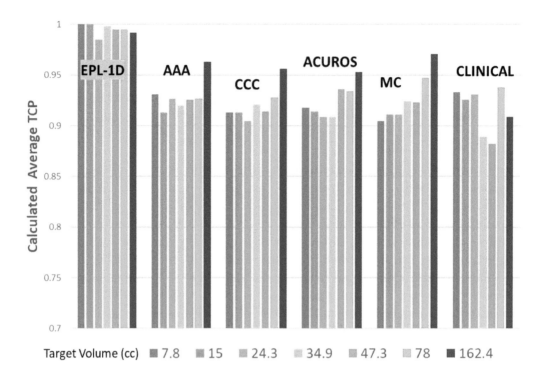

Figure 1.7: Variations in calculated TCP values for ablative radiosurgery of lung lesions for different PTV sizes (cm^3), listed below the figure. The dose algorithms include EPL-1D = Effective Path Length (BrainLab); AAA = Analytical Anisotropic Algorithm (Varian); CCC = Collapsed Cone Convolution (Philips); Acuros XB (Varian); MC = Monte Carlo (BrainLab). The actual clinical rates of local control at a 2-year follow-up are also shown (CLINICAL) (Chetty et al. 2013).

The authors concluded that NTCP differences *could* be clinically significant, but that they were more strongly influenced by the choice of NTCP model and input radiobiological parameters.

1.2   REQUIRED ACCURACY IN DELIVERED DOSE

The term *accuracy* must be contrasted against similar terms such as uncertainty and tolerance (Van Dyk et al. 2018; IAEA 2016). *Accuracy* refers to the proximity of a measured or computed dose value to the accepted true value. In absolute dose calibrations (in units of Gy), the accuracy of a field instrument must be traceable to a national standards laboratory that establishes the true value according to international protocols. *Precision*, on the other hand, refers to the reproducibility of a value obtained from repeated experimental measurements or computational runs. The distribution of results is characterized by a standard deviation of the recorded samples (σ). For a Gaussian distribution of data, 95% of sampled values will normally lie within $\pm 2\sigma$. *Uncertainty* is a composite of the accuracy and precision. *Tolerance* refers to an acceptable level of uncertainty above which there is cause for alarm. If the level of uncertainty significantly exceeds the tolerance value, corrective action is taken to minimize risk.

A delivered dose may be compromised by an incorrect calibration of the treatment unit and imprecision in repositioning the patient. Using a large number of dose fractions has a beneficial averaging effect beyond the radiobiological reasons for fractionation. In lung radiosurgery, however, large doses are administered in much fewer fractions, and this averaging effect is forfeited. Extraordinary patient immobilization techniques, and target tracking or beam gating techniques are required. In cranial radiosurgery, the issue of target motion is diminished but head positioning does remain critical.

In Monte Carlo simulations of radiation, uncertainty in the average dose to a tissue voxel can be caused by inaccuracy and imprecision. Inaccuracy is introduced when using invalid input physical data governing x-ray interactions. Imprecision is reflective of Poisson statistics associated with variable energy deposits from a finite number of simulated particles. The level of these fluctuations can be assayed by running independent batches of particle histories, and determining the standard deviation of the scored results. Reduction of uncertainty to an acceptable level is achieved by running more particle histories and using variance reduction strategies, as will be described in Chapter 5.

1.2.1 Using a Systems Model

Any demands on dose accuracy must be placed within context of uncertainties in the overall clinical procedure, including diagnosis and staging, treatment goal (i.e. curative or palliative), dose prescription, delineation of targets and organs, status of irradiated tissue, and physical limitations. Patient-specific conditions such as human papilloma virus (HPV) and epidermal growth factor receptor (EGFR) status in head and neck tumours, tumour micro-environment such as oxygen supply (Göttgens et al. 2018) and prior therapy require treatment variations. In terms of radiation physics, there are limits to experimental or computational dosimetry that can trump the *achievable* dose accuracy. Setting a tight uncertainty goal of 0.1% for the absolute dose to all voxels in a patient is highly unrealistic and unachievable if dose calibration is intrinsically subject to 1% systematic accuracy. A reasonable compromise must be struck between clinical requirements and achievable accuracy in dose delivery.

Radiotherapy is a multi-stage procedure involving a multi-disciplinary team of professionals. Figure 1.8 depicts the steps for an adaptive radiotherapy procedure. An *in silico* dose distribution displayed on a computer screen will not necessarily materialize *in tissue* if uncertainties are not controlled, risking under-dosing of

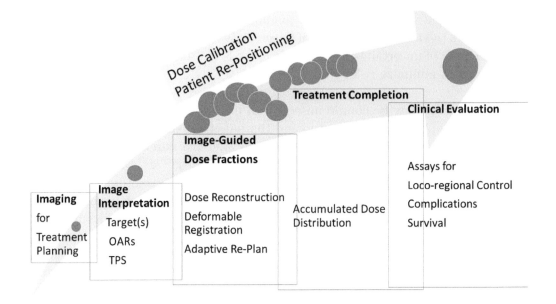

Figure 1.8: Steps in an adaptive radiation treatment process. Each sub-process over-laps and can introduce uncertainties which can propagate right through to radiation delivery. OARs = Organs-at-Risk. TPS = Treatment Planning System.

the tumour or over-exposure of normal tissues. Miscommunication in transferring tasks across workstations and staff members can propagate uncertainties that persist throughout treatment. A systems analysis is appropriate because of the range in possible permutations and combinations of uncertainties (Wong et al. 1997; Killoran et al. 1997).

The numerical specification of tolerable uncertainty has been a topic of debate during different epochs of radiation technology (Van Dyk et al. 2013, 2018). In the era of 2D treatment planning, an international panel (ICRU 1976) concluded that "available evidence for certain types of tumour points to the need for an accuracy of ±5% in the delivery of an absorbed dose to a target volume. Some clinicians have requested even closer limits within 2%, but at the present time it is virtually impossible to achieve such a standard." At the start of 3D treatment planning, a reduction to ±3.5% was suggested as a guideline, recognizing that "... in many cases larger values are acceptable and in a few special cases an even smaller value should be aimed at."(Mijnheer et al. 1987), in agreement with an earlier analysis (Brahme 1984). With consideration of the sigmoidal shape of dose response curves, this uncertainty level maintains the tumour control probability to within approximately 10% of the intended goal. Bentzen et al. (Bentzen et al. 2000) used variations from a dose calibration audit of European cancer centres and showed that reducing dose uncertainty from 10% to 2% could improve tumour control rates by ≈ 17%. A more recent study also maintained a goal of 2% accuracy in delivered dose across patients (Bentzen 2004). With ongoing evolution towards IMRT, the International Commission on Radiation Units and Measurements (ICRU) introduced a composite specification that relates dose uncertainty to positional uncertainty and hence acknowledged the influence of dose gradients (ICRU 2010; Gregoire and Mackie 2011). The Commission recommended that "for a low-gradient (<20%/cm) region, the difference between the measured (or independently computed) absorbed dose and the treatment planning absorbed dose, normalized to the absorbed-dose prescription should be no more than 3.5%. For high-gradient (≥20%/cm) regions, the accuracy of distance-to-agreement (DTA) should be 3.5 mm." A review of IMRT dosimetry showed that there was actually a tendency towards *greater uncertainty* due to complexity of the technique (Thwaites 2013). The author concluded that previous goals for delivered dose should still be maintained and would remain achievable with suitable quality assurance checkpoints.

The reality is that there is no straightforward answer to the question *What level of accuracy is needed on the basis of potential clinical impact?*. The most recent IAEA report on uncertainty in radiotherapy is a tome of almost 300 pages including a bibliography with over 160 publications (IAEA 2016; Van der Merwe et al. 2017). The consultants recognized the difficulty of specifying numerical tolerances and introduced an overarching guideline: "All forms of radiation therapy should be applied as *accurately* as reasonably achievable (AAARA), technical and biological factors being taken into account." This principle is clearly borrowed from the operational ALARA guideline used in the practice of radiation protection. It sets the tone for establishing a general culture of patient safety within a risk-benefit framework. Setting of limits for dose uncertainties to the target and normal tissues revolved around three major themes:

1. **Radiobiological** – Slopes of dose-response curves

2. **Physical** – Positional and dosimetric uncertainties

3. **Clinical** – Dose-volume uncertainties, assessment of treatment response

1.2.2 Radiobiological Considerations

Dose uncertainty has its greatest impact when the local dose lies near the most pronounced slope of the sigmoidal response curve ($\approx D_{50}$) of interest. Typical values for clinically-observed γ_{50} are 2 and 4 for tumours and normal tissue, respectively. These γ_{50} values are averages across patient populations. Treatment uncertainties are inherent and preclude accurate measurement of intrinsic *in vivo* slopes. For example, with a 10% random uncertainty in dose, the intrinsic slope of 4 blurs to an observable shallower value of 2.8 (Pettersen et al. 2008). Hence, the use of average values observed across a patient population may underestimate the dose accuracy requirements. When considering an individual patient, it is possible that personal dose response curves will be steeper and a narrower control of dose at the target and organs-at-risk is important.

Radiation treatments with IMRT tend to deliver a spread of doses at the top or bottom portions of the sigmoidal response curves (Figure 1.6B). This generally leads to wider tolerance in dose uncertainty. Furthermore, amortization of dose over numerous fractions also provides statistical averaging which at first glance should minimize change in radiobiological response. However, the *asymmetry* of the response curves produces a net radiobiological effect that is *not* cancelled by an averaging effect. The influence of dose uncertainty on a treatment aimed at a TCP of >79%, with an NTCP of <21% is plotted in Figure 1.9. If we consider a change of 3% in either TCP or TCP acceptable, a tolerance of ±5% in dose delivery seems reasonable.

1.2.3 Physical Considerations

The range of measured uncertainties in spatial positioning and dose accuracy is shown in Figure 1.10. For simplicity, typical uncertainties in dose (as % deviation) and in geometric localization (in mm) are shown independently, recognizing that some quantities are cross-coupled in regions of dose gradients. Generally, patient repositioning and elasticity of tissue are subject to random day-to-day deviations on the order of 2 to 6 mm. For instance, the flexing of the neck in head and neck treatments affects precision. Discrepancy between the treatment isocentre and imaging isocentre for on-board CT and MRI systems can contribute imprecision. The *largest* spatial uncertainties stem from contouring of targets (5 to 50 mm) and organs (5 to 20 mm).

The largest dose uncertainty is caused by dose computation errors in regions of charged particle disequilibrium, especially build-up and lung regions. The treatment couch attenuation effect is potentially large (up to 20%) but it can be incorporated in the treatment plan (Olch et al. 2014). Multi-leaf collimators (MLCs) are mechanical devices with limitations in positional accuracy and speed limits. Any instability of the dose rate from the accelerator during a treatment or across treatments can limit accuracy. Results of end-to-end testing with static phantoms show overall compliance of a delivered dose within an uncertainty band of 10%, significantly greater than goals set for accuracy.

1.2.3.1 *Dose Auditing*

In response to the increasing complexity of treatment techniques, patient-specific quality assurance (Miften et al. 2018) and independent audits have been conducted by national and international agencies. The goal is ensuring general patient safety and treatment consistency across participants in collaborative clinical trials. Real

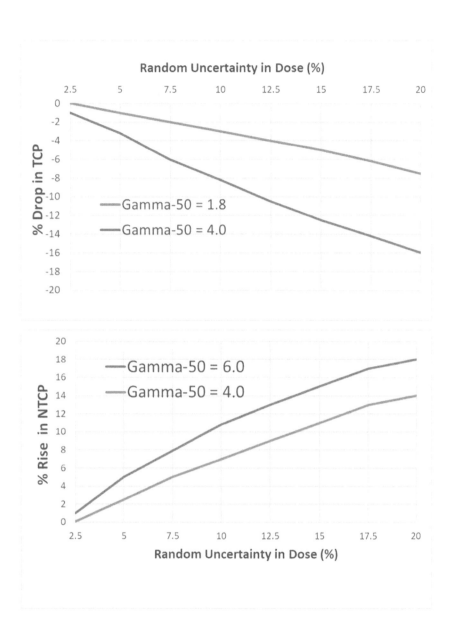

Figure 1.9: Impact of random uncertainties (σ) in dose delivery. (A) Reduction in TCP from a starting value of 79%. (B) Increase in NTCP from a starting value of 21%. Data source (IAEA 2016).

Figure 1.10: A. Range of uncertainties in geometric displacements. B. Range of uncertainties in dose. TPS = Treatment Planning System. Data Sources (IAEA 2016; Van Dyk et al. 2018).

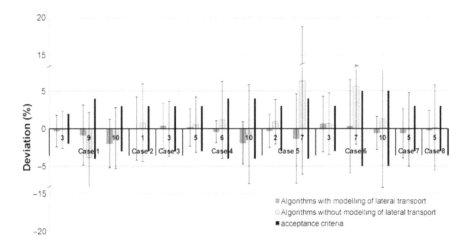

Figure 1.11: Range of uncertainties (2σ) in dose computations. Data sets are grouped for different beam exposure *cases* (1 to 8) and test points (1 to 10) in a static thoracic phantom. The tolerance limits are shown as dark solid lines for each test point. Reproduced with permission (Gershkevitsh et al. 2014).

or virtual human-like phantoms are shipped to participating groups to test protocol compliance with end-to-end testing of the dose delivered. The resultant data provide insight into what level of overall uncertainty exists across a variety of clinical practices. *Significant dose uncertainty is still introduced in the application of dose calculation algorithms.* The European Organization for Research and Treatment of Cancer (EORTC) conducted a survey of 60 radiotherapy centres across 8 European countries. A thoracic anthropomorphic phantom was imaged locally and then treated on site according to a strict protocol. Figure 1.11 shows dose deviations across participants for a set of 8 beam cases and 10 test point locations in the lung phantom. A total of 190 data sets were analysed spanning across a range of dose calculation algorithms marketed by Eleketa, Philips, Siemens, and Varian. Beam energies included cobalt-60 radiation and x-ray beams up to 20 MV. The differences between calculated and measured doses were normalized to dose at a robust nearby point, and plotted to show the range as 2 standard deviations (2σ, 95% confidence interval). The magnitude of systematic offsets and spread was reduced for algorithms based on 3D convolution-superposition methods (mean 0.5%, $2\sigma = 3.3\%$). These models of radiation transport account for laterally-scattered charged particles. Simpler algorithms based on 2D pencil beam or Fourier convolution methods performed more poorly (mean: 0.8%, $2\sigma = 5.6\%$). Differences were pronounced for points lying within the lung-substitute material, with one alarming

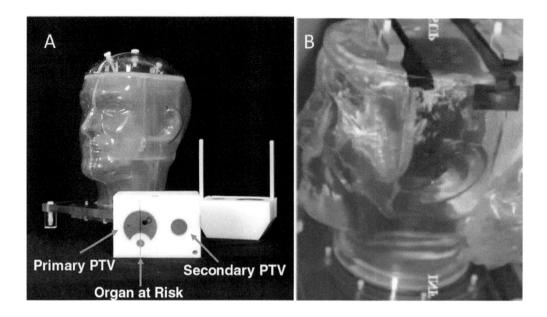

Figure 1.12: A. RPC head and neck phantom with its inserts. Adapted with permission (Molineu et al. 2013). B. Fricke gel inserted in the phantom for 3D dosimetry (Babic et al. 2008), courtesy of Steven Babic and Kevin Jordan, London Regional Cancer Program.

error of +17%. With tolerances set between 2 and 5%, failure rates ranged from 6 to 56%. It is important to note that algorithm performance is not only driven by intrinsic physics modelling but also by proper commissioning of the planning system (Smilowitz et al. 2015; Xue et al. 2017) and CT scanner. Delivery of an absolute dose is further reliant on the beam calibration (Gy per Monitor Unit or time).

A similar survey was conducted in North America by the Imaging and Radiation Oncology Core – Houston (IROC-H) using a head and neck phantom distributed to 763 institutions between 2001 and 2011 (Molineu et al. 2013). This multi-institutional project was commissioned by the National Cancer Institute (USA) to ensure consistency of IMRT procedures across clinical trials of the Radiation Therapy Oncology Group (RTOG). The phantom is shown in Figure 1.12A with inserts used for holding thermoluminescent (TLDs) and film dosimeters. A prototype 3D dose verification procedure using gel dosimetry (Babic et al. 2008) is also shown in Figure 1.12B. The phantom is imaged for treatment planning at the test institution and the dose distribution is delivered in accordance with a clinical protocol. The phantom and embedded dosimeters is returned to IROC-H for unbiased readout, analysis, and feedback to the user. TLD dosimetry was performed at 6,520 test points, and a film recorded the 2D dose distribution. The ratio of computed doses to delivered doses was normally distributed with a standard deviation (1σ) of 3%.

The tolerance was set at $\pm 7\%$ for target doses of 6.6 Gy and 5.4 Gy prescribed to a pair of volumes. In addition, the maximum distance-to-agreement in the dose gradient across the virtual target volumes and organs-at-risk was set at ± 4 mm. These are rather generous settings for defining a passing grade but they probably allow for inherent uncertainty in the IROC-H dosimetry. Results from over 1,100 irradiation procedures showed pass rates for different treatment planning systems, ranging from 75 to 93%. Using CT images of the phantom and similar dose analysis, a trans-Atlantic audit was performed in European cancer centres (Weber et al. 2014). The pass rates for dose points meeting the criteria of 7%/4 mm ranged from 91 to 100% across European institutions. With stricter criteria of $\pm 3\%$/3 mm, the institutional pass rates ranged from to 63 to 97%. A similar study of IMRT for head and neck cancer was conducted across a collaborative network of 13 radiation therapy centres in Ontario, Canada. A virtual anatomical case was distributed and the treatment was delivered locally onto a cylindrical phantom (ArcCHECK, Sun Nuclear Corporation, Melbourne, Florida). The pass rate using criteria of 3%/3 mm was >90% with dose normalization in the high dose region to offset variation in machine dose rate (Létourneau et al. 2013). Pass rates across surveys are difficult to compare because of differences in phantoms and dose normalization procedures.

A recent end-to-end study used optical dosimeters (OSLDs) placed on patient surfaces or inside body cavities, totalling 10,244 point measurements. Treatment techniques included 3D conformal radiotherapy and IMRT procedures with fixed or rotational beams (i.e. VMAT), for a wide variety of tumour sites. Accuracy was determined to be 3% (σ), with a precision ranging from 6.9 and 13.4% across treatment techniques (Riegel et al. 2017). The dosimetric methods used for this type of *in situ* dosimetry have been reviewed elsewhere (Mijnheer et al. 2013).

1.2.4 Clinical Considerations

1.2.4.1 Contouring Uncertainty

Dose prescriptions vary according to treatment objectives and interpretation of anatomical volumes seen in digital images. The numerical dose prescriptions (Gy) can generally be reconciled on the basis of equivalent BED (review Section 1.1.3.2). However, contouring of anatomy is more subjective and produces a cascade of downstream consequences. Contours indirectly impact dose-volume constraints and hence affect the dose optimization during treatment planning. The prediction of radiobiological effects, including TCP and NTCP, also depends on the overlap of targets and organs-at-risk with the computed dose distribution. The IAEA report (IAEA 2016) cites variations between 5 and 50 mm in target volumes (Gahbauer et al. 2004), while normal tissue compartments are subject to smaller variations between 5 and 20 mm. Contour variations stem from visual perception of gross target volume (GTV) edges, often limited by poor image contrast or confounding structures (Jameson et al. 2010; Louie et al. 2010). Planning target volumes (PTV) are further created by adding margins to cover invisible microscopic disease, plus expected uncertainties in targeting during irradiation. There have been technological developments that help reduce margin sizes, including temporary breath-holding (Boda-Heggemann et al. 2016), beam gating, and motion tracking (Onishi 2015; Keall et al. 2006).

Figure 1.13 illustrates difficulties in contouring lung tumours from CT scans. The internal GTV (IGTV) was formed by the union of GTVs each contoured at 10 respiratory phases. Six radiation oncologists participated in a study of 10 clinical cases (Louie et al. 2010). The greatest variability occurred when the target was near a partial lung collapse (i.e. atelectasis) as shown in Figure 1.13A. The volume overlap index (VOI) is plotted in Figure 1.13B, where a VOI of 1.00 indicates perfect consistency of compared volumes. There is clearly a wide range of GTVs drawn from the same set of images due to confounding image features. The improved VOI values for 4D CT scans *versus* a 3D CT scan at mid-respiration indicates that temporal snapshots of lung anatomy lead to more consistent target contouring. Note that under conditions of charged particle disequilibrium (explained later in Chapter 2), the definition of target volume for a small tumour surrounded by lung parenchyma requires careful interpretation (Bibault et al. 2015; Seuntjens et al. 2014). Equilibrium can automatically re-establish itself in a lesion under certain conditions, *even if the target moves* (Disher et al. 2013).

Beyond achieving consensus in target definition, there remains a general need to improve image contrast and consistency of interpretation. Magnetic resonance imaging (MRI) and nuclear imaging (e.g. PET scanning) provide additional insights for identifying active tumour regions, but these methods also operate in different acquisition modes that alter the visual appearance of lesions. For example, variations in pulse sequences for MRI (e.g. T1, T2, diffusion) and miscalibration of nuclear scans (e.g. SUV thresholds for PET-identified lesions) can lead to ambiguity and inconsistency across imaging facilities. Hybrid imaging (e.g. PET-CT, PET-MRI) can help resolve some of these issues.

The international community has developed standardized definitions to harmonize the reporting of volumes used in radiation oncology (ICRU 2010). Metrics, such as DICE coefficient and Jaccard index, characterize the congruence of contours delineated by different observers (Taha and Hanbury 2015). Uncertainty in delineated volumes can be minimized through strict imaging and contouring protocols (www.rtog.org/CoreLab/ContouringAtlases.aspx) (Gregoire and Mackie 2011), specialized training (Vinod et al. 2016; Hallock et al. 2012), and peer-review (Hoopes et al. 2015; Rodrigues et al. 2012).

1.2.4.2 *Dose-Volume Histograms*

Dose-volume histograms (DVHs) are widely used for purposes of dose prescription, dose optimization, and judging clinical acceptability of a treatment plan (see Figure

A

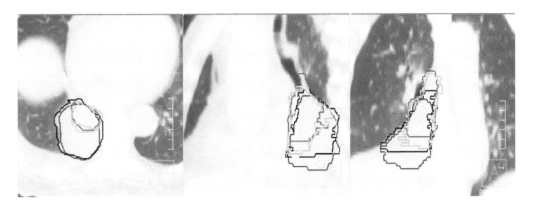

B

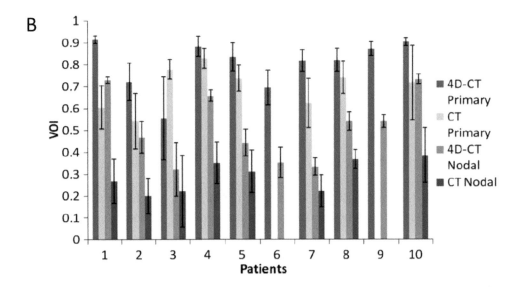

Figure 1.13: A. Inter-observer variability in GTV volumes drawn by 6 radiation oncologists in axial, sagittal and coronal CT views at mid-respiratory phase. B. Volume overlap index (VOI) across 10 clinical cases for primary and nodal tumour IGTV volumes, imaged by 4D-CT or CT at mid-respiratory phase. Reproduced with permission (Louie et al. 2010).

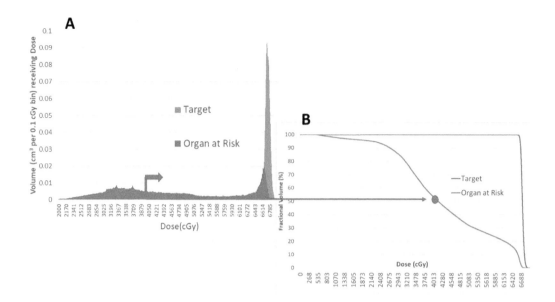

Figure 1.14: A. Differential dose-volume histogram. B. Cumulative dose-volume histogram for a typical prostate treatment plan. The cumulative data point for the organ-at-risk (blue dot) is the right-sided area under the blue differential curve (A) demarcated by the arrow.

1.14). DVH curves, however, are a distillation of 3D information and therefore blind to the spatial layout of the dose. A defect in a dose-volume histogram can only flag a dosimetric anomaly that will require further investigation through detailed scrolling through the 3D dose matrix, $D(x, y, z)$. A more efficient approach was proposed involving the display of "images of regret" that maps regions of clinically-relevant dose violations but it was not widely adopted (Shalev et al. 1988). Note that whatever dose-volume assay is used for driving plan optimization or judging plan quality, it will indirectly be influenced by the accuracy of delineated anatomical regions.

The *differential* histogram counts the number of dose voxels, N_D, having a dose value in the interval $[D, D + dD]$ within a contoured volume (V) of anatomy. It essentially displays the *spectrum* of dose values within the specified volume. Using the known voxel size, dV, this plot normally has units of absolute volume ($N_D \, dV$ [cm^3]) per unit dose increment, dD [cGy]. Figure 1.14A shows a sample of a differential DVH with a small dose bin size of 0.1 cGy. The histogram profile is typical of an IMRT plan for prostate cancer.

The *integral or cumulative* DVH is used more often clinically and is shown in Figure 1.14B. It is a companion plot normally showing the *fractional volume* receiving a dose, D_{min} *or more*. Points on this histogram represent the right-sided partial area under the differential DVH curve, past the threshold dose value (blue arrow), as follows:

$$\text{DVH}_{cum}(D_{min}) = \frac{1}{N_{tot}} \int_{D_{min}}^{D_{max}} N_D(D)\,\mathrm{d}D \qquad \forall\,(x, y, z) \in V \qquad (1.6)$$

where D_{min} is the threshold dose expressed either in absolute units (e.g. Gy) or as a percentage of the prescription dose. N_{tot} is the total number of voxels in the entire volume of interest (V). In these plots, the optimal shape for target coverage approaches a flat-topped box extending to the upper right, up to the prescribed dose value. The optimal organ-sparing shape is an L-shaped curve, notched at the origin. Some key DVH parameters of clinical interest are specific points on the integral DVH curves. They represent dose-volume or volume-dose pairs such as D95% – the minimum dose received by 95% of the target volume or $V_{20\text{Gy}}$ – the volume of lung receiving 20 Gy *or more*.

Figure 1.15 shows a pair of cumulative histograms for an IMRT treatment of a head and neck cancer. Two dose optimization techniques were being compared in this study. There is minimal difference in the dose coverage of the target volume (PTV70), but better sparing of normal organs with fast-inverse direct aperture optimization (FIDAO) (solid lines) *versus* Interior Point Optimization (IPOPT) (dotted lines).

1.2.4.3 Clinical Detection of Dose Inaccuracy

A study in 1967 at the Institut Gustave Roussy of Paris (Dutreix 1984; Wambersie et al. 1969) showed that early treatments with megavoltage electrons seemed less effective than with cobalt-60 treatments at the same tumour dose. After investigation, a calibration problem was uncovered, one that produced a 7% reduction in electron treatment dose. A second incident using 25 MV x-rays for pelvic irradiation produced an overdose of $\approx 10\%$ and oncologists soon reported negative side effects. Similar anecdotal observations had occurred at other institutions when the unit of radiation was changed from Roentgens to *rads* without changing the numerical prescriptions (i.e. 5000 R was simply replaced by 5000 rads), causing a 7% overdose.

The aim of clinical trials is to assess safety and efficacy of new techniques. The number of patients required in a study depends on the magnitude of expected effect (i.e. the signal). This effect must surpass the statistical uncertainty (i.e. noise) in the results. Tight control of dose uncertainty reduces the number of patients required. The impact of dose accuracy on detecting various levels of response (ΔR), for a range of clinically-determined slopes (γ_{clin}) is shown in Figure 1.16. With poorer quality control of the dosimetry during a trial, the effective value of γ_{clin} is reduced and the number of patients required to demonstrate an effect increases significantly.

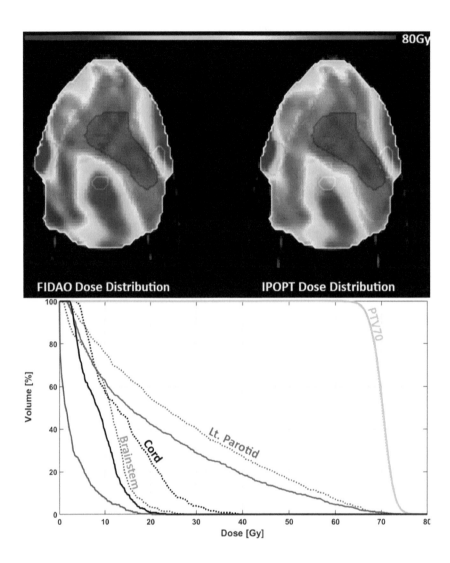

Figure 1.15: Comparison of two dose distributions (upper panel) for IMRT of head and neck cancer, obtained by different optimization methods: fast-inverse direct aperture optimization (FIDAO) and interior point optimization (IPOPT). The cumulative DVH curves (lower panel) are for FIDAO (solid lines) and IPOPT (dotted lines). Data and graphics courtesy of Michael MacFarlane, graduate student at the University of Western Ontario.

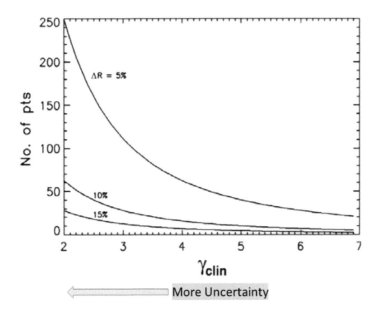

Figure 1.16: Number of patients required in a clinical trial to measure a response ΔR for clinically-determined slopes γ_{clin} of dose response curves. The slope becomes shallower (i.e. decreases in value) with increasing uncertainties as indicated by the arrow. Adapted with permission (Pettersen et al. 2008).

In addition to improving the efficiency of trials (Pettersen et al. 2008; Ibbott et al. 2013), quality assurance of dose distributions leads to unambiguous interpretation of clinical outcome results. It improves effectiveness in answering the intended questions without an ambiguity due to uncertainty in confounding factors such as dose distribution. An historical example from the 1980's demonstrates this concept. Tissue inhomogeneity corrections revealed that inter-patient variations of up to 20% in dose delivery could occur at a target in lung. Yet incorporation of *any* lung corrections was initially discouraged because clinical experience had been based on dose prescriptions assuming an all-water patient (Orton et al. 1984). It was subsequently shown that dose uncertainty across patients could be reduced to 5% by using a suitable inhomogeneity correction method and minimize the risk of radiation pneumonitis. Without individual lung dose corrections, 60% more patients would have been required to detect clinical benefit from a dose escalation (Orton et al. 1998). To recap, better standardization of dose can improve the overall efficiency of clinical trials, reducing socio-economic costs and the chance of drawing erroneous conclusions regarding the safety and efficacy of new treatments.

There is evidence that quality assurance is correlated with achieving better clinical outcomes (Fairchild et al. 2013). A study of locally advanced carcinoma of the head and neck (Peters et al. 2010) conducted by the Trans-Tasman Radiation On-

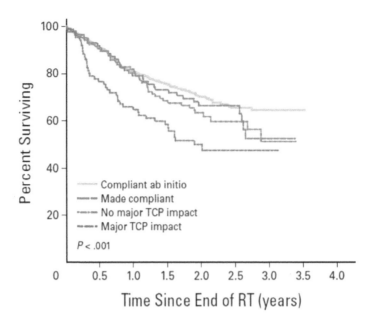

Figure 1.17: Effect of non-compliance with a dose delivery protocol on patient survival in a clinical trial of head and neck tumours. Reproduced with permission (Peters et al. 2010).

cology Group (TROG) revealed that non-compliance with the protocol caused a greater rate of loco-regional failures. This study enrolled 861 patients at 82 cancer centres in 16 countries during the period 2002-2005. Retrospective review of treatment plans revealed 97 protocol violations of 818 cases analysed (i.e. 12%). A subset of 780 patients who had received a dose of at least 60 Gy could be analysed and demonstrated an inferior survival rate at two years after radiotherapy, i.e. 50% *versus* 70% for patients treated in full compliance, as shown in Figure 1.17. This negative effect was twice as large as the beneficial effect expected from adding a hypoxic cell toxin (i.e. tirapazamine, TPZ). The trial was therefore spoiled by poor quality assurance and the clinical question could not be resolved.

Another retrospective analysis of 8 pre-IMRT clinical trials (Ohri et al. 2013) re-affirmed that violations of treatment protocols led to clinically-detectable effects in patients across a wider variety of tumour sites. Deviations included geometric discrepancies in target volumes such as improper margin sizes and dose infractions in the range of 5-10% at critical points in the dose distribution. The impact of infractions on overall survival was assessed as a hazard ratio of 1.74 for all protocol infractions combined. This ratio can be interpreted as an almost 2-fold reduction in median survival time of patients due to substandard treatment in the trial. For more complex multi-step procedures, such as IMRT, the authors cautioned that quality assurance would become even more critical. Collaborative clinical trials groups have now imposed peer reviews before and during trial participation to ensure better compliance across all participating institutions (Molineu et al. 2013; Gershkevitsh et al. 2014).

1.2.4.4 Loco-Regional Control and Patient Survival

The above sections have focused on achieving accuracy in geometric and dosimetric quantities. The tacit hypothesis is that achieving loco-regional control will automatically translate into longer term survival in patients (Suit 1992; Leibel and Fuks 1993). In other words, it is assumed that distant metastases are not pre-established at the time of initiating therapy and that irradiation may prevent metastatic seeding. Important questions were raised about this issue two decades ago when 3D conformal radiotherapy first emerged (Tubiana and Eschwege 2000):

- Does a higher rate of local control of the primary tumour improve the survival rate?

- Does conformal radiotherapy allow the delivery of higher doses by decreasing the incidence of late effects?

- Do higher doses increase tumour control and long-term survival?

The answers to these questions are multi-faceted because of variations in treatment success according to tumour types and the stage of disease at start of treatment. Patients with early stage localized tumours (Stages I and II) are most likely to benefit tangibly from aggressive radiotherapy, while those with advanced disease (Stages III and IV) may benefit from palliation with a limited extension of life. In the latter case, advances in multi-modality therapy are needed to control the localized and disseminated disease. The proportion of early stage tumours *at diagnosis* varies widely across tumour type. According to data from the SEER database of the US National Cancer Institute (progressreport.cancer.gov/diagnosis/stage), disease is localized in 66%, 22%, and 89% for breast, lung, and prostate tumours at diagnosis. Recent data from the Canadian Cancer Society (Canadian Cancer Society 2018) shows a similar pattern. The full spectrum of staging at time of diagnosis is charted in Figure 1.18. The fractions of diagnosed patients with early Stages I and II disease are 82%, 32%, and 74%, for breast, non-small cell lung, and prostate lesions. Although the frequency of localized non-small-cell lung cancer appears low, survival rates from stereotactic ablative radiosurgery are approaching those achieved with surgical resection. This is an important option for inoperable patients. Screening programs including improved imaging techniques and genetic profiling continue to advance the time course of detection and thereby enhance the importance of precision radiotherapy.

A review of major randomized trials in surgical and radiation oncology (Schmidt-Ullrich 2000) shows evidence of achieving better survival for carcinomas with a low

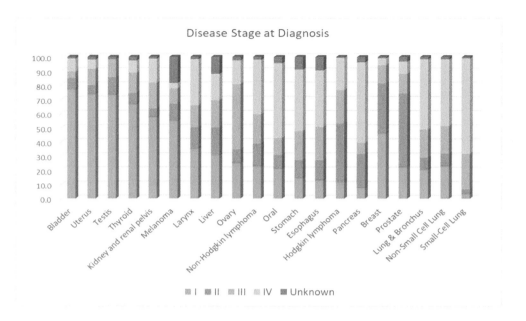

Figure 1.18: Distribution of cancer stage at time of diagnosis in Canadian population. Stages I and II cases are generally considered more favourable for positive response to local radiotherapy. Data source (Canadian Cancer Society 2018).

metastatic potential (e.g. squamous cell tumours of head and neck and cervix, and gliomas), as expected. For tumours with a greater tendency to spread (e.g. breast, prostate, and non-small-cell carcinoma), control of the primary tumour *at an early stage* has influence in limiting the spread of distant metastases. For prostate tumours treated with 3D conformal radiotherapy, post-treatment biopsies at a median follow-up of 10 years confirmed a clear link between early local tumour control and long term survival (Zelefsky et al. 2008). The study demonstrated the effect of dose escalation for intermediate risk patients (42% positive biopsies for target doses <70 Gy *versus* 24% for doses >76 Gy). The authors concluded firmly that "... these data provide evidence for the fundamental rationale for dose escalation in the treatment of clinically localized prostate cancer, namely that effective local treatment can decrease metastases and cancer related deaths." A more recent analysis of dose escalation trials across a wider range of solid tumours showed improvements in local control but not always in survival (Yamoah et al. 2015). For some tumours including pancreatic, ovarian, and glioblastoma tumours, results have been persistently disappointing over many decades (Allen et al. 2013). Controlling these tumours will require development of multi-pronged approaches including possibly molecularly-directed imaging and therapy. Fundamental research on metastases is revealing a complex process involving circulating cells, cell dormancy, immunology, and communication from the primary tumour site (Galmarini et al. 2014; Demicheli et al. 2008; Chiarella et al. 2012). These studies should eventually identify critical molecular pathways that can be disrupted for synergistic combinations of local and systemic therapy (Baumann et al. 2016).

1.3 EVALUATION OF ALGORITHM PERFORMANCE

The dose computation algorithm *per se* can be isolated from other steps in radiotherapy delivery and its accuracy thus verified independently. Accuracy is assessed by a comparison between a pair of dose distributions, one of which is considered the gold standard. Benchmark dose results are obtained from experimental or computational dosimetry (e.g. Monte Carlo simulation). Techniques for 3D dosimetry in phantoms and patients are rapidly evolving (Mijnheer 2018). Experimentally, diode arrays can sample beam fluences in planar or cylindrical surfaces during mock treatments and these measurements can be compared with those expected from a treatment plan applied to the phantom. These devices do *not* actually measure the complete 3D dose distribution *per se*. Volumetric gels offer a more comprehensive test in tissue-equivalent absorbers of simple geometry. Taking this approach to the next logical step, the addition of organ motion and deformation of gels during or between radiation exposures can simulate dose deposition in moving tissue voxels, achieving four-dimensional (4D) verification of dynamic processes. Transit dosimetry can also be performed during radiation treatments of patients using on-line portal imaging devices (EPIDs) (Mijnheer et al. 2015, 2017). Backprojection of the exit photon fluence through CT image data obtained in the same treatment session determines the fluence incident on the patient (explained later in Chapter 7). A secondary forward dose calculation is then used to reconstruct *in vivo* dose distributions. *The quality of this dose reconstruction is reliant on the dose calculation algorithm* used in the forward re-projection. Alternatively, incident beam fluences recorded by the treatment machine in log files can be replayed through the CT data set (Katsuta et al. 2017). Once again, the quality of the result is dependent on the quality of the dose algorithm used.

1.3.1 Gamma Testing

As the dimensionality of dosimetry has escalated up to 4D, more sophisticated ways of comparing dose distributions have evolved. The simplest comparison would subtract the dose values at common locations in a pair of dose distributions, assuming a static situation with perfect alignment. The difficulty of precisely aligning two spatial distributions obtained by different computational or experimental methods complicates this approach. Data sets measured with different dosimeters introduce a variety of coordinate systems, spatial resolutions, and detector noise characteristics. In regions of low dose gradients, a direct comparison between numerical dose values is relatively simpler. However, in regions of high dose gradients, mismatched spatial resolution and misalignment will confound the inter-comparison. A small spatial offset in a dose distribution with an intense gradient can result in an overly-alarming discrepancy. In regions with steep gradients, the proximity of the nearest voxel with a closely-matching dose value merits consideration. In regions with shallow dose gradients, the local dose discrepancy is more important.

A two-component vector metric was devised to combine disparities in local dose values (ΔD) and *distance-to-agreement* (DTA) (Δd) (Low 2010). Comparative analysis of *reference* and *evaluation* dose distribution matrices yields a matrix of γ values. Tolerances are set by the user ($\Delta D, \Delta d$) for normalization of discrepancies. Typical tolerance values are 3% for dose difference and 3 mm for proximity to agreement (DTA). The γ value is calculated as follows for each pair of voxels:

$$\gamma(\mathbf{r_e}, \mathbf{r_r}, \Delta D, \Delta d) = min\sqrt{\frac{|D_e(\mathbf{r_e}) - D_r(\mathbf{r_r})|^2}{(\Delta D)^2} + \frac{|\mathbf{r_e} - \mathbf{r_r}|^2}{(\Delta d)^2}} \qquad (1.7)$$

where the dose $D_e(\mathbf{r_e})$ is the dose value being evaluated in comparison with the reference value $D_r(\mathbf{r_r})$ at 3D locations $\mathbf{r_e}$ and $\mathbf{r_r}$, respectively. The discrepancy between voxel pairs, Γ, can be interpreted as a vector distance from the origin to a point in virtual dose-distance space. A set of Γ values is calculated within a restricted neighbourhood search region and the *minimum value* is selected as the γ index of matching quality. Note that γ is a composite index and it does *not* assure compliance with each local dose or DTA tolerances specified by the user. The fraction of all evaluated points with a value $\gamma \leq 1.0$ is the *pass rate*, describing overall matching between two dose distributions. The γ values are often displayed as a colour map with dimensions of the dose matrices under consideration. This map will flag regions that require further dosimetric attention, much like the fire alarm panel at the entrance of a large building, but true and false alarms can occur. The gamma comparison assay can be affected by spatial resolution and noise in dose, determining the DTA distance in 2D *versus* 3D (Pulliam et al. 2014), and lack of reciprocity between the reference and test data sets (Schreiner et al. 2013). Refinements have been devised to improve the sensitivity and specificity of the test for quality assurance purposes (Steers and Fraass 2016; Sumida et al. 2015; Moodie et al. 2014; Bakai et al. 2003; Harms et al. 1998). However, limitations of gamma testing for detecting changes in clinically-relevant parameters (e.g. DVH constraints) have been identified (Kry et al. 2014; Caivano et al. 2014; Cozzolino et al. 2014; Nelms et al. 2013; Stasi et al. 2012; Nelms et al. 2011; Zhen et al. 2011; Waghorn et al. 2011). Any intercomparison assay should be checked with a range of input thresholds and artificially-imposed discrepancies to ensure detection of clinically-important variances for specific clinical techniques. In other words, quality assurance measurement tools need to be properly commissioned to assure patient safety.

Figure 1.19 shows an example of a γ–map for dose distributions obtained by using three algorithms for 18 MV x-ray beams of different field size incident on a layered phantom (Han et al. 2011). The tested dose algorithms included the triple-A (AAA, Eclipse TPS, Varian Medical Systems) and collapsed cone convolution (CCC, Pinnacle TPS, Philips Medical Systems) compared with the reference distribution

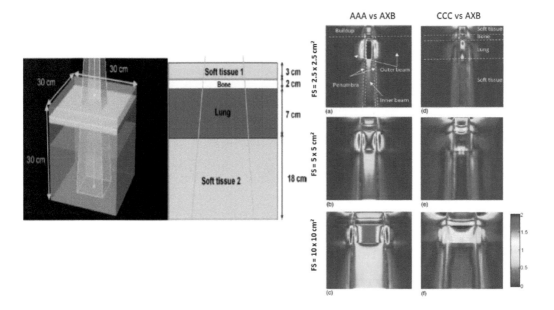

Figure 1.19: Gamma maps for tolerances of (2%/2 mm) for Anisotropic Analytical Algorithm (AAA) and Collapsed Cone Convolution (CCC) versus benchmark results from the Acuros XB Boltzmann equation solver. The multi-layered heterogeneous phantom is shown on the left side, with soft tissue, bone, and lung substitute materials. 18 MV x-ray beams of different field sizes (2.5, 5.0, 10.0 cm^2) were applied. Reproduced with permission (Han et al. 2011).

computed with a Boltzmann equation solver (Acuros XB, Eclipse TPS, Varian Medical Systems). The AAA algorithm (left panels) has limitations in predicting lateral disequilibrium effects. The CCC algorithm is superior in this regard, but there are some discrepancies lateral to the bone slab and at lung interfaces. This type of visual map is effective for pre-screening regions where inadequate commissioning or intrinsic limitations of a dose algorithm require deeper investigation.

1.3.2 Computational Speed

There is great potential for improving the accuracy of computed dose distributions by executing more advanced algorithms on high-performance cost-effective workstations. This is not just expressing a technological need for speed but rather encourages optimization of treatment plans beyond a "good enough" state. Looking ahead, dose re-computation of daily dose distributions will also demand more significantly more computational resources. Exponential gains in computational power have gone unabated for many decades and will bring Monte Carlo and Boltzmann methods to the forefront of many clinical applications. It is therefore instructive to review advances in computing power achieved over past decades and anticipate future applications. The performance *per unit cost* over a 40-year period (1970-

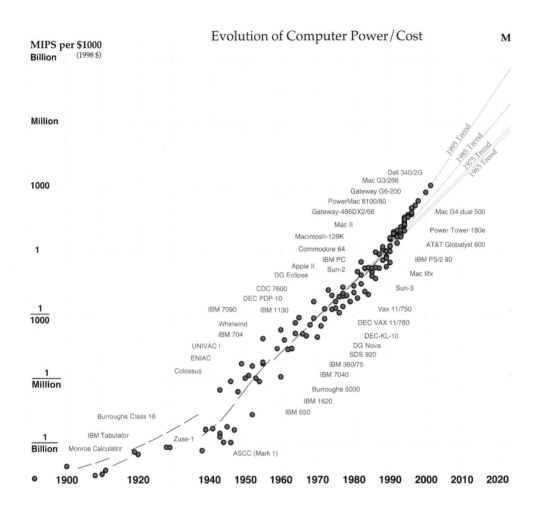

Figure 1.20: Computational power expressed in millions of instructions per second (MIPS) normalized per 1,000 dollars (1998 USD), reproduced with permission of Hans Moravec (Moravec 1998).

2010) has escalated enormously. Figure 1.20 shows exponential growth, expressed in millions of instructions per second (i.e. MIPS) normalized per cost of 1,000 US dollars (Moravec 1998). The MIPS rating is a general baseline indicator of intrinsic computing power. Over successive decades, manufacturers have sustained **5 doublings in performance per decade** – a remarkable 32-fold gain per decade. The observed annual trend backprojects to Moore's Law originally describing transistor packing per integrated circuit, with a typical doubling time of 1.5 years. Over the period (1974-2014), the transistor packing rose from five thousand to five billion elements per circuit – a gain factor of one million!

The late Hans Meuer of the University of Mannheim developed a standardized way of assessing the speed of supercomputers more specifically for scientific and engineering applications. On an annual basis (www.top500.org), high performance computers are ranked in terms of their FLOP rate (FLOP/s – floating point operations per second) while solving a system of linear equations (i.e. Linpack benchmark, Rmax parameter). The term *floating point* refers to arithmetic operations on real numbers with fractional values. In the top-500 competition, these numbers are computed with double-precision with approximately 15 significant digits – ample accuracy for radiotherapy applications! Supercomputers broke through the PFLOP/s barrier in 2008 where P denotes peta or 10^{15}. The fastest machine as of November 2017 is the Sunway TaihuLight installed at the National Supercomputing centre in Wuxi, China; it boasts a benchmark performance of almost 100 PFLOP/s. However, supercomputers costs several hundreds of million dollars (USD) and consume significant electrical power. More down-to-earth examples highlight gains in computational power of affordable consumer products (https://pages.experts-exchange.com/processing-power-compared/). Early Nintendo game consoles (NES, circa 1983) had a clocking speed comparable to the Apollo guidance computer that landed humans on the moon for the first time (1969). An Apple IPhone 4 (2010) has the calculation rate of a Cray-2 supercomputer of 1985, i.e. 1.6 GFLOP/s where G denotes giga or 10^9. An Apple watch doubles this rate to 3 GFLOP/s. An updated Nintendo Wii console yields 12 GFLOP/s, matched by a Sony Smartwatch 3. We will further explore the specific consequences of such advances on dose computation in Chapter 7.

1.4 SUMMARY AND SCOPE OF BOOK

Dose calculations applied to radiation oncology have evolved to become the indispensable critical element in treatment planning systems. While this chapter has concentrated on megavoltage x-rays used in external beam radiation therapy, many of the concepts are reflected in other types of radiation therapy with heavy particles or brachytherapy. Inaccurate dose distributions can misrepresent the therapeutic window and risk possible optimal outcomes in individual patients. An over-prediction of TCP values creates a false positive expectation, potentially blocking an opportunity for dose intensification. Conversely, an artificially low value of NTCP may lead

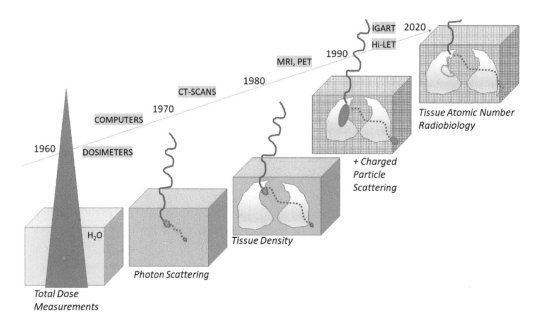

Figure 1.21: Progress in dose calculation algorithms, presented in a chronological order, along with major technological achievements. IGART = Image guided adaptive radiotherapy; Hi-LET = heavy particle beam accelerators.

to toxic dose escalation. Speed enhancements due to rapid advances in computer technology will enable greater accuracy and make Monte Carlo simulation or Boltzmann solutions practical. The scope of applications of dose algorithms will expand to include computations of adaptive and end-of-treatment dose distributions (Park et al. 2016). Vendors that implement platform-independent codes will be able to ride the technological wave of faster processors. This will bring long-awaited products to market featuring dose tracking and instant re-optimization during image-guided therapy (Raaymakers et al. 2017).The present practice of using a faster algorithm during dose optimization cycles and a slower more accurate algorithm for the final results should be gradually abandoned. The initial optimization obtained with an inferior algorithms can, in fact, yield a sub-optimal solution (Zhou et al. 2017). The long term strategy should be aimed at unification – using a *single* high-performance algorithm for consistent treatment planning, optimization, and dose reconstruction.

Figure 1.21 presents an overview of progress achieved in dose calculation algorithms and associated techniques. The notable gain in algorithms has been the distinct modelling of scattered photons and charged particles in heterogeneous tissue. Assumptions about the atomic compositions will gradually fade to further improve the accuracy at interfaces of different tissue with implanted foreign materials and smart devices. The eventual fusion of physical and functional-molecular imaging during treatment planning will lead to comprehensive biophysical optimization.

The purpose of this book is to prepare aspiring medical physicists and allied radiation specialists for the expected wave of new dose calculation algorithm developments. Chapter 1 introduced basic principles of radiation oncology from the perspective of the basic sciences of biology and physics, with emphasis on the need for accurate and consistent dosimetry. The advances made in cost-effective computing power were also reviewed as they will lead to more comprehensive computational models of radiation transport with fewer assumptions. Calculation methods that were dismissed as too slow or costly just a few decades ago are now being implemented. Chapter 2 will present fundamental principles of radiation interactions of megavoltage x-rays with matter. Chapter 3 will present a historical backdrop for algorithmic developments and introduce the key concepts that underpin contemporary algorithms including convolution-superposition, Monte Carlo simulation, and Boltzmann solution of radiation transport equations. Chapters 4, 5, and 6 constitute the core of this book, providing details of these more advanced algorithms. These three chapters need not be studied in their order of presentation. The concluding Chapter 7 provides a forecast of anticipated future developments.

BIBLIOGRAPHY

Ahnesjö, A., P. Andreo, and A. Brahme (1987). Calculation and application of point spread functions for treatment planning with high energy photon beams. *Acta Oncologica 26*(1), 49–56.

Allen, B., E. Bezak, and L. Marcu (2013). Quo vadis radiotherapy? Technological advances and the rising problems in cancer management. *BioMed Research International ID2013*, 1–9.

Andreo, P. (1991). Monte Carlo techniques in medical radiation physics. *Physics in Medicine and Biology 36*(7), 861–920.

Andreo, P., D. Burns, A. Nahum, J. Seuntjens, and F. Attix (2017). Macroscopic aspects of the transport of radiation through matter. In P. Andreo, D. Burns, A. Nahum, J. Seuntjens, and F. Attix (Eds.), *Fundamentals of Ionizing Radiation Dosimetry*, Chapter 6, pp. 279–313. Weinheim, Germany: Wiley-VCH.

Babic, S., J. Battista, and K. Jordan (2008). Three-dimensional dose verification for intensity-modulated radiation therapy in the radiological physics centre head-and-neck phantom using optical computed tomography scans of ferrous xylenol-orange gel dosimeters. *International Journal of Radiation Oncology, Biology, and Physics 70*(4), 1281–1291.

Bakai, A., M. Alber, and F. Nüsslin (2003). A revision of the gamma-evaluation concept for the comparison of dose distributions. *Physics in Medicine and Biology 48*(21), 3543–3553.

Battista, J., W. Rider, and J. Van Dyk (1980). Computed tomography for radiotherapy planning. *International Journal of Radiation Oncology, Biology, and Physics 6*(1), 99–107.

Baumann, M., M. Krause, J. Overgaard, J. Debus, S. Bentzen, J. Daartz, C. Richter, D. Zips, and T. Bortfeld (2016). Radiation oncology in the era of precision medicine. *Nature Reviews: Cancer 16*(4), 234.

Bentzen, S. (2004). High-tech in radiation oncology: should there be a ceiling? *International Journal of Radiation Oncology, Biology, and Physics 58*(2), 320–330.

Bentzen, S., J. Bernier, J. Davis, J. Horiot, G. Garavaglia, J. Chavaudra, K. Johansson, and M. Bolla (2000). Clinical impact of dosimetry quality assurance programmes assessed by radiobiological modelling of data from the thermoluminescent dosimetry study of the European Organization for Research and Treatment of Cancer. *European Journal of Cancer 36*(5), 615–620.

Bentzen, S., L. Constine, J. Deasy, A. Eisbruch, A. Jackson, L. Marks, R. Ten Haken, and E. Yorke (2010). Quantitative analyses of normal tissue effects in the clinic (QUANTEC): An introduction to the scientific issues. *International Journal of Radiation Oncology, Biology, and Physics 76*(Supplement 3), 3–9.

Bentzen, S. and M. Joiner (2009). The linear-quadratic approach in clinical practice. In M. Joiner and A. Vander Kogel (Eds.), *Basic Clinical Radiobiology* (4th ed.)., Chapter 4, pp. 120–134. London, England: Hodder Arnold.

Bibault, J., X. Mirabel, T. Lacornerie, E. Tresch, N. Reynaert, and E. Lartigau (2015). Adapted prescription dose for Monte Carlo algorithm in lung SBRT: Clinical outcome on 205 patients. *PLOS ONE 10*(7), 1–10.

Boda-Heggemann, J., A. Knopf, A. Simeonova-Chergou, H. Wertz, F. Stieler, A. Jahnke, L. Jahnke, J. Fleckenstein, L. Vogel, and A. Arns (2016). Deep inspiration breath hold based radiation therapy: a clinical review. *International Journal of Radiation Oncology, Biology, and Physics 94*(3), 478–492.

Bortfeld, T., J. Bürkelbach, R. Boesecke, and W. Schlegel (1990). Methods of image reconstruction from projections applied to conformation radiotherapy. *Physics in Medicine and Biology 35*(10), 1423–1434.

Brahme, A. (1984). Dosimetric precision requirements in radiation therapy. *Acta Radiologica: Oncology 23*(5), 379–391.

Bufacchi, A., B. Nardiello, R. Capparella, and L. Begnozzi (2013). Clinical implications in the use of the PBC algorithm versus the AAA by comparison of different NTCP models/parameters. *Radiation Oncology 8*(1), 1–13.

Busch, M. (1968). Approximation and variation of human body surface for 3-dimensional dose calculation in teletherapy. In J. Cunningham (Ed.), *Role of Computers in Radiotherapy: Report of a Panel held in Vienna, 1-14 July 1967,* Chapter 8, pp. 77–92. Vienna: International Atomic Energy Agency.

Caivano, R., G. Califano, A. Fiorentino, M. Cozzolino, C. Oliviero, P. Pedicini, S. Clemente, C. Chiumento, and V. Fusco (2014). Clinically relevant quality assurance for intensity modulated radiotherapy plans: gamma maps and DVH-based evaluation. *Cancer Investigation 32*(3), 85–91.

Canadian Cancer Society (2018). Canadian Cancer Statistics: A 2018 Special Report on Cancer Incidence by Stage. Report ISSN 0835-2976, Canadian Cancer Society, Toronto, Canada.

Chen, W. (2014). Impact of dose calculation algorithm on radiation therapy. *World Journal of Radiology 6*(11), 874.

Chetty, I., H. Li, N. Wen, and S. Kumar (2013). Correlation of dose computed using different algorithms with local control following stereotactic ablative radiotherapy (SABR)-based treatment of non-small-cell lung cancer. *Radiotherapy and Oncology 109*, 498–504.

Chiarella, P., J. Bruzzo, R. Meiss, and R. Ruggiero (2012). Concomitant tumor resistance. *Cancer Letters 324*(2), 133–141.

Cozzolino, M., C. Oliviero, G. Califano, S. Clemente, P. Pedicini, R. Caivano, C. Chiumento, A. Fiorentino, and V. Fusco (2014). Clinically relevant quality assurance (QA) for prostate RapidArc plans: gamma maps and DVH-based evaluation. *Physica Medica 30*(4), 462–472.

Cunningham, J. (1989). Keynote address: development of computer algorithms for radiation treatment planning. *International Journal of Radiation Oncology, Biology, and Physics 16*(6), 1367–1376.

Demicheli, R., M. Retsky, W. Hrushesky, M. Baum, and I. Gukas (2008). The effects of surgery on tumor growth: a century of investigations. *Annals of oncology 19*(11), 1821–1828.

Disher, B., G. Hajdok, S. Gaede, M. Mulligan, and J. Battista (2013). Forcing lateral electron disequilibrium to spare lung tissue: A novel technique for stereotactic body radiation therapy of lung cancer. *Physics in Medicine and Biology 58*(19), 6641–6662.

Dutreix, A. (1984). When and how can we improve precision in radiotherapy? *Radiotherapy and Oncology 2*(4), 275–292.

Fairchild, A., W. Straube, F. Laurie, and D. Followill (2013). Does quality of radiation therapy predict outcomes of multicenter cooperative group trials? A literature review. *International Journal of Radiation Oncology, Biology, and Physics 87*(2), 246–260.

Fowler, J. (2010). Twenty-one years of biologically effective dose. *British Journal of Radiology 83*(991), 554–568.

Fowler, J., A. Dasu, and I. Toma-Dasu (2015). *Optimum Overall Treatment Time in Radiation Oncology*. Madison, Wisconsin: Medical Physics Publishing.

Gahbauer, R., T. Landberg, J. Chavaudra, J. Dobbs, N. Gupta, G. Hanks, J. Horiot, K. Johansson, T. Möller, S. Naudy, J. Purdy, I. Santenac, N. Suntharalingam, and H. Svensson (2004). Prescribing, recording, and reporting electron beam therapy. *Journal of the International Commission on Radiation Units and Measurements 4*(1), 2.

Galmarini, C., O. Tredan, and F. Galmarini (2014). Concomitant resistance and early-breast cancer: should we change treatment strategies? *Cancer and Metastasis Reviews 33*(1), 271–283.

Gershkevitsh, E., C. Pesznyak, B. Petrovic, J. Grezdo, K. Chelminski, M. Do Carmo Lopes, J. Izewska, and J. Van Dyk (2014). Dosimetric inter-institutional comparison in European radiotherapy centres: Results of IAEA supported treatment planning system audit. *Acta Oncologica 53*(5), 628–636.

Göttgens, E., C. Ostheimer, P. Span, J. Bussink, and E. Hammond (2018). HPV, hypoxia and radiation response in head and neck cancer. *The British Journal of Radiology 91*, 20180047–20180047.

Gregoire, V. and T. Mackie (2011). Dose Prescription, Reporting and Recording in Intensity-Modulated Radiation Therapy: A Digest of the ICRU Report 83. *Imaging in Medicine 3*(3), 367–373.

Guckenberger, M., R. Klement, M. Allgäuer, N. Andratschke, O. Blanck, J. Boda-Heggemann, K. Dieckmann, M. Duma, I. Ernst, U. Ganswindt, P. Hass, C. Henkenberens, R. Holy, D. Imhoff, H. Kahl, R. Krempien, F. Lohaus, U. Nestle, M. Nevinny-Stickel, C. Petersen, S. Semrau, J. Streblow, T. Wendt, A. Wittig, M. Flentje, and F. Sterzing (2016). Local tumor control probability modeling of primary and secondary lung tumors in stereotactic body radiotherapy. *Radiotherapy and Oncology 118*(3), 485–491.

Hall, E. (2006). *Radiobiology for the Radiologist* (6th ed.). Philadelphia: J.B. Lippincott Company.

Hallock, A., G. Bauman, N. Read, D. D'Souza, F. Perera, I. Aivas, L. Best, J. Cao, A. Louie, E. Wiebe, T. Sexton, S. Gaede, J. Battista, and G. Rodrigues (2012). Assessment and improvement of radiation oncology trainee contouring ability utilizing consensus-based penalty metrics. *Journal of Medical Imaging and Radiation Oncology 56*(6), 679–688.

Han, T., J. Mikell, M. Salehpour, and F. Mourtada (2011). Dosimetric comparison of Acuros XB deterministic radiation transport method with Monte Carlo and model based convolution methods in heterogeneous media. *Medical Physics 38*(5), 2651–2664.

Harms, W., D. Low, J. Wong, and J. Purdy (1998). A software tool for the quantitative evaluation of 3D dose calculation algorithms. *Medical Physics 25*(10), 1830–1836.

Hoopes, D., P. Johnstone, P. Chapin, C. Kabban, W. Lee, A. Chen, B. Fraass, W. Skinner, and L. Marks (2015). Practice patterns for peer review in radiation oncology. *Practical Radiation Oncology 5*(1), 32–38.

IAEA (1966). Computer Calculation of Dose Distributions in Radiotherapy. Report TRS No.57, International Atomic Energy Agency, Vienna.

IAEA (1968). The Role of Computers in Radiotherapy: Report of a Panel held in Vienna 1-14 July 1967. Report STI/PUB/203, International Atomic Energy Agency, Vienna.

IAEA (2016). Accuracy Requirements and Uncertainties in Radiotherapy: A Report of the International Atomic Energy Agency. Report Human Health Series No. 31, International Atomic Energy Agency, Vienna.

Ibbott, G., A. Haworth, and D. Followill (2013). Quality Assurance for Clinical Trials. *Frontiers in Oncology 3*, 311.

ICRU (1976). Determination of Absorbed Dose in a Patient Irradiated by Beams of X or Gamma Rays in Radiotherapy. Report 24, International Commission on Radiation Units and Measurements, Bethesda, Maryland.

ICRU (2010). Prescribing, Recording and Reporting Photon-Beam Intensity-Modulated Radiation Therapy. Report 83, International Commission on Radiation Units and Measurements, Bethesda, Maryland.

Jaffray, D., S. Das, P. Jacobs, R. Jeraj, and P. Lambin (2018). How advances in imaging will affect precision radiation oncology. *International Journal of Radiation Oncology, Biology, and Physics 101*(2), 292–298.

Jameson, M., L. Holloway, P. Vial, S. Vinod, and P. Metcalfe (2010). A review of methods of analysis in contouring studies for radiation oncology. *Journal of Medical Imaging and Radiation Oncology 54*(5), 401–410.

Joiner, M. and A. Van der Kogel (2016). *Basic Clinical Radiobiology*, Volume 2. CRC Press, Taylor & Francis Group.

Katsuta, Y., N. Kadoya, Y. Fujita, E. Shimizu, K. Matsunaga, K. Sawada, H. Matsushita, K. Majima, and K. Jingu (2017). Patient-Specific Quality Assurance Using Monte Carlo Dose Calculation and Elekta Log Files for Prostate Volumetric-Modulated Arc Therapy. *Technology in Cancer Research and Treatment 16*(6), 1220–1225.

Keall, P., G. Mageras, J. Balter, R. Emery, K. Forster, S. Jiang, J. Kapatoes, D. Low, M. Murphy, and B. Murray (2006). The Management of Respiratory Motion in Radiation Oncology: Report of AAPM Task Group 76. *Medical Physics 33*(10), 3874–3900.

Killoran, J., H. Kooy, D. Gladstone, F. Welte, and C. Beard (1997). A numerical simulation of organ motion and daily setup uncertainties: implications for radiation therapy. *International Journal of Radiation Oncology, Biology, and Physics 37*(1), 213–221.

Kry, S., A. Molineu, J. Kerns, A. Faught, J. Huang, K. Pulliam, J. Tonigan, P. Alvarez, F. Stingo, and D. Followill (2014). Institutional patient-specific IMRT QA does not predict unacceptable plan delivery. *International Journal of Radiation Oncology, Biology, and Physics 90*(5), 1195–1201.

Lagendijk, J., B. Raaymakers, C. Van Den Berg, M. Moerland, M. Philippens, and M. Van Vulpen (2014). MR guidance in radiotherapy. *Physics in Medicine and Biology 59*(21), R349–R369.

Leibel, S. and Z. Fuks (1993). The impact of local tumor control on the outcome in human cancer. In H. Beck-Bornholdt (Ed.), *Current Topics in Clinical Radiobiology of Tumors*, Chapter 10, pp. 113–127. Berlin: Springer-Verlag.

Létourneau, D., A. McNiven, and D. Jaffray (2013). Multicenter collaborative quality assurance program for the province of Ontario, Canada: First-year results. *International Journal of Radiation Oncology, Biology, and Physics 86*(1), 164–169.

Liang, X., J. Penagaricano, D. Zheng, S. Morrill, X. Zhang, P. Corry, R. Griffin, E. Han, M. Hardee, and V. Ratanatharathom (2016). Radiobiological impact of dose calculation algorithms on biologically-optimized IMRT lung stereotactic body radiation therapy plans. *Radiation Oncology 11*(1), 10.

Lindblom, E., A. Dasu, and I. Toma-Dasu (2015). Optimal fractionation in radiotherapy for non-small cell lung cancer: A modelling approach. *Acta Oncologica 54*(9), 1592–1598.

Louie, A., G. Rodrigues, J. Olsthoorn, D. Palma, E. Yu, B. Yaremko, B. Ahmad, I. Aivas, and S. Gaede (2010). Inter-observer and intra-observer reliability for lung cancer target volume delineation in the 4D-CT era. *Radiotherapy and Oncology 95*(2), 166–171.

Low, D. (2010). Gamma Dose Distribution Evaluation Tool. *Journal of Physics: Conference Series 250*(1), 12071.

MacDermott, P. (2016). Deterministic radiation transport: a rival to monte carlo methods. In *Tutorials in Radiotherapy Physics* (First ed.)., Chapter 4, pp. 170–216. Boca Raton, Florida: CRC Press, Taylor & Francis Group.

MacDonald, J. (1965). A New Method for Obtaining Multi-Field and Rotational Dose Distributions in Cobalt-60 Teletherapy. *Radiology 85*(4), 716–724.

Mackie, T., A. Bielajew, D. Rogers, and J. Battista (1988). Generation of photon energy deposition kernels using the EGS Monte Carlo code. *Physics in Medicine and Biology 33*(1), 1–20.

Mackie, T., J. Scrimger, and J. Battista (1984). A convolution method of calculating dose for 15 MV x rays. *Medical Physics 12*(2), 188–96.

Miften, M., A. Olch, D. Mihailidis, J. Moran, T. Pawlicki, A. Molineu, H. Li, K. Wijesooriya, J. Shi, P. Xia, N. Papanikolaou, and D. Low (2018). Tolerance limits and methodologies for IMRT measurement-based verification QA: Recommendations of AAPM Task Group 218. *Medical Physics 45*(4), e53–e83.

Mijnheer, B. (2018). *Clinical 3D Dosimetry in Modern Radiation Therapy*. Boca Raton, Florida: CRC Press, Taylor & Francis Group.

Mijnheer, B., J. Battermann, and A. Wambersie (1987). What degree of accuracy is required and can be achieved in photon and neutron therapy? *Radiotherapy and Oncology 8*(3), 237–252.

Mijnheer, B., S. Beddar, J. Izewska, and C. Reft (2013). In vivo dosimetry in external beam radiotherapy. *Medical Physics 40*(7), 1–19.

Mijnheer, B., P. González, I. Olaciregui-Ruiz, R. Rozendaal, M. van Herk, and A. Mans (2015). Overview of 3-year experience with large scale electronic portal imaging device based three dimensional transit dosimetry. *Practical Radiation Oncology 5*(6), e679–e687.

Mijnheer, B., R. Rozendaal, I. Olaciregui Ruiz, P. González, R. Van Oers, and A. Mans (2017). New developments in EPID-based 3D dosimetry in the Netherlands Cancer Institute. *Journal of Physics: Conference Series 847*(1), 012033.

Moiseenko, V., J. Deasy, and J. Van Dyk (2005). Radiobiological Modelling for Treatment Planning. In *The Modern Technology of Radiation Oncology, Vol.2*, Chapter 5, pp. 185–220. Madison, WI: Medical Physics Publishing Corporation.

Molineu, A., N. Hernandez, T. Nguyen, G. Ibbott, and D. Followill (2013). Credentialing results from IMRT irradiations of an anthropomorphic head and neck phantom. *Medical Physics 40*(2), 0221011–0221018.

Moodie, T., J. Sykes, and R. Gajewski (2014). A revision of the γ-evaluation concept for the comparison of dose distributions. *Physics in Medicine and Biology 59*(23), 7557.

Moravec, H. (1998). When will computer hardware match the human brain? *Journal of Evolution and Technology 1*(1), 10.

Nelms, B., M. Chan, G. Jarry, M. Lemire, J. Lowden, C. Hampton, and V. Feygelman (2013). Evaluating IMRT and VMAT dose accuracy: practical examples of failure to detect systematic errors when applying a commonly used metric and action levels. *Medical Physics 40*(11), 111722.

Nelms, B., H. Zhen, and W. Tomé (2011). Perbeam, planar IMRT QA passing rates do not predict clinically relevant patient dose errors. *Medical Physics 38*(2), 1037–1044.

Ohri, N., X. Shen, A. Dicker, L. Doyle, A. Harrison, and T. Showalter (2013). Radiotherapy protocol deviations and clinical outcomes: A meta-analysis of cooperative group clinical trials. *Journal of the National Cancer Institute 105*(6), 387–393.

Olch, A., L. Gerig, H. Li, I. Mihaylov, and A. Morgan (2014). Dosimetric effects caused by couch tops and immobilization devices: Report of AAPM Task Group 176. *Medical Physics 41*(6), 061501–30.

Onishi, H. (2015). Respiratory motion management. In Y. Nagata (Ed.), *Stereotactic Body Radiation Therapy: Principles and Practices*, Chapter 7, pp. 91–102. Tokyo, Japan: Springer.

Orton, C., S. Chungbin, E. Klein, M. Gillin, T. Schultheiss, and W. Sause (1998). Study of lung density corrections in a clinical trial (RTOG 88-08). *International Journal of Radiation Oncology, Biology, and Physics 41*(4), 787–794.

Orton, C., P. Mondalek, J. Spicka, D. Herron, and L. Andres (1984). Lung corrections in photon beam treatment planning: Are we ready? *International Journal of Radiation Oncology, Biology, and Physics 10*(12), 2191–2199.

Park, S., T. McNutt, W. Plishker, H. Quon, J. Wong, R. Shekhar, and J. Lee (2016). SCUDA: A software platform for cumulative dose assessment. *Medical Physics 43*(10), 5339–5346.

Peters, L., B. O'Sullivan, J. Giralt, T. Fitzgerald, A. Trotti, J. Bernier, J. Bourhis, K. Yuen, R. Fisher, and D. Rischin (2010). Critical impact of radiotherapy protocol compliance and quality in the treatment of advanced head and neck cancer: Results from RTOG 0202. *Journal of Clinical Oncology 28*(18), 2996–3001.

Pettersen, M., E. Aird, and D. Olsen (2008). Quality assurance of dosimetry and the impact on sample size in randomized clinical trials. *Radiotherapy and Oncology 86*(2), 195–199.

Pulliam, K., J. Huang, R. Howell, D. Followill, R. Bosca, J. O'Daniel, and S. Kry (2014). Comparison of 2D and 3D gamma analyses. *Medical Physics 41*(2), 021710.

Raaymakers, B., I. Jürgenliemk-Schulz, G. Bol, M. Glitzner, A. Kotte, B. van Asselen, J. de Boer, J. Bluemink, S. Hackett, and M. Moerland (2017). First patients treated with a 1.5 T MRI-Linac: clinical proof of concept of a high-precision, high-field MRI guided radiotherapy treatment. *Physics in Medicine and Biology 62*(23), L41.

Riegel, A., Y. Chen, A. Kapur, L. Apicello, A. Kuruvilla, A. Rea, A. Jamshidi, and L. Potters (2017). In vivo dosimetry with optically stimulated luminescent dosimeters for conformal and intensity-modulated radiation therapy: A 2-year multicenter cohort study. *Practical Radiation Oncology 7*(2), e135–e144.

Rinks, M. (2012). Computerised treatment planning systems: where did it all begin? *Journal of Medical Radiation Sciences 59*(1), 1–4.

Rodrigues, G., A. Louie, G. Videtic, L. Best, N. Patil, A. Hallock, S. Gaede, J. Kempe, J. Battista, P. De Haan, and G. Bauman (2012). Categorizing segmentation quality using a quantitative quality assurance algorithm. *Journal of Medical Imaging and Radiation Oncology 56*(6), 668–679.

Rogers, D., B. Faddegon, G. Ding, C. Ma, J. We, and T. Mackie (1995). BEAM: a Monte Carlo code to simulate radiotherapy treatment units. *Medical Physics 22*(5), 503–524.

Rogers, D., B. Walters, and I. Kawrakow (2009). BEAMnrc User's Manual. Manual NRC509, National Research Council of Canada, Ottawa, Canada.

Schmidt-Ullrich, R. (2000). Local tumor control and survival: clinical evidence and tumor biologic basis. *Surgical Oncology Clinics of North America 9*(3), 401–414.

Schreiner, L., O. Holmes, and G. Salomons (2013). Analysis and evaluation of planned and delivered dose distributions: Practical concerns with γ- and χ- evaluations. *Journal of Physics: Conference Series 444*(ID012016), 1–9.

Seuntjens, J., E. Lartigau, S. Cora, G. Ding, S. Goetsch, and J. Nuyttens (2014). Prescribing, Recording, and Reporting of Stereotactic Treatments with Small Photon Beams. *Journal of the ICRU 14*, 1–160.

Shalev, S., D. Viggars, M. Carey, and M. Stewart (1988). Treatment planning using images of regret. *Medical Dosimetry 13*(2), 57–61.

Smilowitz, J., I. Das, V. Feygelman, B. Fraass, S. Kry, I. Marshall, D. Mihailidis, Z. Ouhib, T. Ritter, M. Snyder, and L. Fairobent (2015). AAPM Medical Physics Practice Guideline 5.a.: Commissioning and QA of Treatment Planning Dose Calculations - Megavoltage Photon and Electron Beams. *Journal of Applied Clinical Medical Physics 16*(5), 14–34.

Smith, I., J. Fetterly, J. Lott, J. MacDonald, L. Myers, P. Pfalzner, and D. Thomson (1964). *Cobalt 60 teletherapy*. Harper & Row.

Sontag, M. and J. Cunningham (1978). The equivalent tissue-air ratio method for making absorbed dose calculations in a heterogeneous medium. *Radiology 129*(3), 787–794.

Stasi, M., S. Bresciani, A. Miranti, A. Maggio, V. Sapino, and P. Gabriele (2012). Pretreatment patient-specific IMRT quality assurance: A correlation study. *Medical Physics 39*(12), 7626–7634.

Steers, J. and B. Fraass (2016). IMRT QA: Selecting gamma criteria based on error detection sensitivity. *Medical Physics 43*(4), 1982–1994.

Suit, H. (1992). Local control and patient survival. *International Journal of Radiation Oncology, Biology, and Physics 23*(3), 653–660.

Sumida, I., H. Yamaguchi, H. Kizaki, K. Aboshi, M. Tsujii, N. Yoshikawa, Y. Yamada, O. Suzuki, Y. Seo, and F. Isohashi (2015). Novel radiobiological gamma index for evaluation of 3-dimensional predicted dose distribution. *International Journal of Radiation Oncology, Biology, and Physics 92*(4), 779–786.

Taha, A. and A. Hanbury (2015). Metrics for evaluating 3D medical image segmentation: analysis, selection, and tool. *BMC Medical Imaging 15*(1), 29.

Thwaites, D. (2013). Accuracy required and achievable in radiotherapy dosimetry: have modern technology and techniques changed our views? *Journal of Physics: Conference Series 444*(1), 012006.

Tommasino, F., A. Nahum, and L. Cella (2017). Increasing the power of tumour control and normal tissue complication probability modelling in radiotherapy: recent trends and current issues. *Translational Cancer Research 6*(S5), S807–S821.

Tsien, K. (1955). The application of automatic computing machines to radiation treatment planning. *The British Journal of Radiology 28*(332), 432–439.

Tubiana, M. and F. Eschwege (2000). Conformal radiotherapy and intensity-modulated radiotherapy: Clinical data. *Acta Oncologica 39*(5), 555–567.

Van der Merwe, D., J. Van Dyk, B. Healy, E. Zubizarreta, J. Izewska, B. Mijnheer, and A. Meghzifene (2017). Accuracy requirements and uncertainties in radiotherapy: a report of the International Atomic Energy Agency. *Acta Oncologica 56*(1), 1–6.

Van Dyk, J. and J. Battista (2014). Has the use of computers in radiation therapy improved the accuracy in radiation dose delivery? *Journal of Physics: Conference Series 489*(1), 012098.

Van Dyk, J., J. Battista, and G. Bauman (2013). Accuracy and uncertainty considerations in modern radiation oncology. In J. Van Dyk (Ed.), *The Modern Technology of Radiation Oncology, Vol.3* (First ed.)., Chapter 11, pp. 361–405. Madison, Wisconsin: Medical Physics Publishing.

Van Dyk, J., J. Battista, and G. Bauman (2018). Accuracy requirements for 3d dosimetry in contemporary radiation therapy. In B. Mijnheer (Ed.), *Clinical 3D Dosimetry in Modern Radiation Therapy*, Chapter 2, pp. 15–61. Boca Raton, Florida: CRC press, Taylor & Francis Group.

Vinod, S., M. Min, M. Jameson, and L. Holloway (2016). A review of interventions to reduce inter-observer variability in volume delineation in radiation oncology. *Journal of Medical Imaging and Radiation Oncology 60*(3), 393–406.

Vogelius, I. and S. Bentzen (2013). Meta-analysis of the alpha/beta ratio for prostate cancer in the presence of an overall time factor: bad news, good news, or no news? *International Journal of Radiation Oncology, Biology, and Physics 85*(1), 89–94.

Waghorn, B., S. Meeks, and K. Langen (2011). Analyzing the impact of intrafraction motion: Correlation of different dose metrics with changes in target D95%. *Medical Physics 38*(8), 4505–4511.

Wambersie, A., J. Dutreix, and A. Dutreix (1969). Dosimetric precision required in radiotherapy. Consequences of the choice and performance required of detectors. *Journal Belge de Radiologie 52*(2), 94.

Weber, D., V. Vallet, A. Molineu, C. Melidis, V. Teglas, S. Naudy, R. Moeckli, D. Followill, and C. Hurkmans (2014). IMRT credentialing for prospective trials using institutional virtual phantoms: Results of a joint European Organization for the Research and Treatment of Cancer and Radiological Physics Center project. *Radiation Oncology 9*(1), 1–8.

Wong, E., J. Van Dyk, J. Battista, R. Barnett, and P. Munro (1997). Uncertainty analysis: A guide to optimization in radiation treatment planning. *Radiotherapy and Oncology 48*(1), S151.

Wong, J., E. Slessinger, F. Rosenberger, K. Krippner, and J. Purdy (1984). The delta-volume method for 3-dimensional photon dose calculations. In *Proceedings of the 8th International Conference on the Use of Computers in Radiation Therapy*, pp. 26–30. Computer Society Press, IEEE, Silver Spring, Maryland.

Wong, O., A. McNiven, B. Chan, J. Moseley, J. Lee, L. Le, C. Ren, J. Waldron, J. Bissonnette, and M. Giuliani (2018). Evaluation of differences between estimated delivered dose and planned dose in nasopharynx patients using deformable image registration and dose accumulation. *Journal of Medical Imaging and Radiation Sciences 49*(1), S2–S3.

Xue, J., J. Ohrt, J. Fan, P. Balter, J. Park, L. Kim, S. Kirsner, and G. Ibbott (2017). Validation of treatment planning dose calculations: Experience working with medical physics practice guideline 5.a. *International Journal of Medical Physics, Clinical Engineering and Radiation Oncology 06*(01), 57–72.

Yamoah, K., T. Showalter, and N. Ohri (2015). Radiation therapy intensification for solid tumors: A systematic review of randomized trials. *International Journal of Radiation Oncology, Biology, and Physics 93*(4), 737–745.

Yu, C., T. Mackie, and J. Wong (1995). Photon dose calculation incorporating explicit electron transport. *Medical Physics 22*(7), 1157–1165.

Zelefsky, M., V. Reuter, Z. Fuks, P. Scardino, and A. Shippy (2008). Influence of local tumor control on distant metastases and cancer related mortality after external beam radiotherapy for prostate cancer. *The Journal of Urology 179*(4), 1368–1373.

Zhen, H., B. Nelms, and W. Tomé (2011). Moving from gamma passing rates to patient DVH-based QA metrics in pretreatment dose QA. *Medical Physics 38*(10), 5477–5489.

Zhou, C., N. Bennion, R. Ma, X. Liang, S. Wang, K. Zvolanek, M. Hyun, X. Li, S. Zhou, W. Zhen, C. Lin, A. Wahl, and D. Zheng (2017). A comprehensive dosimetric study on switching from a Type-B to a Type-C dose algorithm for modern lung SBRT. *Radiation Oncology 12*(1), 80.

CHAPTER 2

X-Ray Interactions and Energy Deposition

Jerry J. Battista and George Hajdok

London Health Sciences Centre and University of Western Ontario

CONTENTS

2.1 INTRODUCTION

IN this chapter, we review fundamental elements of radiological physics relevant to the understanding of megavoltage x-ray dose calculations. We describe need-to-know concepts that influence the strengths and limitations of dose calculation algorithms when applied to contemporary radiation oncology.

The learning objective of this chapter is to understand how x-rays interact in human tissue and ultimately deposit dose through secondary particles. Photons carry energy and interact sporadically in the human body until they are either fully absorbed or escape the patient. Every point in the patient is therefore a potential site of interaction between an x-ray photon and a host atom constituting the tissue. For a therapeutic beam, a large number of coupled interactions occur and deposit energy packets throughout the tissues, leading to atomic excitation, ionization, and radiobiological damage to DNA. Figure 2.1 illustrates the energy dissipation of a *single* primary photon through a shower of secondary particles. Detailed histories of individual particles are modelled through Monte Carlo simulations addressed in Chapter 5. A cascade of energy results in mutual exchanges of energy between photons and charged particles, predominantly atomic electrons. The various particle entities in Figure 2.1 will be explained later in this chapter. For the moment, the key point to remember is that while megavoltage photons teleport energy across relatively large distances in the human body, charged particles launched by these photons have a much more limited range. They are directly responsible for the local deposition of energy in tissue volume elements (i.e. voxels). Because the number of interactions is random and enormous, most algorithms aim to predict the *expectation value* of the energy deposited in a voxel.

A dose calculation algorithm must model the energy transferred from x-rays that is rapidly disseminated to a three-dimensional matrix of surrounding tissue voxels. X-ray interactions occur randomly with atomic targets on the order of 10^{-27} cm^2 in effective area and each occurs very swiftly. Transit times through a water molecule (0.3 nm) are on the order of $< 10^{-18}$ seconds. The interaction probabilities are governed by quantum electrodynamics and relativistic physics (Evans 1955; Andreo et al. 2017; Podgorsak 2016; ICRU 2011). However, for our purposes, we adopt a simpler particle collision model, acknowledging that some underlying details and intermediate steps of the interaction process may be omitted. The condensed summary presented in this chapter is intended to be a quick-start learning guide on the subject of radiological physics.

2.2 PHOTON INTERACTIONS

X-rays travel undetected through a medium until they interact and release charged particles that can either excite or ionize atoms. The three most important modes of interaction in the megavoltage energy range are Compton scattering, photoelectric effect, and pair production. The processes are sketched in Figure 2.2. The reader is encouraged to use an interactive software tool (RadSim) that shows animations of ricocheting particles involved in these mechanisms and logs the kinematic data, including scattering angles and energy losses (Verhaegen et al. 2006).

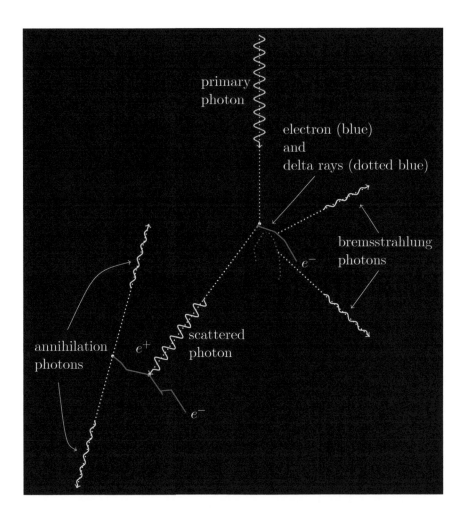

Figure 2.1: Evolution of the history of a single primary megavoltage photon (top). The family tree includes bidirectional energy exchanges between photons (yellow) and charged particles. Energy deposits occur along electron (e^- blue) or positron (e^+ red) tracks.

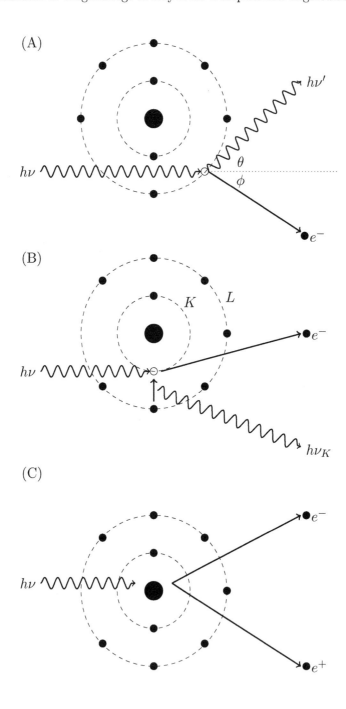

Figure 2.2: Schematic of major photon interactions and emerging particles.
(A) Compton scattering (B) Photoelectric effect with subsequent emission of a K-fluorescence x-ray ($h\nu_K$). (C) Positron-electron pair production near the nucleus. For simplicity, the photon icon does not show changes in wavelength for scattered and secondary photons.

Compton events occur with individual orbital electrons with a binding energy well below the energy of the incoming photon. If the binding energy is negligible, the atomic ownership of the electron becomes irrelevant, and the probability of interaction *per electron* is independent of atomic number (Z). The outcome of a Compton interaction is a scattered photon and a recoil electron sharing the incoming energy, and emerging with correlated directions.

The photoelectric interaction is an x-ray absorption event occurring in the inner orbital space. If the incoming photon energy exceeds the atomic binding energy of an inner electron, this electron is ejected with some minor recoiling of the atom as a whole. For inbound photon energies of only a few MeV, the launch direction of the photoelectrons is tightly forward-peaked. After the atom is ionized by the initial event, the subsequent stabilization leads to spontaneous tertiary emissions of Auger electrons or fluorescent x-rays (shown in Figure 2.2). The probability of photoelectric absorption *per atom* depends strongly on atomic number (Z^n) where $n \approx 4.5$ for elements constituting human tissue ($Z < 15$). The probability of occurrence drops off by orders of magnitude as the photon energy is increased beyond a few hundred keV.

Pair production causes the incoming photon to be fully absorbed in the vicinity of the nucleus. For this event, the minimum or threshold energy for the incoming photon is 1.022 MeV – the combined rest mass energy of the electron-positron created out of this process. Pair production can also occur with orbital electrons instead of the nucleus with a greater threshold of 2.04 MeV, yielding a triplet of emerging particles (2 electrons, 1 positron). While triplet events can involve any one of the Z orbital electrons, these events are less prevalent than nuclear events because of the higher energy threshold and screening of the Coulomb field. In either pair or triplet production, the energy exceeding threshold energy is shared as kinetic energy among the emerging charged particles. The probability of nuclear pair production events per atom is quadratically dependent on atomic number (Z^2). The emerging positron has a very high likelihood of annihilation in a sea of numerous host electrons particularly as the positron slows down. Upon annihilation, a pair of gamma rays will be emitted. If the positron exhausts all of its kinetic energy and recombines with a quasi-stationary electron, the gamma emissions will each inherit characteristic energies of 0.511 MeV and launch in directly opposite directions; otherwise the gamma rays will split a share of the residual kinetic energy and emerge at oblique angles.

Photonuclear interactions, such as (γ, n) or (γ, p) reactions, occur with megavoltage energies exceeding the threshold values for nuclear activation. While photonuclear events may be of concern for radiation protection if radioisotopes are induced in components of a linear accelerator head, they are much rarer in tissue (Podgorsak 2016). Nuclear reactions are usually ignored in dose calculations.

2.2.1 Atomic Target Theory

The occurrence of x-ray interactions is random and the interaction types are mutually exclusive; only one of these interactions can occur per inbound photon. The probabilities of occurrence are expressed in terms of effective target size for each type of interaction in units of *barns* (10^{-24} cm^2). This rather peculiar terminology is traced back to American scientists at Purdue University collaborating on the Manhattan nuclear project during World War II. The term comes from the description of baseball pitchers with limited talent who "couldn't hit the side of a barn door"! The word may have been used to camouflage sensitive nuclear data during the war. The cross-sectional area represents the effective size of the bull's eye presented to an incoming fluence of projectiles (per cm^2). This effective area is *not the actual* cross-sectional area of the target, but is rather an indicator of probability. In the International System (SI) of units (ICRU 2011), a barn equals 100 fm^2.

Imagine a solitary electron that is isolated in a rarefied voxel at a moment in time, as shown in Figure 2.3. For 6 MeV photons, the Compton cross section is approximately 73 milli-barns (i.e. 73×10^{-27} cm^2) per electron target. Now consider a uniform incoming fluence of photons (10^{27} per cm^2). Then, only 73 photons would, on average, be scattered by a solitary electron. In other words, the odds of a single photon being scattered by a single electron is indeed minuscule, only 73 chances in 10^{27} attempts. Consider a typical voxel size of a dose distribution matrix (0.25 cm × 0.25 cm × 0.25 cm). Water-like tissue has an electron density (ρ_e) of 3.34 × 10^{23} electrons per cm^3. The number of interactions occurring in a radiologically thin voxel *via* a specific type of interaction (i) is given by

$$N_{\text{int}} = \Phi \, \sigma_i \, \rho_t \, \mathrm{d}A \, \mathrm{d}x \tag{2.1}$$

where:

$$N_{\text{int}} \equiv \text{number of interactions}$$
$$\Phi \equiv \text{incident photon fluence } [\text{cm}^{-2}]$$
$$\sigma_i \equiv \text{cross-section } [\text{cm}^2 \text{ per target}] \text{ for interactions of type } i$$
$$\rho_t \equiv \text{density of targets in voxel } [\text{targets per cm}^3]$$
$$\mathrm{d}A \equiv \text{area of the voxel face } [\text{cm}^2]$$
$$\mathrm{d}x \equiv \text{thickness of the voxel } [\text{cm}]$$

We assumed negligible attenuation of fluence across the voxel depth ($\mathrm{d}x$) and minimal overlap of targets. Then approximately 0.6% of the incoming stream of photons (i.e. $N_{\text{int}}/\Phi \, \mathrm{d}A$) would be deflected by Compton events. The number and type of interactions depend on the atomic content and density of targets in each voxel. In a body region where the atomic composition of soft tissues is quasi-constant, density (ρ_t) generally plays the dominant role. Bone or foreign elements embedded in tissue have a greater density and higher effective atomic number (Z). They trigger more interaction events than soft tissue for two reasons: more interaction probability per target (σ_i) and greater density of targets (ρ_t). For dose calculation

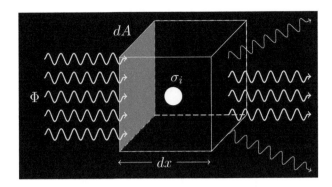

Figure 2.3: A photon fluence Φ (photons per cm^2) is incident on a face of a voxel (dA) with thickness dx occupied by a solitary target presenting a cross section σ_i for a specific type of interaction (i).

Table 2.1: Conversion of microscopic cross sections to macroscopic quantities

Quantity	Units	Symbol	Relationships
Electronic cross section	cm^2/electron	σ_e	σ_a/Z
Atomic cross section	cm^2/atom	σ_a	$Z\,\sigma_e$
Atomic density	atoms/cm^3	ρ_a	$N_A\,(1/A)\,\rho$
Electron density	electrons/cm^3	ρ_e	$N_A\,(Z/A)\,\rho$
Mass attenuation coefficient	cm^2/g	μ/ρ	$(\sigma_a\,\rho_a)/\rho$ or $(\sigma_e\,\rho_e)/\rho$
Linear attenuation coefficient	cm^{-1}	μ	$\sigma_a\rho_a$ or $\sigma_e\rho_e$

where:

$N_A \equiv$ Avogadro's number [6.02×10^{23} atoms per mole]
$Z/A \equiv$ ratio of atomic number (Z) to atomic mass (A)
$\rho \equiv$ gravimetric density [g/cm^3] of tissue in voxel

purposes, *microscopic* cross sections (σ_i) are converted to *macroscopic* quantities such as mass attenuation coefficient (μ/ρ) (see Table 2.1). This conversion yields the interaction probabilities *per gram* of material. The mass density (ρ) clearly plays an important role for all macroscopic quantities. The linear attenuation coefficient (μ) is the probability of interaction *per unit thickness* of tissue traversed. Note that (Z/A) is the number of electrons per gram with a value of approximately 0.5 for all stable elements commonly found in tissues, except for hydrogen; (Z/A) is unity for its most abundant isotope $(^1\text{H}_1)$.

2.2.2 Prevalence of Interactions

Figure 2.4 shows the probability map of each type of interaction (i) as a function of photon energy and absorber specified by atomic number. The probability was calculated from mass attenuation coefficients for each element as follows:

$$P_i = \frac{(\mu_i/\rho)}{(\mu/\rho)} = \frac{\mu_i}{\mu} \tag{2.2}$$

with the total probability sum

$$(\mu/\rho) = \frac{(\mu_{pe} + \mu_c + \mu_{pt})}{\rho} \tag{2.3}$$

The total mass attenuation coefficient (μ/ρ) represents the probability of occurrence by any one of the alternative mechanisms *per unit mass* of attenuating material (Hubbell and Seltzer 1999). The subscripts specify Compton (c), photoelectric (pe), or pair-triplet production (pt) interactions. When normalized in this way, the probabilities from Equation 2.2 serve as *branching ratios* for Monte Carlo simulations in Chapter 5. They are used to randomly select the types of interactions occurring at an interaction site by drawing random numbers in the interval [0,1]. Each type of interaction is assigned a proportional range within this unit interval and is selected when the random number value lands in the allocated range. Note that branching ratios change dynamically as different tissues are traversed and photon energy degrades.

Compton interactions are clearly dominant for soft tissues as seen in the bottom of the Compton volcano of Figure 2.4B. At lower energy, competition with photoelectric events is observed (Figure 2.4A) but only in higher atomic number (Z) materials. At higher photon energy and for absorbers of high atomic number, the onset of pair production events occurs at 1.02 MeV (Figure 2.4C). Exposure of cortical bone, teeth fillings, prosthetic implants or cardiac pacemaker hardware can trigger these events locally. The high-Z regions are relevant to hardware components found in a linear accelerator head. Note that a radiation beam undergoes a sequence of interactions in different materials with progressive degradation of energy. Hence, a photon history traces a complex trajectory throughout the coloured spaces on the right-side landscapes of Figure 2.4.

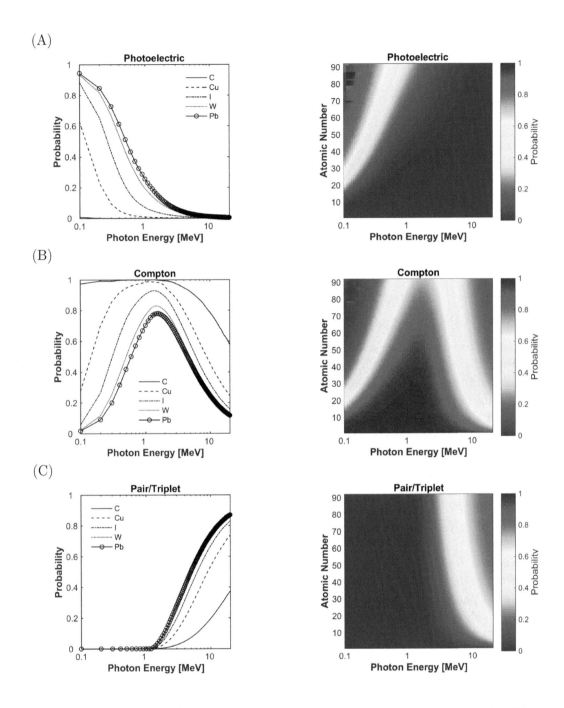

Figure 2.4: Relative probability plots and maps of photon interactions. (A) Photoelectric absorption. (B) Compton scattering. (C) Pair-triplet production. Data source (Berger et al. 1999).

(A) (B)

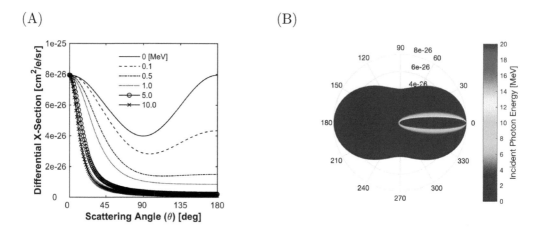

Figure 2.5: Angular distributions of photons Compton-scattered per unit *solid* angle. (A) Linear plot and (B) polar plot for each energy.

2.2.3 Angular Distribution of Compton-Scattered Particles

It is instructive to visualize a *differential* cross section in a way that exhibits the anisotropic distribution of Compton-scattered photons. This is well understood theoretically and it is described analytically by an equation derived originally by Klein and Nishina (1929) for the probability of scattering of an unpolarized photon by an ubound electron, $(\mathrm{d}\sigma/\mathrm{d}\Omega)_{\mathrm{KN}}$. The expression is found in (too) many textbooks on radiological physics (Podgorsak 2016; Andreo et al. 2017) and will be described in Chapter 5. Figure 2.5 shows plots of the Klein-Nishina cross section, in units of cm^2 per electron per steradian, *versus* the photon scattering angle, for a range of incoming photon energies. The *gedanken* interpretation of this display is as follows. Imagine a thin scattering foil placed in vacuum and irradiated with a horizontal stream of photons from the left side. A small particle counter will subtend a fixed solid angle $(\mathrm{d}\Omega)$ that intercepts photons emerging at different scattering angles (θ) as in Figure 2.5A. In the accompanying polar plot (B) on the right, the radial distance from the origin reflects the number of detected scattered photons at each angle for each photon energy. For megavoltage energies, scattered photons are tightly bunched in the forward direction. The total area enclosed by each coloured region is the total Compton probability $(\mathrm{cm}^2$ per electron) for scattering in all possible directions. For increasing energy, this area is in the forward direction but reduced in overall size, reflecting the decrease in probability of Compton events.

An alternative format for exhibiting angular scattering uses the probability *per unit planar angle* (radians) instead of per unit solid angle (steradians). Now imagine *ring-shaped detectors* of variable circumference instead of the fixed detector size that we previously considered. A large ring detector will intercept particles scattered into the wedge region defined by an angular increment ($\mathrm{d}\alpha$ in Figure 2.6).

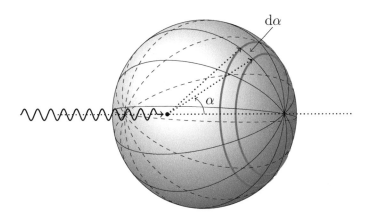

Figure 2.6: Spherical geometry for scattering into a ring-like detector subtending a fixed wedge gap ($d\alpha$) at angle α.

The distribution profile retains some of the core features of the original distribution per unit solid angle (Figure 2.5), but it becomes modulated by the sine wave function (i.e. $\sin\alpha$). Referring to Figure 2.6, the solid angle subtended in between two adjacent cones (α), encompassing the full range of azimuthal angles (2π) is $d\Omega = 2\pi \sin\alpha \, d\alpha$. The detected count rate in this *gedanken* experiment exhibits the plain angular pattern seen in Figure 2.7, for both Compton-scattered photons ($d\sigma/d\theta$), and recoil electrons ($d\sigma/d\phi$). This representation may give the false impression that very few particles are launched in the forward direction. This is indeed not the case! The plots of Figure 2.7 rather reflect a detector that changes its size with angle. For a given scattering angle, the ring-like detector will subtend a larger capture area near the equator ($\alpha = 90$ degrees in Figure 2.6) because the circumference approaches a maximum. Conversely, a ring detector shrinks to a very small circumference near polar angles (i.e. $\alpha = 0$ or 180 degrees). Hence, particle counts plummet because of a vanishingly small detector, rather than a scarcity of emitted particles.

An analogy using the earth's globe may be helpful in this regard, courtesy of "Rock" Mackie. The surface area encompassed by a fixed angular increment in latitude is maximum near the equator and decreases near the poles because the circumference shrinks. In a Mercator projection of the earth onto a cylindrical flat surface, Greenland's area is greatly exaggerated because the circumference has been *artificially* stretched. Its actual surface area (in units of km^2) is much smaller.

2.2.4   Energy Spectrum of Compton Recoil Electrons

The energy spectrum of the recoil electrons is of great interest because these particles are directly responsible for depositing most of the dose in tissue voxels. This spectrum, denoted by ($d\sigma/dE_k$), can be derived by the chain rule applied to the

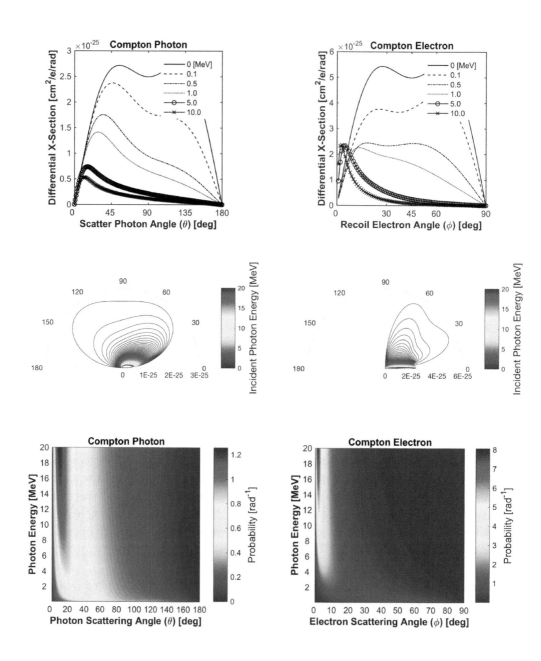

Figure 2.7: Angular distributions of Compton-scattered photons (left) and recoil electrons (right) per unit *planar* angle in units of radians (rad). Three presentation formats are shown: linear plot (top row), polar plot (middle row), and probability colour map (bottom row).

(A)

(B)

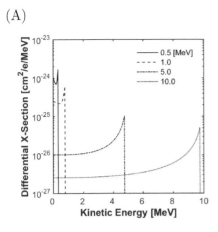

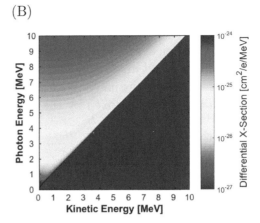

Figure 2.8: A. Energy spectra of Compton recoil electrons B. Probability map. The maximum allowable energy transfer limit ($E_{k,\mathrm{max}}$) is sharply visible as a quasi-diagonal line in the probability map. For photon energies exceeding a few MeV, $E_{k,\mathrm{max}}$ approaches the theoretical limit of ($h\nu - 0.256$ MeV).

differential photon cross section

$$\frac{\mathrm{d}\sigma}{\mathrm{d}E_k} = \left(\frac{\mathrm{d}\sigma}{\mathrm{d}\Omega}\right)_{\mathrm{KN}} \left(\frac{\mathrm{d}\Omega}{\mathrm{d}\theta}\right) \left(\frac{\mathrm{d}\theta}{\mathrm{d}E_k}\right) = \left(\frac{\mathrm{d}\sigma}{\mathrm{d}\Omega}\right)_{\mathrm{KN}} \times 2\pi \sin\theta \times \left(\frac{\mathrm{d}E_k}{\mathrm{d}\theta}\right)^{-1} \qquad (2.4)$$

where E_k is the kinetic energy of the recoil electron, $(\mathrm{d}\sigma/\mathrm{d}\Omega)_{KN}$ is the differential Klein-Nishina cross section, and θ is the *photon* scattering angle. The combination of the first two terms is the photon angular distribution shown in Figure 2.7 (left side). The kinematics expression that couples the energies and angles of the scattered photon and electron can be easily derived (Podgorsak 2016). The electron spectra are shown in Figure 2.8 for a wide range of mono-energetic incident photons. The last term in equation 2.4 causes the upward curl in the spectra. Photons that are side-scattered or back-scattered over a wide angular range ($\theta > 90$ degrees) result in similar electron energy piling up near $E_{k,max} = h\nu - 0.256$ MeV. In other words, the slope $\mathrm{d}E_k/\mathrm{d}\theta$ is very shallow for these photon scattering angles. The reciprocal slope is therefore very steep and causes spectral bunching around $E_{k,\mathrm{max}}$. The shape of the Compton recoil continuum and Compton edge is easily observed experimentally using radiation spectrometers (Knoll 2010).

2.3 PRIMARY PHOTON BEAM ATTENUATION

Two major calculation steps describe energy deposition in an absorbing medium:

1. transfer of energy from the incident primary photons to secondary particles

2. spreading of transferred energy throughout the irradiated volume

In this section, we focus on the first step. Attenuation of a primary x-ray beam fluence occurs through interactions that remove photons from the incoming stream. Particle removal mechanisms include absorption by photoelectric (pe) or pair-triplet production (ptp) and angular deflections by Compton (c) events. The probability that any one of these mutually exclusive events occurs per unit distance of travel in tissue with density (ρ) is given by the total *linear* attenuation coefficient (cm^{-1})

$$\mu = \rho \left(\frac{\mu_{pe}}{\rho} + \frac{\mu_c}{\rho} + \frac{\mu_{pt}}{\rho} \right) = \mu_{pe} + \mu_c + \mu_{pt} \tag{2.5}$$

Consider a narrow pencil beam of mono-energetic x-rays incident on an absorber and observed with a collimated co-linear detector. The number of photons surviving without interaction after traversal of a thickness (L) in a homogeneous medium follows the classical Beer-Lambert equation, originally describing the transmission of light through a semi-opaque medium

$$N(L) = N_0 \exp(-\mu L) \tag{2.6}$$

where N_0 is the number of incident particles (i.e. at $L = 0$ cm). $N(L)$ denotes the number reaching a depth, L, without any prior interaction at shallower depths ($< L$). The probability, P, that particles will have interacted at any shallower depth ($< L$) and the probability, $P(L)$, that a photon will have interacted within a thin layer dL at a specific depth, L, are given by the following pair of equations:

$$P(< L) = \frac{[N_0 - N(L)]}{N_0} = 1 - \exp(-\mu L) \tag{2.7}$$

$$P(L) = \mathrm{d}P/\mathrm{d}L = \mu \exp(-\mu L) \tag{2.8}$$

The probability of a photon interacting at L is the same as the probability *density* of having a photon travel freely without interaction up to depth L. Figure 2.9 shows the probability density distribution of free paths (L) for photons with 1, 5, or 10 MeV incident on a water absorber. The ordinate can be interpreted as the relative number of interactions occurring at each depth. While the probability of interaction per cm (i.e. μ) is constant, fewer photons have opportunity to reach deeper locations due to possible upstream interactions. In Monte Carlo simulations, this frequency distribution is randomly sampled to determine the distance between photon interaction sites (Chapter 5).

The *mean* free path (mfp) is exactly what its name implies. It is the average value of free path lengths, L, in units of length [cm]

$$\mathrm{mfp} = \frac{\int L \mu \exp(-\mu L) \, \mathrm{d}L}{\int \mu \exp(-\mu L) \, \mathrm{d}L} = \frac{1}{\mu} \tag{2.9}$$

Recall that μ is energy-dependent so that mean free paths change as photon energy progressively degrades through successive scattering events. The dimensionless

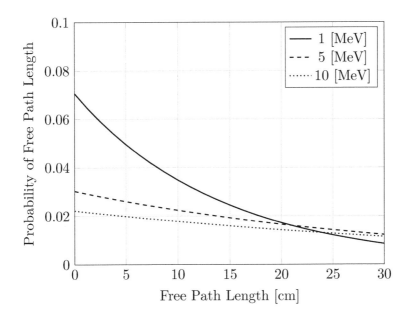

Figure 2.9: Free path distributions, L[cm], in a water absorber exposed to a narrow beam of photons of 1, 5, or 10 MeV.

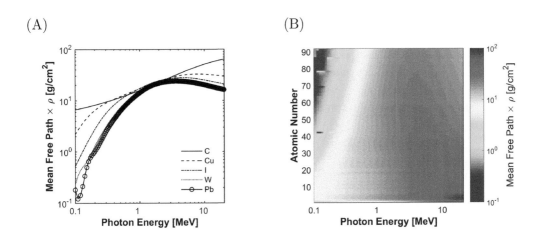

Figure 2.10: Mean free path lengths multiplied by physical density (mfp $\times \rho$ in units of g per cm^2) for atomic elements (Z) and photon energy range of interest. (A) logarithmic plot. (B) colour map.

product $\mu L = L/\text{mfp}$ is the radiological thickness of an absorber expressed in units of mean free paths. For example, if $\mu = 0.05$ cm^{-1}, the mean free path is 20 cm. For a physical thickness of 10 cm, μL equals $\frac{1}{2}$ of a mean free path (mfp). Figure 2.10 displays the mean free path lengths re-scaled by the natural density of elements

$$\text{mfp} \times \rho = \left(\frac{\rho}{\mu}\right) = \left(\frac{\mu}{\rho}\right)^{-1} \tag{2.10}$$

2.3.1 TERMA, KERMA, KERMA$_c$ and SCERMA

The cohort of photons reaching any point inside a patient is characterized by the differential photon fluence Φ_E equal to $d\Phi(E)/dE$ (photons cm^{-2} per MeV). The energy fluence is a spatially-varying function throughout the patient space ($\mathbf{r}$) because of differential attenuation in tissue and possible inclusion of previously-scattered photons. For applications to dose calculations, it is advantageous to only include *primary* photons emerging from the radiation source in the fluence spectrum. In this context, *primary* designates photons that have not yet interacted *in the patient*. However, they may have pre-scattered within the head components of the accelerator. The energy fluence of each primary spectral component reaching a location $\mathbf{r}(x, y, z)$ in the patient is given by

$$\Psi_E(\mathbf{r}) = E\,\Phi_E(\mathbf{r_0})\left(\frac{\mathbf{r_0}}{\mathbf{r}}\right)^2 \exp\left[-\int \mu(E, l)\,dl\right] \tag{2.11}$$

where:

$\Psi_E(\mathbf{r}) = \frac{d\Psi(E)}{dE} \equiv$ energy fluence of photons having energy in the interval between $[E, E + dE]$ at $\mathbf{r}$

$E \equiv$ quantum energy ($h\nu$) of photons comprising the photon fluence

$\mu(E, l) \equiv$ linear attenuation coefficient along a beam ray at position l

$\mathbf{r_0} \equiv$ reference location from the source

$dl \equiv$ pathlength increment along a beam ray

The link between energy fluence and dose (defined in Section 2.4) requires several intermediate quantities. Conversion coefficients include: mass attenuation (μ/ρ), mass energy transfer (μ_{tr}/ρ), or mass energy absorption coefficients (μ_{en}/ρ), respectively (Table 2.2). These coefficients are expansions of microscopic interaction cross-sections (Table 2.1), amplified by the average fraction of the incident energy that is either released or transferred to specific particles (Podgorsak 2016). If each mass coefficient is expressed in units of cm^2 per g, and energy fluence is expressed in MeV per cm^2, then *all* quantities ending in *ERMA* will be in units of MeV per unit mass (g).

Table 2.2: Quantities ending in ERMA, as used in radiation dosimetry

Concept	Quantity	Definition
Total Energy transferred to charged and uncharged particles	TERMA Total energy released per unit mass	$\text{TERMA} = \int (\mu/\rho)\, \Psi_E \, \mathrm{d}E$ $\text{TERMA} = \text{KERMA}_c + \text{SCERMA}$
Energy transferred to charged particles only	KERMA Kinetic energy released per unit mass	$\text{KERMA} = \int (\mu_{tr}/\rho)\, \Psi_E \, \mathrm{d}E$
Net energy absorbed close to charged particle tracks	KERMA_c Kinetic energy released per unit mass and deposited locally	$\text{KERMA}_c = \int (\mu_{en}/\rho)\, \Psi_E \, \mathrm{d}E$
Photon energy radiated away by charged particles	KERMA_{rad} Kinetic energy released to radiative photons per unit mass	$\text{KERMA}_{rad} = \text{KERMA} - \text{KERMA}_c$
Scattered and radiated photon energy (i.e. non-primary)	SCERMA Scattered energy released per unit mass	$\text{SCERMA} = \text{TERMA} - \text{KERMA}_c$

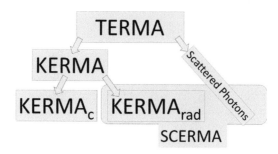

Figure 2.11: Summary of quantities ending in *ERMA*, showing the partitioning of energy released by incoming primary photons.

All ERMA-ending quantities are segmented in Figure 2.11 to illustrate the partitioning and conservation of energy. The total TERMA (total energy released per unit mass) focuses on *all* of the energy liberated, on average, from the point of impact. The KERMA (kinetic energy released per unit mass) focuses on the average amount of energy transferred *to charged particles only*, setting them in motion (hence the word kinetic). KERMA is further subdivided into energy that will eventually be deposited in: (a) the vicinity of the charged particle tracks, i.e. the collisional KERMA$_c$, and (b) *radiative* KERMA$_{rad}$, which escapes the immediate track neighbourhood *via* bremsstrahlung x-rays or annihilation γ-rays radiated during the slowing down of liberated charged particles. Note that KERMA$_c$ reflects *net local absorption* of energy near tracks and it is closely related to primary absorbed dose under equilibrium conditions as will be described in Section 2.4.1.

SCERMA (scattered energy released per unit mass) includes non-primary energy obtained simply by subtracting KERMA$_c$ from TERMA (Russell and Ahnesjö 1996; Ahnesjö et al. 2005, 2017). It captures all *secondary photon* energy that is dispersed at remote distances from the primary interaction site. This quantity includes energy dispersed not only by scattered photons but also by radiated photons. If radiative events are rare, this quantity reflects energy from Compton-scattered photons and the acronym seems more precise; otherwise the word *scatter* in the acronym must be interpreted in a broader sense to include radiated photon energy.

The energy spectrum TERMA$_E$ is an abbreviated form of $\frac{\mathrm{dTERMA}}{\mathrm{d}E}(E)$. This spectrum changes at every spatial position ($\mathbf{r}$) in the patient because of changes in the energy fluence and the local mass attenuation coefficient (Ahnesjö et al. 1987), as follows:

$$\mathrm{TERMA}_E(\mathbf{r}) = \frac{\mu}{\rho}(\mathbf{r}, E)\,\Psi_E(\mathbf{r}) \qquad (2.12)$$

where μ/ρ is the mass attenuation coefficient at location ($\mathbf{r}$) and $\Psi_E(\mathbf{r})$ is the energy fluence spectrum (Equation 2.11). Similar expressions are used to calculate KERMA

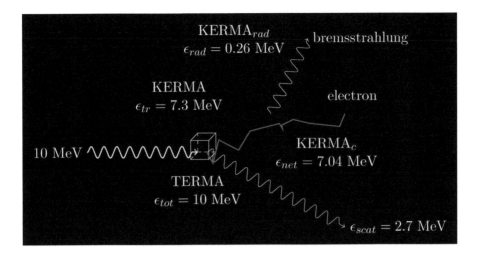

Figure 2.12: Sketch of a single 10 MeV photon interaction illustrating total energy transferred (10 MeV), energy released to an electron (7.3 MeV), and net electron energy that is absorbed near the electron track (7.04 MeV). These events contribute to KERMA and KERMA$_c$, respectively. The SCERMA contribution comes from all secondary photon energies (i.e. 2.7 + 0.26 = 2.96 MeV). This value is in agreement with SCERMA = TERMA − KERMA$_c$ = 10.0 - 7.04 = 2.96 MeV.

and collision KERMA$_c$ spectra substituting mass energy transfer and mass energy absorption coefficients *in lieu* of mass attenuation coefficients, respectively.

Figure 2.12 shows a numerical example of discrete energy contributions to TERMA, KERMA, collision KERMA$_c$, radiative KERMA$_{rad}$, and SCERMA. We consider a single Compton interaction that releases a 7.3 MeV recoil electron and scattered photon that emerges from the interaction site with an energy of 2.7 MeV. The recoil electron subsequently slows down mainly through soft electron-electron collisions close to the electron track, but also radiates a bremsstrahlung burst of 0.26 MeV, leaving behind a net energy of 7.04 MeV available for primary local energy deposition along the electron track. The distal energy associated with SCERMA is derived from the 2 emerging photons (i.e. 0.26 + 2.7 MeV = 2.96 MeV). This energy will be deposited well away from the primary charged particle track.

2.4 ENERGY IMPARTED TO A VOXEL

Generally, a voxel of tissue will be subjected to a mixture of photons and secondary charged particles. The basic equation for energy imparted by all these particles can be expressed as (ICRU 2011)

$$\epsilon = (R_{in} - R_{out})_u + (R_{in} - R_{out})_c + \Sigma Q \tag{2.13}$$

where bracketed quantities are radiant energies (R) of uncharged (u) and charged (c) particles that either enter (in) or leave (out) a voxel through any of its facets. The radiant energy is the product of particle number and kinetic energy specified at points of voxel entry or exit. Note that energy imparted is *specific to the type of absorbing tissue in the voxel* as this will affect the energy extraction in the voxel. For an identical incoming fluence of particles, different amounts of energy will be imparted to soft tissue or to bone. The lagging term ΣQ designates the sum of mass-energy conversion quantities, occurring when radiant energy transforms to mass or *vice versa*. By convention, conversion from mass to photon energy has a positive Q-value and vice versa for photon energy conversion to particulate mass. This term plays a role in nuclear pair production, for example, ($\Delta Q = -1.022$ MeV) and annihilation ($\Delta Q = +1.022$ MeV) events only when they occur *inside* the voxel. Application of equation 2.13 requires monitoring of all energy entering or leaving a voxel, as can be done by Monte Carlo simulations.

The concept of energy imparted is further demonstrated by a simple numerical example (Figure 2.13). A 10 MeV photon enters the voxel of unit mass (1 g) and soon undergoes a nuclear pair production event. Of the available energy of 8.978 MeV (i.e. $10 - 1.022$ MeV), the positron receives the majority of the energy, 8 MeV, leaving only 0.978 MeV for the electron. The electron loses all of this kinetic energy within the voxel. The positron proceed to expend only 0.4 MeV inside the voxel, before being annihilated in-flight. This event releases a pair of gamma rays each with equal energy (4.311 MeV) that escapes the voxel. The energy imparted by this single event is $\epsilon = [10 - 2(4.311)] + (0 - 0) + (-1.022 + 1.022) = 1.378$ MeV. Intuitively, this is the energy deposited inside the voxel by charged particles only – the electron (0.978 MeV) and positron (0.4 MeV) combined.

An analogy may be helpful in understanding the deposition of energy in a voxel. Consider the task of tracking the economic impact of tourism in a country. Assume that customs officers can verify all cash amounts that are carried in or out of the country by tourists crossing at all border crossings. Electronic transfers of currency to or from the country are tracked by banks (i.e. radiative transfers). Banks also impose a service charge (i.e. ΣQ) for this service and for currency exchanges, leaving the tourist with less liquid cash to spend locally. The net cash expenditures in the country with local currency dollars (i.e. deposited dose) can then be determined using a monetary form of equation 2.13.

2.4.1 Absorbed Dose

Absorbed dose is the quantity related to the total energy imparted by all particles entering the voxel of interest, $d\bar{\epsilon}$, per unit mass, dm, of the voxel. The voxel mass must be small enough to represent local dose but sufficiently large to dampen statistical fluctuations in the energy imparted over the radiation exposure time. Thus dose is intended to denote an average non-stochastic value of the *concentration of*

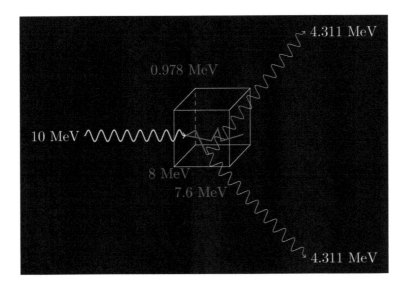

Figure 2.13: Sketch of a single photon entering a voxel. The energies shown refer to incident photon energy and unequal kinetic energies imparted to an electron-positron pair. The positron starts with a lion's share of 8 MeV but only loses 0.4 MeV along its track within the voxel before prematurely annihilating in-flight, producing two escaping gammas of 4.311 MeV each.

energy deposited, on average, in a voxel for a sufficient radiation exposure level

$$D = \frac{\mathrm{d}\bar{\epsilon}}{\mathrm{d}m} \tag{2.14}$$

Absorbed dose, TERMA, KERMA, and KERMA$_c$ are inter-related quantities all expressed in units of Joules per kg or Gray (1 Gy = 1 J kg^{-1}) in honour of Louis Harold Gray, a British physicist who established the physical foundation for understanding biological effects of radiation. If we consider a dose calculation algorithm that only tracks photons and assume that charged particles drop all of their kinetic energy on-the-spot at photon interaction sites, absorbed dose becomes *synonymous* with the collision KERMA$_c$ liberated inside a voxel. The dose can then be calculated by *only* tracking inbound and outbound *photons*, as follows:

$$D \stackrel{\mathrm{CPE}}{=} \frac{[(R_{in} - R_{out})_u + \Sigma Q]}{\mathrm{d}m} = \mathrm{KERMA}_c \tag{2.15}$$

At megavoltage energies, charged particles actually have an elongated range comparable to voxel dimensions, and energy is potentially transported across many voxels. With a large number of charged particles bombarding a voxel, the charged particle equilibrium (CPE) state can develop that mimics the full collapse of particle tracks onto photon interaction sites within a single voxel. This equilibrium state requires a uniform supply of secondary charged particles that can balance the incoming and outgoing fluences. In other words, $(R_{in} \overset{CPE}{=} R_{out})_c$ and ignoring the mass-energy exchange events, the dose and collision KERMA$_c$ become identical; energy released on-the-spot equals energy deposited. This will be explained further in Section 2.7.

Outside the special realm of CPE, energy is really deposited along tracks of charged particles stopping in, starting in, or passing through a voxel. This deposition can be determined directly by tracking charged particle trajectories *inside* the voxel space. Alternatively, the energy fluence of all particles $\vec{\Psi}^{(\gamma,cp)}$, including both photons (γ) and charged particles (cp), can be monitored at all *borders* of the voxel (Equation 2.13). All incoming and outgoing energy can be tallied by using a surface integral, as follows:

$$D = \frac{d\bar{\epsilon}}{dm} = \frac{\left[-\oiint \vec{\Psi}^{(\gamma,cp)} \cdot d\vec{S} + \Sigma Q \right]}{dm} = \frac{\left[-\oiint \vec{\Psi}^{(\gamma,cp)} \cdot d\vec{S} + \Sigma Q \right]}{\rho \, dV} \qquad (2.16)$$

The negative sign simply denotes that energy was removed from the incoming energy and dumped in the voxel. Applying the divergence identity to the surface integral, and counting all types of particles in the fluence, we obtain the following:

$$D = -\frac{\nabla \cdot \vec{\Psi}^{(\gamma,cp)}}{\rho} + \frac{\Sigma Q}{\rho \, dV} \qquad (2.17)$$

If we assume mass-energy conversions are rare, the absorbed dose at a voxel is simply the divergence of the vector form of energy fluence. Vector fluences reflect the flow of particles along different directions (Kempe and Brahme 2010) and will be explained in more detail in Chapters 3 and 6. The magnitude of divergence corresponds to the amount of energy *extracted* by a voxel through all interactions by all types of particles including photons (γ) and charged particles (cp). This is essentially Gauss' law for radiation physics, reflecting the deposition of charged particle energy within the confines of a voxel!

It can also be shown that if we focus our attention only on photons with the incoming quantum energy, $\vec{\Psi}^{(\gamma_0)}$, the TERMA can also be expressed in the form of a divergence operator (Ahnesjö et al. 1987)

$$\text{TERMA} = \left(\frac{1}{\rho} \right) \left[\frac{-\oiint \vec{\Psi}^{(\gamma_0)} \cdot d\vec{S}}{dV} \right] = -\frac{\nabla \cdot \vec{\Psi}^{(\gamma_0)}}{\rho} \qquad (2.18)$$

The energy released in the voxel is related to the divergence of the photon fluence vector at the incident energy; this complements Equation 2.12.

Similarly, the collision KERMA can be calculated from the overall energy fluence of *all* incoming and outgoing photons (γ) regardless of their origin and energy, $\vec{\Psi}^{(\gamma)}$.

This photon fluence includes primary, scattered, and radiated photons incident on a voxel

$$\text{KERMA}_c = \left(\frac{1}{\rho}\right)\left[\frac{-\oiint \vec{\Psi}^{(\gamma)} \cdot d\vec{S}}{dV}\right] + \frac{\Sigma Q}{\rho\, dV} = -\frac{\nabla \cdot \vec{\Psi}^{(\gamma)}}{\rho} + \frac{\Sigma Q}{\rho\, dV}$$

$$(2.19)$$

Charged particle equilibrium (CPE) is established when the divergence of charged particles is nil, i.e. $\nabla \cdot \vec{\Psi}^{(cp)} \overset{\text{CPE}}{=} 0$. This is implicit in Equation 2.19 because the divergence term only includes photons. It follows from Equation 2.17 that $D \overset{\text{CPE}}{=} \text{KERMA}_c$. We will show in Section 2.7 that this situation is *as if* all the tracks of secondary charged particles were fully compacted at their launch sites. It becomes intuitively clear that dose would then be equal to energy released because of highly localized on-the-spot absorption.

The total KERMA can also be calculated from the overall incident photon energy fluence, $\vec{\Psi}^{(\gamma)}$, but photons that are radiated by charged particles slowing down *within the voxel* must be excluded from the *exit* photon fluence count. This differential treatment of incoming and outgoing particles is notated as $\gamma^{\text{non}-\text{rad}}$ in the divergence term

$$\text{KERMA} = \left(\frac{1}{\rho}\right)\left[\frac{-\oiint \vec{\Psi}^{(\gamma^{\text{non}-\text{rad}})} \cdot d\vec{S}}{dV}\right] + \frac{\Sigma Q}{\rho\, dV}$$

$$= -\frac{\nabla \cdot \vec{\Psi}^{(\gamma^{\text{non}-\text{rad}})}}{\rho} + \frac{\Sigma Q}{\rho\, dV} \qquad (2.20)$$

In summary, the divergence operator is a powerful mathematical tool that allows the determination of several important inter-related dosimetric quantities. The key point to remember is that the *correct type of incoming and outgoing particles must be tallied* in the energy fluence to obtain each quantity.

2.5 INTERACTIONS OF CHARGED PARTICLES

It should now be evident to the reader that the most important consequence of photon interactions is the spawning of fast charged particles:

Principle #3: "All x-ray energy is deposited by charged particles liberated in the absorber"

The energy transferred from x-rays, whether they are primary or secondary photons, is deposited relatively close to the interaction sites. Charged particles experience continuous Coulomb forces from an array of atoms lying closely along their path. Trajectories end when all of their kinetic energy has been exhausted and they consequently reach their terminus. Charged particles have a range that is much shorter than the mean free path of photons of comparable energy. At megavoltage energies, the mean free path of photons is on the order of 10 cm in water, while the free path of secondary charged particles is orders of magnitude shorter in the same medium. For example, a 5 MeV electron will produce, on average, over 100,000 individual atomic ionizations before reaching its end of track in water (i.e. 2.5 cm). For this reason, the *continuous slowing down approximation* – CSDA (Andreo et al. 2017) is often applied to describe quasi-continuous energy loss as the particle slows down. Individual interactions of charged particles are not generally modelled as was the case for photons. Multiple scattering events are often lumped together or *condensed* to accelerate Monte Carlo simulations (Chapter 5).

Figure 2.14 summarizes Coulomb interactions experienced by fast charged particles, according to their proximity (i.e. impact parameter b) to an atom of classical radius a. The ratio of b/a conveniently categorizes Coulomb interactions that occur with either distant orbital electrons or closer to the nucleus. In the far zone ($b/a \gg 1$), the net charge experienced by the incoming particles is neutral and a soft collision occurs, resulting mainly in atomic excitation in the form of distortion of the atomic electron cloud as a whole. In the intermediate zone ($b/a \sim 1$), one-on-one encounters with orbital electrons ionize the atom. The ejected electrons, renamed delta rays (δ-rays), develop side tracks of their own. As is the case for photoelectric interactions, characteristic x-rays or Auger electrons may be emitted when the atom relaxes back to its ground energy state through shuffling of electrons across orbitals. The interaction probabilities of electron-electron or positron-electron collisions are described by Möller and Bhabha cross sections, respectively. These cross-sections are orders of magnitude greater than photon cross-sections at comparable energy. Charged particle interactions are therefore often treated as being continuous, with miniscule free paths between ionization events. Energy transfer to tissue is thereby viewed as a multitude of soft collisions punctuated by occasional harder energy transfers.

Coulomb scattering without any loss of energy (i.e. an elastic collision) is also possible wherever the incoming particles sense a net charge from the atom, near the nucleus. These elastic encounters are responsible for shaping the tortuous path of electrons and positrons. In the ultra-near zone ($b/a \ll 1$) a strong Coulomb interaction occurs directly with the nucleus. If the charged particle is not captured by the nucleus, it will be redirected elastically without any energy loss or with a radiative loss (i.e. bremsstrahlung event). The emission of bremsstrahlung x-rays transmits energy away from the main charged particle track. Table 2.3 summarizes the salient features of each type of interaction.

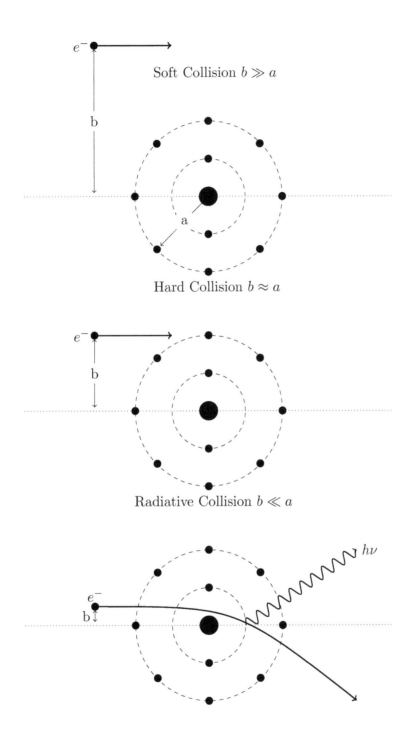

Figure 2.14: Charged particle encounters at different impact distances (b) relative to the classical radius of an atom (a).

Table 2.3: Summary of charged particle interactions

Type of Interaction	Ratio (b/a)	Interaction Target	Probability Cross-Section	Energy Transfer	Angular Deflection	Consequence
Soft Collision	> 1	Electron Cloud	Bethe	Small	Minimal	Atomic Excitation
Hard Collision	< 1	Orbital Electron	Möller Bhabha	Large	Yes	Atomic Ionization δ-rays
Elastic Coulomb Scattering	< 1	Atom	Screened Rutherford	None	Yes	Tortuous Path
Nuclear Collision	$\ll 1$	Nucleus (Radiative)	Bethe-Heitler	Large	Yes	Bremsstrahlung x-rays
		Nucleus (Elastic)	Rutherford	None	Yes	Tortuous Path

2.5.1 Stopping Power of an Absorber

As mentioned previously, numerous interactions occur along a particle track and the CSDA concept can be applied to determine the average energy loss per step. The mass stopping power, (S/ρ), describes the *average* rate of energy loss per increment of *radiological* path length ($\rho \, \mathrm{d}l$) by a charged particle with a given (instantaneous) kinetic energy E_k at the start of the incremental step

$$\left(\frac{S}{\rho}\right) = \left(\frac{1}{\rho}\right)\left(\frac{\mathrm{d}E_k}{\mathrm{d}l}\right) = \frac{\mathrm{d}E_k}{\rho \, \mathrm{d}l} \tag{2.21}$$

where a decrement of energy ($\mathrm{d}E_k$) is lost to the stopping material of density ρ along a track increment $\mathrm{d}l$. Note that as the kinetic energy is lost, the stopping power will be progressively changing along the particle track.

The division by ρ in equation 2.21 may give the impression that the mass stopping power is independent of density, but, this is not rigorously correct. The mass collisional stopping power for a condensed medium must allow for inter-atomic effects; individual atoms no longer interact as independent entities. An electron or positron with a kinetic energy of only a few MeV travels at relativistic speed and the associated electric field is intensified by Lorentz contraction. This accentuates the distortion of the electron cloud of atoms near the track and suppresses the Coulomb interaction with more distant screened atoms. Tabulated stopping powers (Berger et al. 2000) include this polarization correction. As an example, the adjustment is

(A)

(B)

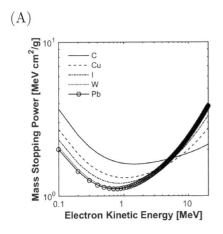
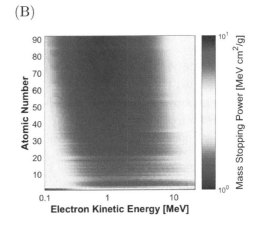

Figure 2.15: Mass stopping powers for electrons as a function of energy and atomic number of absorber. (A) Logarithmic plot. (B) colour map.

on the order of a 10% reduction in carbon for electrons with energy of 10 MeV.

Figure 2.15 shows a map of mass stopping powers for electrons as a function of atomic number and electron energy. The parabolic trend (A) is due to collisional encounters prevailing at lower energy and the rise in the number of radiative encounters as energy increases.

The linear stopping power describes the average energy loss rate of charged particle per increment of geometric distance

$$S = \left(\frac{S}{\rho}\right) \times \rho \tag{2.22}$$

By way of analogy, this linear stopping power is akin to a vehicle's fuel consumption rate per unit of distance travelled in metric units (e.g. litres per 100 km). The stopping power depends on the charge (i.e. type of vehicle) and speed (i.e. highway or city driving) of the particle and the nature of the stopping medium (i.e. road conditions). As the charged particle slows down, the stopping power escalates sharply because the Coulomb force exerts itself for a longer duration on passing by an atom. At a starting energy of 5 MeV, the mass stopping power for an electron in carbon ($Z = 6$) is 1.72 MeV per g cm^{-2} and the linear stopping power is 3.44 MeV per cm of travel using a density of $\rho = 2$ g cm^{-3}. When the particle has slowed to a kinetic energy of 100 keV, the linear stopping power has more than doubled to 7.32 MeV per cm. The stopping power includes energy losses due to collisions with atomic electrons (i.e. collisional stopping power) and with the nucleus (i.e. radiative stopping power). The ratio of radiative to collisional stopping power can be estimated by the ratio ($\approx E_k \times Z/800$) where E_k is the particle's kinetic energy and Z is the atomic number of the absorber. For instance, for an electron with energy of 5 MeV interacting in carbon ($Z = 6$), bremsstrahlung energy losses amount to only $\sim 4\%$ of collisional losses.

The dose deposited by charged particles, D_{cp}, along a short track segment can be calculated *via* the collisional stopping power, $(S/\rho)_c$. This quantity is a subset of the stopping power and excludes radiative energy losses that deposit their energy remotely away from a track. Imagine a fluence of charged particles, $\Phi^{(cp)}$, entering a voxel. The energy deposited per unit mass can be calculated simply as follows:

$$D_{cp} = \frac{\Phi^{(cp)} (S/\rho)_c \, \rho \, dA \, dl}{\rho \, dA \, dl} = \Phi^{(cp)} (S/\rho)_c \tag{2.23}$$

where:

$$\Phi^{(cp)} \equiv \text{incident fluence of charged particles } [e \text{ per cm}^2]$$
$$(S/\rho)_c \equiv \text{mass collisional stopping power } [\text{MeV cm}^2/\text{g}]$$
$$\rho \equiv \text{voxel density } [\text{g/cm}^3]$$
$$dA \equiv \text{entrance area to a voxel } [\text{cm}^2]$$
$$dl \equiv \text{track segment within voxel } [\text{cm}]$$

2.5.2 Range

Charged particles lose discrete packets of energy (dE_K) as they undergo multiple collisions, culminating in the termination of their track. The CSDA concept also applies to the range R_{CSDA} of charged particles, reflecting the continuously changing energy loss rate as the particle slows down. The range of the particle is its overall string length including its tortuous path due to nuclear scattering. It is derived from the reciprocal mass stopping power and inherits the unit of radiological distance (g cm^{-2})

$$R_{\text{CSDA}} = \int_0^{E_k} \left[\frac{S}{\rho}(E) \right]^{-1} dE \tag{2.24}$$

Figure 2.16 shows a map of R_{CSDA} values for different materials and electron energies of interest.

To obtain the geometric range, R'_{CSDA} in units of geometric distance (cm), the R_{CSDA} is divided by the macroscopic density of the stopping material

$$R'_{\text{CSDA}} = (1/\rho) \times R_{\text{CSDA}} \tag{2.25}$$

(A)

(B)

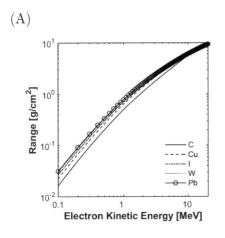
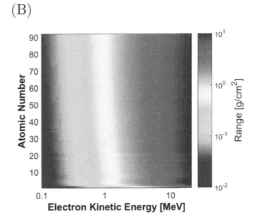

Figure 2.16: Electron CSDA ranges for different elements of the periodic table and electron energy. (A) Logarithmic plot. (B) colour map. Data source (Berger et al. 2000).

2.5.3 Angular Deflections: Scattering Power

Nuclear encounters cause a multitude of deflections of charged particles (ICRU 2007). The mass scattering power reflects their angular dispersion per unit of radiological distance travelled through an absorber. It describes the statistical expectation value of the cumulative scattering angle after an increment of travel in a material of density ρ. The mass scattering power, (T/ρ), is given by

$$(T/\rho) = \left(\frac{1}{\rho}\right)\left(\frac{d\Theta^2}{dl}\right) \tag{2.26}$$

where $d\Theta^2$ is the change in mean-square angle (i.e. a forced positive entity), relative to the starting angular distribution per increment of distance, dl. Figure 2.17 shows values of mass scattering powers of electrons as a function of atomic number and electron energy. The linear scattering power describes the angular dispersion per increment of geometric distance (cm) instead of radiological distance

$$T = (T/\rho) \times \rho \tag{2.27}$$

In summary, charged particles undergo a multitude of energy losses and deflections. These statistical processes are at the core of dose calculation algorithms used for electron beams. In developing x-ray dose algorithms, the modelling details can be relaxed because of significant pre-blurring of energy transport caused by polyenergetic x-rays, release of charged particles from isolated launch sites, and the polyenergetic nature of knock-on charged particles. Early x-ray dose algorithms ignored charged particle ranges altogether and assumed that all transferred energy was deposited on-the-spot. With clinical introduction of higher energy x-ray beams,

(A)   (B)

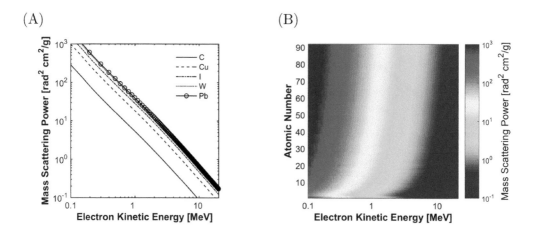

Figure 2.17: Mass scattering power of electrons as a function of energy and atomic elements. (A) Logarithmic plot and (B) colour map.

the consequences of this over simplification soon became apparent. This subject was studied by analytical (Wang and Jette 1999) and Monte Carlo methods (Woo and Cunningham 1990); refinements were then made to x-ray dose algorithms (Yu et al. 1995). The coupling of x-ray and charged particle transport is introduced in the following section using the concept of convolution.

2.6 SHOWER OF SECONDARY PARTICLES

The primary fluence of megavoltage photons required to deliver a typical daily radiotherapy dose (2 Gray) is on the order of 10^{11} photons per cm^2. The number of atoms is on the order of 10^{23} atoms per cm^3 in soft tissue. This large statistical ensemble of projectiles and targets leads to bursts of released energy throughout the irradiated volume of the patient. One approach to describing the spreading and deposition of energy from distributed x-ray interaction sites uses a Green's function – a concept used in the study of electrodynamics, for example (Jackson 2007). This method forms the basis of the convolution dose calculation algorithm (Chapter 4).

2.6.1 Energy Deposition Kernels

In a Monte Carlo simulation, photons can be forced to interact at an isolated spot in an absorber. The free path of photons can be fixed to a constant depth, rather than being randomly selected as actually happens in nature. Energy that emerges from the interaction site is then deposited in surrounding voxels as depicted in Figure 2.18. Comet-shaped maps portray the repercussions of an impulse of many interactions at the origin (i.e. ground zero). These point spread functions (Ahnesjö et al. 1987) have different names such as *dose spread arrays* (Mackie et al. 1984), *differential pencil beams* (Mohan et al. 1986), or *energy deposition kernels* (Mackie

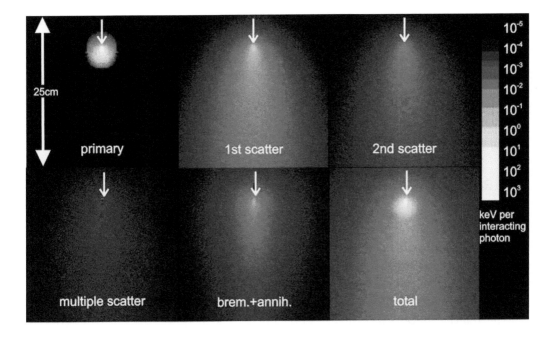

Figure 2.18: The energy deposition kernels for mono-energetic 6 MeV photons launching secondary radiation from a launch point in a large uniform water phantom. The map values are energy deposited (keV) in surrounding voxels per photon interaction at the origin (tip of white arrow). These maps were computed using Monte Carlo simulations (Mackie et al. 1988) and reproduced under the condition of *fair dealing* from a Ph.D. thesis submitted to the University of Western Ontario (Sharpe 1997).

et al. 1988); the reader is cautioned that they may also be normalized differently. The kernel values in Figure 2.18 are normalized per single photon impulse (e.g. keV deposited in the surrounding voxels per single interaction event at the origin). The total kernel exhibits a hot spot due to short-range primary charged particles. A more diffuse radiation pattern is due to longer range secondary photons. The sum of all voxel values in the total energy kernel is related to total energy released, i.e. TERMA and (μ/ρ) of Table 2.2. By tagging particles, the total kernel can be split according to the type of interactions that give rise to energy deposits. The primary kernel reflects energy absorbed locally from electron or positron tracks *only*. The sum of all its voxel values is related to KERMA$_c$ and hence (u_{en}/ρ) (Boyer 1988; Mackie et al. 1988). Remote dose deposition by scattered and radiated photons is accounted for separately in kernels dedicated to each type of interaction with long distance implications. Figure 2.18 shows the energy distributed by single Compton scattering events (first), dual scattering events (second), higher-order scattering events (multiple), and radiated photons (bremsstrahlung plus annihilation). A composite sum of these long-range kernels reflects the SCERMA distribution (not shown).

2.6.2 A Reciprocal Point of View

In their normal orientation, the kernels reflect the incremental energy distributed by a sending voxel (S) exposed to unit fluence to all surrounding receiving voxels (R). In a large *homogeneous* absorber, interaction and dose deposition sites can be interchanged without affecting the radiation transmitted between pairs of voxels applying the reciprocity theorem (Andreo et al. 2017). By inverting the kernel orientation (see Figure 2.19), we obtain an iso-influence kernel. The kernel values then reflect the relative importance of neighbouring voxels to dose at a receiver voxel. Using a baseball analogy, the same information is reflected from either a pitcher's or catcher's point of view. Reciprocity is maintained in a homogeneous medium because scattering, attenuation, and energy losses along all pathways are invariant to particle direction.

Primary kernels can be measured with specialized equipment (O'Connor and Malone 1989). An ion chamber was constructed with a series of build-up caps and was centred within a hollow spherical shell made of interlocking wooden segments. The chamber detected the charged particles released by each segment and iso-influence curves were converted to primary kernels using the reciprocity concept. Experimental measurement of scattered photon energy kernels is much more difficult, however, if not impossible. Compton-scattered radiation reaching a detector has a broad spectrum of energies and flight paths; isolation singly and multiply scattered photons, for example, would require detectors with extreme energy or temporal resolution.

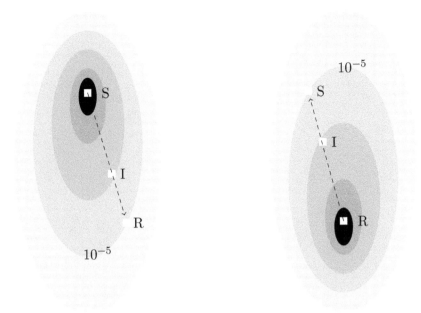

Figure 2.19: Illustration of reciprocity in a homogeneous absorber. The energy spread kernel (left) is inverted to produce the iso-influence kernel (right). S = Sender voxel. R = Receiver voxel. I = Intervening voxel along a single scattering pathway.

We now consider the analytic calculation of iso-effect kernels for singly and doubly-scattered photons in a uniform water absorber from first principles. These kernels have been derived using Compton cross sections and attenuations along scattering ray paths (Boyer and Mok 1985) (Wong et al. 1981). Figure 2.20A shows scattering voxels dV_1 or dV_2 with routes that lead to the voxel of interest at P. Numerical results are shown in Figure 2.20B for a beam of cobalt-60 radiation irradiating a large water phantom. Data are presented as the fraction of total dose contributed to a target point P by individual surrounding voxels. The contributions of dV_1 along dual-scatter pathways, of dV_2 along dual-scattering pathways, and of dV_1 along direct first-scatter paths are shown in order (left to right) in the figure. Near to point P, the first-scatter dose (right panel) from nearby dV_1 voxels is much greater than its contributions *via* dual-scattering (left panel); in the near zone, only a limited number of dV_2 voxels that can participate. However, when dV_1 is situated further away, as sketched in Figure 2.20A, first-scatter and second-scatter contributions become comparable in magnitude, on the order of 10^{-5}; in the far zone, there is a large number of dV_2 voxels that can tip singly-scattered photons towards point P. This clearly demonstrates the interplay of voxels in delivering scattered energy. The situation is more complex when scattering occurs across multiple voxels.

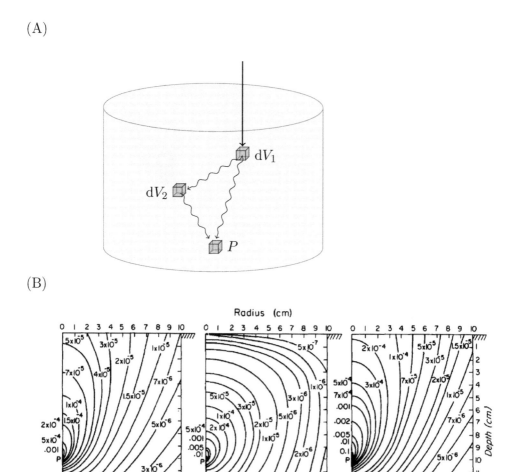

Figure 2.20: (A) Scattered photons reaching point P via voxels dV_1 and dV_2. (B) From left to right, contributions of dV_1 along dual-scattering path; contributions of dV_2 along dual-scattering path; direct contributions of dV_1 along single-scattering path. Iso-effect lines are expressed in units of fraction of total dose reaching point P per unit volume of scatterer. Figure (B) is reproduced with permission (Wong 1981).

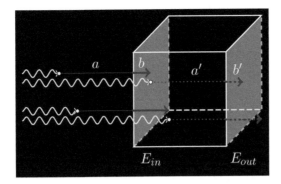

Figure 2.21: Concept of complementary track segments in longitudinal charged particle equilibrium.

2.7 CHARGED PARTICLE EQUILIBRIUM (CPE)

The release of recoil electrons and positrons from numerous interaction sites results in omni- directional particles that bombard voxels with a wide range of energies. Where layers of adjoining voxels are sufficient, these particles can establish an equilibrium state; the energy fluence spectrum entering a voxel is precisely balanced by the exiting fluence. This condition mimics on-the-spot absorption of charged particles released by interactions inside the dose voxel.

2.7.1 Longitudinal Equilibrium

Equilibrium is a 4-dimensional phenomenon. Particle tracks reach a voxel from all directions over a period of time. However, to simplify the explanation, we consider longitudinal (Figure 2.21) and oblique (Figure 2.23) tracks separately, and assume temporal equilibration over a finite radiation exposure period. Consider a subset of mono-energetic electrons approaching a voxel. A build-up layer located ahead of the voxel entrance has a thickness equal to the range of launched electrons $(a + b)$. For each inbound electron there is a corresponding electron launched inside the voxel and leaving with the same energy as determined at the exit surface; the energy is balanced. This occurs because the energy lost along the external portion of a track (a) is the same as that of energy lost inside the voxel by another electron launched inside the voxel (a').This analysis applies to other pairs of tracks and satisfies the energy balance conditions required to achieve CPE. The residual ranges (b and b') are also matched in length. Now imagine that the top two tracks are abutted together. The amalgamated energy losses along the hybrid track $(b + a')$ will be equivalent to the energy *released* inside the voxel. Hence primary dose becomes synonymous with $KERMA_c$, where the electron fluence equilibrates.

We now establish a convolution framework for calculating depth-dose curves analytically, concentrating only on primary radiation for now (Papanikolaou et al. 2004). For a hypothetical monoenergetic non-divergent beam incident on a homogeneous medium, the basic shape of the depth-dose curve (Figure 2.22) has a build-up portion at shallow depths due to gradual overlapping of charged particle tracks released upstream (Figure 2.21). The shift in the depth-dose curve from the $KERMA_c$ curve is due to the downstream push of energy from primary launch sites. The upper and lower panels illustrate the calculation procedure in forward and reciprocal directions, respectively. In the *pitcher's* view, the energy released from charged particles accumulates at multiple downstream dose points as the kernel slides down the $KERMA_c$ curve. In the *catcher's* view, the kernels are reversed and reflect a collection of weighted upstream voxel contributions at a single dose point. The maximum dose occurs at a depth (d_{max}), approximately equal to the range of liberated charged particles; dose equals $KERMA_c$ for this pure equilibrium state. For shallower depths ($< d_{max}$), disequilibrium is observed and the dose curve drops well below the $KERMA_c$ curve. For deeper locations photon attenuation causes the release of fewer charged particles and the dose eventually drops off exponentially at the rate of the primary attenuation coefficient (μ). The deeper region is said to be in transient charged particle equilibrium (TCPE) because the dose remains linearly proportional to the $KERMA_c$. The resultant dose curve is calculated using standard mathematics (Johns et al. 1949) with a primary kernel shaped by a stronger decay constant, $\beta \gg \mu$. Ignoring a small backscattering component (Iwasaki 1985), the primary depth-dose curve is given by a *convolution* of attenuated $KERMA_c$ and the primary kernel. The two terms are bracketed explicitly in the integral below.

$$D_p(x) = N \int_0^x \left[\text{KERMA}_{c,0} \, \exp(-\mu x') \right] \left[\beta \exp(-\beta(x - x')) \right] \mathrm{d}x' \qquad (2.28)$$

$$= N \times \text{KERMA}_{c,0} \times \frac{\beta}{(\beta - \mu)} \left[\exp(-\mu x) - \exp(-\beta x) \right] \qquad (2.29)$$

The curve reaches its peak value where:

$$\mathrm{d}D_p/\mathrm{d}x = 0 \qquad (2.30)$$

from which we can solve for the depth where dose is at its maximum ($d_{\max}$)

$$d_{\max} = \frac{1}{(\beta - \mu)} \ln(\beta/\mu) \qquad (2.31)$$

where:

$D_p \equiv$ primary dose as a function of depth (x)

$N \equiv$ Normalization factor

$\text{KERMA}_{c,0} \equiv$ incident collision $KERMA_c$

$\mu \equiv$ linear attenuation coefficient [cm^{-1}] for photon fluence

$\beta \equiv$ exponential coefficient that characterizes the energy deposition from divergent charged particles

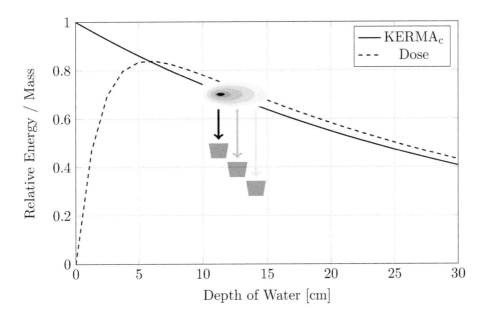

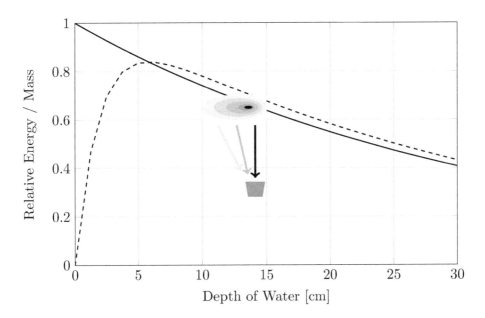

Figure 2.22: Collisional KERMA$_c$ and depth-dose curves for monoenergetic photons incident on water. The top panel shows a forward calculation using the energy deposition kernel. The bottom panel shows the reciprocal calculation with an inverted kernel. The bucket icons represent voxels that accumulate energy deposits.

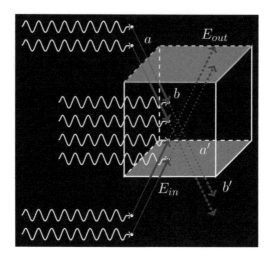

Figure 2.23: Concept of lateral build-up of charged particle equilibrium. Incoming (E_{in}) and outgoing (E_{out}) electron energies match at the top and bottom of the voxel. Abutted oblique tracks are also complementary as in longitudinal equilibrium.

For deeper locations (i.e. $x \gg d_{\max}$), transient charged particle equilibrium (TCPE) is established because $\beta \gg \mu$, and Equation 2.29 reduces to a simpler form

$$D_p(x) = \exp(\mu\,\bar{x})\,\mathrm{KERMA}_c(x) \approx (1 + \mu\,\bar{x})\,\mathrm{KERMA}_c(x) \qquad (2.32)$$

where $\bar{x} = 1/\beta$ is the downstream shift of charged particle energy deposition; this concept is similar to that of the mean free path used for photons. The proportionality constant, $(1 + \mu\,\bar{x})$, linking dose to collision KERMA_c is greater than unity for all depths $x > d_{\max}$ where TCPE is maintained.

Principle #4: "In regions of charged particle equilibrium, dose is closely related to KERMA$_c$. Otherwise, these quantities become uncorrelated"

2.7.2 Lateral Equilibrium

Disequilibrium is a multi-dimensional phenomenon. We now consider side-scattered electrons released by Compton interactions. These electrons have less energy and range than forward-directed electrons (as in Figure 2.21) but the principles of complementary adjoining tracks still apply, as illustrated in Figure 2.23. Lateral tracks will abut with each other to mimic full tracks and deposit a full quota of energy as if released in the voxel. This effect now occurs symmetrically from both sides of the voxel, as long as there are enough surrounding exposed voxels to achieve build-up. For a maximum bilateral effect at the centre of the voxel, the field size must exceed twice the projected lateral range of side-scattered electrons. Otherwise there is diminished opportunity to match outgoing and incoming energies and this causes a

Figure 2.24: Two-dimensional convolution process. The energy deposition kernel (centre) is applied at each photon interaction site and blurs the TERMA distribution (left). The result of the convolution process is a fuzzy-edged dose distribution (right).

precipitous drop in central axis dose. For voxels lying near to a beam edge, lateral disequilibrium is inevitable because of the lack of compensatory fluence from unexposed side regions. This phenomenon produces the radiological penumbra of the beam in an absorber. In low density absorbers where the range of charged particles is elongated, the penumbral width becomes significantly inflated.

Figure 2.24 shows convolution blurring in 2D whereby each photon impulse site (left) is smudged by the kernel footprint (centre). The kernels are applied across the beam in a grid pattern with their intensity modulated by the attenuated local TERMA. Longitudinal disequilibrium effects produce dose build-up and build-down near the entrance and exit surfaces; lateral disequilibrium affects the shape and width of the beam penumbra.

We explore a simple analogy, courtesy of "Rock" Mackie. Imagine watering a lawn with a hose and nozzle with a selected lateral spray pattern. If the nozzle is directed laterally towards each part of the lawn area uniformly, most of the inner lawn will also be evenly watered. However, near the fringes of the lawn, some water will have sprayed outside the lawn area, with no water being sprayed back in from the dry periphery. The edge of the lawn will therefore receive one-half of the water received by the inner lawn. The width of the watering penumbra is governed by the width of the scattering pattern (i.e. nozzle setting). A narrower spray pattern (i.e. lower beam energy) can produce a more sharply-defined water coverage.

Figure 2.25 shows calculated *primary dose* curves for an 18 MV x-ray beam for a variety of circular field sizes to illustrate the 3D nature of charged particle equilibrium (Papanikolaou et al. 2004). Disequilibrium occurs at the central axis for smaller fields. When the field size is small, the bilateral beam penumbrae can "collide" and overlap; the central axis dose plummets. In addition to the geometric effects of the radiation source size, lateral disequilibrium is responsible for the radiological beam penumbra in tissue for all field sizes.

Note that the term *convolution* applies strictly to the situation where kernel shapes are applied identically to every point of photon interaction in the patient. This assumes non-divergent mono-energetic photon beams incident on an infinitely large homogeneous absorber. In practice, the kernels require a re-orientation along divergent beam rays. The x-ray spectrum also requires that either a separate convolution be performed for each spectral component or that kernels be modified as the beam penetrates tissue. Bigger effects on the kernel shapes are caused by tissue inhomogeneities, including the interface of the patient with surrounding air. When applying a variable kernel at each photon launch site, the term convolution is no longer applicable and the correct terminology is *superposition*. Details of this method will be presented in Chapter 4.

2.8 SUMMARY

X-ray interactions occur randomly throughout the irradiated regions and launch secondary particles that spread energy away from interaction loci. The key concepts are summarized in Figure 2.26. The TERMA from all photons is divided into its collisional component due to charged particles that deposit dose locally and secondary photons that deposit dose more remotely. If charged particle equilibrium fails, dose calculation algorithms must explicitly consider the transport of charged particle energy because the equivalence of dose and collision KERMA is no longer tenable. The required algorithms are introduced in Chapter 3.

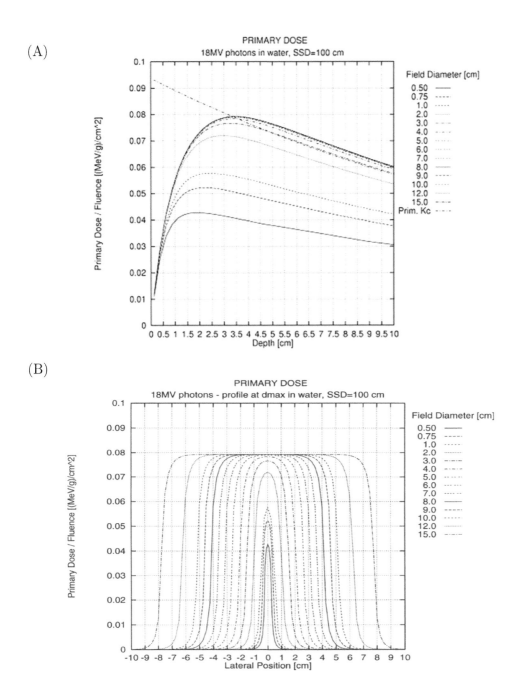

Figure 2.25: (A) Depth-dose curves. (B) Lateral beam profiles in water for 18 MV x-rays as a function of field size. Lateral disequilibrium causes a large reduction in central dose for small fields and a wide radiological penumbra for *all* field sizes. Reproduced with permission (Papanikolaou et al. 2004).

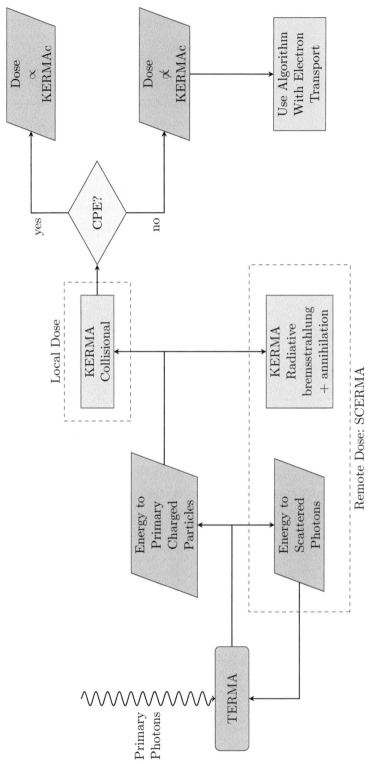

Figure 2.26: Relationship between dosimetric quantities with and without charged particle equilibrium (CPE).

BIBLIOGRAPHY

Ahnesjö, A., P. Andreo, and A. Brahme (1987). Calculation and application of point spread functions for treatment planning with high energy photon beams. *Acta Oncologica 26*(1), 49–56.

Ahnesjö, A., B. van Veelen, and A. Tedgren (2017). Collapsed cone dose calculations for heterogeneous tissues in brachytherapy using primary and scatter separation source data. *Computer Methods and Programs in Biomedicine 139*, 17–29.

Ahnesjö, A., L. Weber, A. Murman, M. Saxner, I. Thorslund, and E. Traneus (2005). Beam modeling and verification of a photon beam multisource model. *Medical Physics 32*(6), 1722–1737.

Andreo, P., D. Burns, A. Nahum, J. Seuntjens, and F. Attix (2017). *Fundamentals of Ionizing Radiation Dosimetry*. Wiley-VCH, Weinheim, Germany.

Berger, M., J. Coursey, and M. Zucker (2000). ESTAR: Stopping-power and range tables for electrons. Report NISTIR 4999, National Institute of Standards and Technology, Gaithersburg, Maryland.

Berger, M., J. Hubbell, and S. Seltzer (1999). XCOM: Photon cross sections database. Report NBSIR 87-3597, National Institute of Standards and Technology, Gaithersburg, Maryland.

Boyer, A. (1988). Relationship between attenuation coefficients and dose-spread kernels. *Radiation Research 113*(2), 235–242.

Boyer, A. and E. Mok (1985). A photon dose distribution model employing convolution calculations. *Medical Physics 12*(2), 169–177.

Evans, R. (1955). *The Atomic Nucleus*. McGraw-Hill, New York, NY.

Hubbell, M. and S. Seltzer (1999). Tables of x-ray mass attenuation coefficients and mass energy-absorption coefficients. Report NISTIR 5632, National Institute of Standards and Technology, Gaithersburg, Maryland.

ICRU (2007). Elastic scattering of electrons and positrons. *Journal of the ICRU 7*(1), 1–162.

ICRU (2011). Fundamental Quantities And Units For Ionizing Radiation (Revised). *Journal of the ICRU 11*(1), 1–35.

Iwasaki, A. (1985). A method of calculating high energy photon primary absorbed dose in water using forward and backward spread dose distribution functions. *Medical Physics 12*(6), 731–737.

Jackson, D. (2007). *Classical Electrodynamics*. John Wiley & Sons.

Johns, H., E. Darby, R. Haslam, L. Katz, and E. Harrington (1949). Depth dose data and isodose distributions for radiation from a 22 MeV betatron. *American Journal of Roentgenology, Radium Therapy, and Nuclear Medicine 62*(2), 257–268.

Kempe, J. and A. Brahme (2010). Analytical theory for the fluence, planar fluence, energy fluence, planar energy fluence and absorbed dose of primary particles and their fragments in broad therapeutic light ion beams. *Physica Medica 26*(1), 6–16.

Knoll, G. (2010). *Radiation Detection and Measurement*. John Wiley & Sons.

Mackie, T., A. Bielajew, D. Rogers, and J. Battista (1988). Generation of photon energy deposition kernels using the EGS Monte Carlo code. *Physics in Medicine and Biology 33*(1), 1–20.

Mackie, T., J. Scrimger, and J. Battista (1984). A convolution method of calculating dose for 15 MV x rays. *Medical Physics 12*(2), 188–96.

Mohan, R., C. Chui, and L. Lidofsky (1986). Differential pencil beam dose computation model for photons. *Medical Physics 13*(1), 64–73.

O'Connor, J. and D. Malone (1989). A cobalt-60 primary dose spread array derived from measurements. *Medical Physics 34*(8), 1029–1042.

Papanikolaou, N., J. Battista, A. Boyer, C. Kappas, E. Klein, and T. Mackie (2004). Tissue inhomogeneity corrections for megavoltage photon beams. Report of the AAPM radiation therapy committee task group 65. Report AAPM 85, American Association of Physicists in Medicine, Alexandria, Virginia.

Podgorsak, E. (2016). *Radiation Physics for Medical Physicists* (3rd ed.). Heidelberg: Springer-Verlag.

Russell, K. and A. Ahnesjö (1996). Dose calculation in brachytherapy for a 192-Iridium source using a primary and scatter dose separation technique. *Physics in Medicine and Biology 41*, 1007–1024.

Sharpe, M. (1997). *A unified method of calculating the dose rate and dose distribution for therapeutic x-ray beams*. Ph. D. thesis, Department of Medical Biophysics, University of Western Ontario.

Verhaegen, F., S. Palefsky, and F. DeBlois (2006). RadSim: a program to simulate individual particle interactions for educational purposes. *Physics in Medicine and Biology 51*(8), N157–N161.

Wang, L. and D. Jette (1999). Photon dose calculation based on electron multiple-scattering theory: primary dose deposition kernels. *Medical Physics 26*(8), 1454–65.

Wong, J. (1981). Second scatter contribution to dose in a cobalt-60 beam. *Medical Physics 8*(6), 775–782.

Wong, J., R. Henkelman, J. Andrew, J. Van Dyk, and H. Johns (1981). Effect of small inhomogeneities on dose in a Cobalt-60 Beam. *Medical Physics 8*(6), 783–791.

Woo, M. and J. Cunningham (1990). The validity of the density scaling method in primary electron transport for photon and electron beams. *Medical Physics 17*(2), 187–194.

Yu, C., T. Mackie, and J. Wong (1995). Photon dose calculation incorporating explicit electron transport. *Medical Physics 22*(7), 1157–1165.

Jack Cunningham holding a traditional isodose chart in transparency format. He pioneered computer programs such as IRREG, CBEAM, IBEAM and ETAR (*circa* 1972-1980), marketed on the *TP-11* system (Atomic Energy of Canada). Isodose chart collection courtesy of J.C.F. MacDonald. Photo courtesy of Jerry Battista (*circa* 2010).

CHAPTER 3

Conceptual Overview of Algorithms

Jerry J. Battista

London Health Sciences Centre and University of Western Ontario

CONTENTS

3.1 SCOPE OF CHAPTER

Tʜᴇ conceptual developments of x-ray dose calculation algorithms followed an evolutionary progression that mirrors what led to a deeper understanding of visible light. The realization that optical interference patterns could not be predicted solely by ray-tracing led to the view that light had electromagnetic wave properties. Similarly, dose patterns with hot and cold spots observed at tissue interfaces, resembling optical fringe patterns, could not be explained solely by beam rays. This type of dose turbulence was eventually attributed to secondary *charged particles* launched by x-rays.

This chapter begins with an overview of two major strategic advances in dose calculations: (1) splitting radiation dose into primary and secondary components, and (2) tissue densitometry using x-ray computed tomography (CT). Early semi-empirical methods of dose computation assumed that the energy transferred from x-rays to electrons was wholly absorbed *on-the-spot*. In the era of cobalt radiation, consideration of a significant amount of scattered photons was more important. Electron ranges of only a few mm's could easily produce charged particle equilibrium (CPE) in most tissue regions away from the patient's surface. For higher megavoltage energies, photon scattering was less prevalent but electron ranges were longer (i.e. cm's) and could disrupt equilibrium inside the body. Algorithms had to be improved, particularly in regions of low density tissue where the ranges were further elongated. Tracking of charged particle migration away from photon interaction sites became much more important. While Chapter 2 established the fundamental physics of x-ray interactions and introduced standard radiation dosimetry quantities, additional vector quantities are added in this chapter in support of the later chapters (Chapters 5 and 6) on general radiation transport theory.

The didactic philosophy adopted for this chapter is as follows (Schey 1997):

> "...we have a deep-seated conviction that mathematics ...is best discussed in a context which is not exclusively mathematical. Thus, we will soft pedal mathematical rigour which we think is an obstacle to learning this subject on first exposure, and appeal as much as possible to physical and geometric intuition."

> H.M. Schey

The main objective of this chapter is to introduce and compare three dose modelling approaches used currently in clinical radiation oncology. These methods all solve the same radiation propagation problem, but view it from a different vantage and invoke different mathematical tools: (1) convolution-superposition of Green's functions, (2) stochastic Monte Carlo simulation, and (3) deterministic solution of Boltzmann equations. These topics will be covered in specialized Chapters 4, 5, and

6, respectively. At this introductory level, the intent is to set a common stage for efficient learning across these mathematical techniques. In the concluding section, the algorithms will be categorized.

Figure 3.1 shows the main particles that must be considered in the design of a dose computation algorithm. The accelerator head has an x-ray target and fixed collimator upstream, followed downstream by variable beam-shaping collimators. The primary collimator confines the x-ray beam to a conical shape. This exiting fluence then optionally passes through a field-flattening filter that makes the downstream fluence uniform for conventional radiotherapy beams. This filter, however, can be retracted when using smaller fields at high dose rate, thereby reducing treatment time for the patient. This manoeuvre, implemented on tomotherapy and modern accelerators, minimizes undesirable extra-focal scattered radiation and also makes the x-ray spectrum less variable across the radiation field. Flat segmented ion chambers monitor the fluence rate and beam symmetry for dose calibration, quality assurance, and patient safety. The secondary collimator jaws are mobile and set the overall field size. Multi-leaf collimators (MLCs) produce irregularly-shaped fields for general organ shielding and for achieving beam intensity modulation (i.e. IMRT) through moving leaves or overlapping fields. These hardware components attenuate primary x-rays in a controlled manner but they also introduce undesirable scattered photon and charged particle contamination into the radiation field reaching the patient. During commissioning of a dose calculation algorithm, this radiation must be modelled to set initial boundary conditions for the source model and bootstrap the dose computation. For the purpose of dose computation, it is important to clarify the terms *primary and scattered radiation*. Primary x-radiation designates all photons emerging from the head of a linear accelerator. It may include photons that have pre-scattered within the accelerator hardware. This primary radiation directly reaches locations in the patient without having prior interactions *in the patient*. On the other hand, scattered radiation reaches locations along multiple routes after interacting in the overlying and underlying patient volume.

3.1.1 Contemporary Algorithms

Numerical techniques that produce practical dose distributions in patients have been published in various articles (Knöös 2017; Thomas 2016a,b; MacDermott 2016a,b; Seuntjens et al. 2014; Lu 2013; Papanikolaou and Stathakis 2009; Reynaert et al. 2007; Oelkfe and Scholz 2006). Theory and clinical aspects are also covered within textbooks (Andreo et al. 2017a; Gibbons 2016; Seco and Verhaegen 2016; Metcalfe et al. 2007a,b; Mayles et al. 2007).

(a) Convolution-Superposition Method - Green's Function Approach

This approach relies on the superposition of impulse-response functions, a technique often used to solve complex integral equations in many fields of physics and

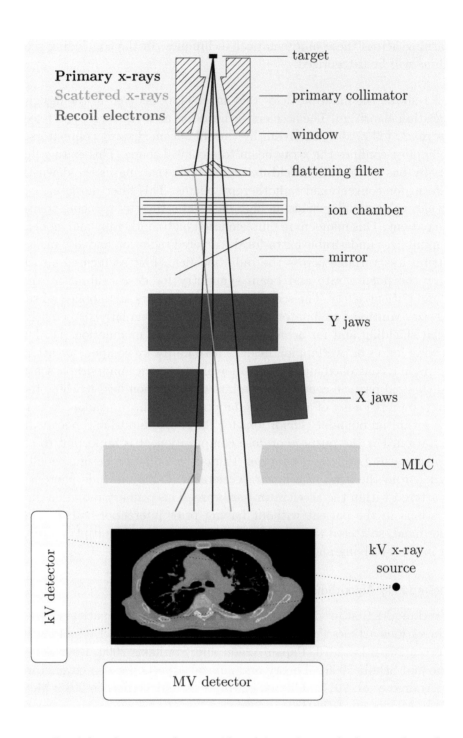

Figure 3.1: Particles that must be considered in a dose calculation algorithm. The image-guidance detectors are used for quality assurance and reconstruction of delivered dose distributions, as will be described in Chapter 7.

engineering. This has become the *most prevalent algorithm* in commercial software for clinical treatment planning (Ahnesjö et al. 1987). The required database is a set of energy deposition kernels that are pre-established by computer simulation of x-ray interactions forced to occur at an isolated point in a large homogeneous absorber. For a beam of x-rays, kernels are each intensity-modulated according to the local primary energy released (i.e. TERMA) and resultant patterns of energy propagation are overlapped to compose the overall dose distribution. In heterogeneous tissue, the situation is more complex. Ray tracing is most often used to determine radiological paths that predict not only the primary beam intensity throughout the patient, but also to reshape the kernel pattern for each pair of interaction and dose deposition sites. The convolution-superposition method offers a good compromise between the realism of radiation modelling and practical computation time.

(b) Monte Carlo Method - Stochastic Simulation

The Monte Carlo method is named after the world famous casino in Monaco. It is used to compute dose distributions resulting from a large number of individual x-ray interactions (Andreo 2017; Seco and Verhaegen 2016; Chetty et al. 2007; Reynaert et al. 2007; Verhaegen and Seuntjens 2003). A random sampling technique is applicable to the simulation of a large number of interactions because they are fundamentally governed by quantum mechanics that describes their probability of occurrence (i.e. cross-sections of Chapter 2).

The integral of any function can be evaluated by random sampling, even when the solution cannot be expressed analytically in a closed-form equation. The concept is quite simple to understand if we consider a plot of a one-dimensional function. With proper normalization, the fraction of random points (i.e. hits) that land under the curve yields the *area under the curve* or value of the integral. This approach will be explained simply in Section 3.4.1 and expanded in Chapter 5. The technique can be extended to multi-dimensional integration as required for dose computations.

The photon fluence of megavoltage x-rays delivering a radiotherapy dose fraction of 2 Gy is on the order of 10^{11} photons cm^{-2}. The number of atoms in soft tissue is also vast, on the order of 10^{23} per cm^3. This large ensemble leads to a long succession of discrete energy releases dispersed randomly throughout the irradiated region (Jenkins et al. 2012; Raeside 1976). The average dose, deposited by charged particles, and its variation in tissue voxels can be scored using the Monte Carlo method. In dose computations, however, achieving a precision of a few percent in high dose regions generally requires several million photon histories, or more in low dose regions. The disadvantage of the Monte Carlo method thus becomes apparent. Long computation times are required to reduce statistical variations that only decay by the inverse square root of the number of histories. The Monte Carlo method is well suited to parallel computing because batches of independent particle histories or interactions can be distributed across different processors. Efforts are ongoing to reduce the computation time using customized codes and multi-core processors (Rodriguez and Brualla 2018; Paudel et al. 2016; Fleckenstein et al. 2013; Jahnke et al. 2012; Hissoiny et al. 2011).

Pure mathematicians sometimes describe Monte Carlo simulation as an inelegant brute-force technique to evaluate complex integrals. Practically speaking, however, this method still has a very strong following not only in scientific fields but also in socio-economic simulations (Kancharla 2013). The technique is also easy to understand intuitively, facilitating trouble-shooting of software at the single event level. In medical physics, the Monte Carlo simulation method has been the *de facto* gold standard technique for many decades although this is now being challenged by proponents of Boltzmann equation solvers described below. It has proven to be highly flexible and an invaluable tool for evaluating the assumptions and performance of weaker energy-propagation models, and for education. The Monte Carlo method reduces systematic uncertainty by using probability functions based on proven fundamental physics (i.e. interaction cross-sections) but it is subject to stochastic random uncertainties. These can be mitigated by increasing the number of radiation particles simulated. A recent review lists over 20 Monte Carlo computer codes specialized for clinical treatment planning or dose quality assurance (Brualla et al. 2017).

(c) Solving Boltzmann Transport Equations - Deterministic Method

The flow of radiation particles can be expressed in a set of coupled equations based on interaction cross-sections that constrain the trajectories and energy exchanges. In our application, we will focus on a two-particle system of equations with mutual energy transfers between x-rays and charged particles (MacDermott 2016b; Varian 2015b; Failla et al. 2015; Gifford et al. 2006). Radiation flow is described by vector quantities introduced later in this chapter (Section 3.4.2). These vectors can be mapped, much like electric fields or wind patterns, with a local intensity and direction. The radiation field can be imagined to permeate the tissue space, constrained by consistency of energy flow across tissue voxels. Readers are strongly encouraged to refresh their knowledge of vectors in multi-dimensional space before reading Chapter 6 (Schey 1997; Spiegel 1959).

Ultimately, the solution to the Boltzmann equations yields the spatial distribution of charged particle fluences, subsequently converted to a dose distribution (via Equation 2.23 of Chapter 2). This method is conceptually different and mathematically more challenging when compared with convolution-superposition or Monte Carlo methods. The Boltzmann method is deterministic in its nature, yielding the *expectation value* of doses without statistical fluctuations, i.e. assuming a high level of radiation exposure. The efficiency of deterministic problem solving, compared to Monte Carlo simulation, has been analysed theoretically (Borgers 1998; Bielajew 2016). The speed of computation is dependent upon the dimensionality of the problem being solved (i.e. the *curse of dimensionality*). A dose distribution calculated by either method is expected to converge for a large number of Monte Carlo histories, provided that the input radiation physics data, patient-related data, and run-time parameters (e.g. voxel size, dose matrix dimensions) are matched.

3.2   FIRST GENERATION ALGORITHMS: HISTORICAL PERSPECTIVE

While this book focuses mainly on contemporary modes of dose computation, this section includes a historical perspective (Metcalfe et al. 2007a; Ahnesjö and A. 1999; Wong and Purdy 1990; Cunningham 1989). Early algorithms developed in the 1960s faithfully reproduced dose distributions that had been previously measured experimentally in a water phantom. Dose distributions for each beam could then be rapidly recalled and added for multi-beam treatment plans, accounting for the external surface of the patient. This automation replaced manual treatment planning performed by summing intersecting dose values on transparencies of isodose charts. Tissue-air ratios (TARs) became the database of choice for dose computation (Johns et al. 1956). These quantities reflect the combination of primary beam penetration and build-up of scattered photons at depth in a homogeneous absorber. Scatter-air ratios (SARs), on the other hand, isolate dose due only to scattered radiation. Similar quantities were defined later when dose calibration was switched from in-air to in-phantom for higher energy x-ray beams. These quantities were renamed accordingly to tissue-phantom (TPR) and scatter-phantom (SPR) ratios, respectively. This divide-and-conquer strategy of isolating scattered from primary radiation greatly improved the predictive capability of dose calculation algorithms. Irregularly-shaped fields could be modelled — well ahead of the availability of multi-leaf collimation. The effects of missing tissue in tangential fields could also be modelled accurately. In retrospect, **scatter functions were key to the development of more versatile and accurate dose calculation methods.** They paved the way to the next generation of algorithms based on convolution principles using scatter kernels.

In parallel with these advances, Monte Carlo simulation of radiation transport was being adapted to medical physics energies and applications (Rogers 2006; Verhaegen and Seuntjens 2003; Andreo 1991; Raeside 1976). The infrastructure codes originated in major national research laboratories, including ETRAN and ITS (National Institute of Standards and Technology, USA), EGS (Stanford Linear Accelerator Center - SLAC) in collaboration with the National Research Council of Canada, PENELOPE (University of Barcelona, Spain), MCNP (Los Alamos National Laboratory), GEANT (European Organization for Nuclear Research - CERN), and PEREGRINE (Lawrence Livermore National Laboratory). The types of simulated particles, their energies, and applications have been reviewed (Andreo 2017; Brualla et al. 2017; Seco and Verhaegen 2016). Figure 3.2 provides a summary of the progress achieved in algorithm capabilities near the end of the 20th century.

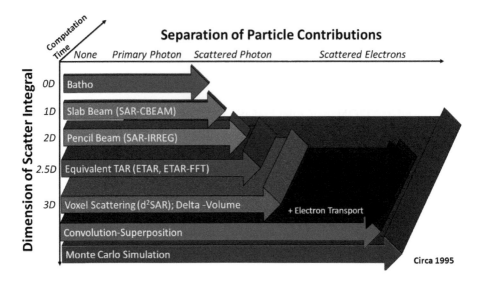

Figure 3.2: Dose algorithms *circa* 1995. Batho = Batho's power-law method; SAR = scatter-air ratio; CBEAM and IRREG = program module names; FFT = fast Fourier transform; d²SAR = doubly-differentiated SAR values with respect to field radius and depth. The projected depth of each arrow indicates the computational burden. This graphic was created by Michael Sharpe during his doctoral research program.

3.2.1 First Giant Leap: Splitting of Primary and Secondary Dose

Knowledge of the relative magnitude of primary and secondary dose contributions is important because it sets priorities for improving the accuracy of dose calculations. Tissue inhomogeneities affect primary and secondary particles in a different way and span different spatial domains. In Chapter 2, we reviewed exponential attenuation of primary rays, and described the complex influence that a single voxel can have on its scattering amplitude and secondary attenuation of photon fluences passing through the voxel (Figure 2.20). Empirically, the primary and secondary dose contributions can be decoupled at depth in a large homogeneous medium as follows:

$$D(d, FS) = D_p(d, FS \approx 0) + D_s(d, FS) \qquad (3.1)$$

where d is the depth and FS is the radiation field size. Figure 3.3 shows the shape of a typical empirical curve for the total dose at a central axis point in a large water phantom, as a function of field size. With increasing field size, the total dose continues to climb because of the build-up of scattered radiation. With decreasing field size, there is a smooth reduction in dose and an extrapolation onto the y-axis helps identify the primary collision KERMA$_{c,prim}$. Note that the ≈ 0 symbol in equation 3.1 is only intended to signify this extrapolation procedure. This quantity equals the saturation value of the primary dose, for field sizes that are wide enough to establish lateral charged particle equilibrium (CPE). For high energy megavoltage beams, the

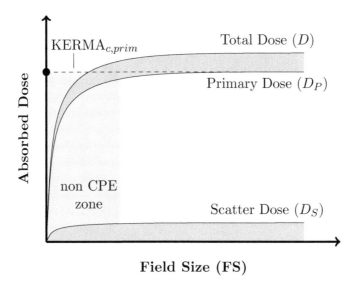

Figure 3.3: Lateral build-up curves at a central axis point for high-energy megavolt-age x-ray beams, including primary and secondary components of dose in a water phantom. The y-intercept identifies the primary collision KERMA, $KERMA_{c,prim}$, which equals the saturation value of primary dose, D_P, under conditions of charged particle equilibrium (CPE).

extended lateral migration of secondary charged particles causes a wider build-up region with inequality of primary KERMA and primary dose. As the beam size narrows to a critical field size, the abrupt decline of primary dose occurs because of loss of lateral charged particle equilibrium. More energy is scattered away from the core of the beam by charged particles than is injected back in from lateral regions. Empirical scatter functions measured at the central axis as a function of energy and field size reflect lateral scattered radiation patterns, including both photons and charged particles (Woo et al. 1990).

Figure 3.4 shows a plot of the scatter-to-total dose ratio (SAR/TAR) at the central axis of a cobalt beam incident on water, as a function of depth and field radius. As a function of depth in an absorber, the dose contribution from scattered radiation increases while primary dose reduces exponentially. The fractional scatter is spatially-variable and builds up to a significant proportion of approximately 50% for deep voxels in large fields of radiation. The consequence for dose calculation algorithms is that variation in scattered radiation must be modelled. Furthermore, this scatter component is strongly affected by tissue densities such that the pattern seen in Figure 3.4 cannot be applied universally to all situations in a patient. This is the core reason why computation of dose distributions requires significant resources and is often so time-consuming.

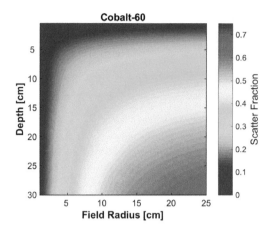

Figure 3.4: Scatter fraction (i.e. SAR/TAR) at the central axis of a cobalt beam incident on a water phantom, as a function of depth and field radius. Data source (Johns and Cunningham 1974).

3.2.1.1 Differential Scatter Functions

The total scatter contribution can be decomposed into the scatter from sub-elements of an absorber using a progressive mathematical peeling process (see Figure 3.5). Differentiation of scatter quantities (e.g. SARs or SPRs) with respect to circular field radius, starting from the largest radius, yields the scatter that comes from shells of the scattering medium. If the beam is non-divergent, the shell is cylindrical in shape. If the beam is divergent, it is conically-shaped like an empty ice cream cone. A segmentation of the shell scatter with respect to azimuthal angle yields the scatter from a divergent column or pencil beam. A further differentiation of the pencil shape with respect to depth, starting at the surface of the absorber, yields the scatter contribution from individual layers of voxels (Cunningham and Beaudoin 1974). The shapes of different scattering elements are illustrated on the right side of Figure 3.5. Note that scattering from these elemental volumes includes a majority of single-scattering events originating within, and some pre-scattered photons originating in inner exposed zones that are then backscattered towards point P by the element under consideration.

3.2.1.2 Iso-Influence Kernels

Extraction of voxel scattering quantities from empirical scatter functions is very difficult because of confounding effects of multiple scattering and noise in differential dose measurements. However, they can be determined by Monte Carlo simulation.

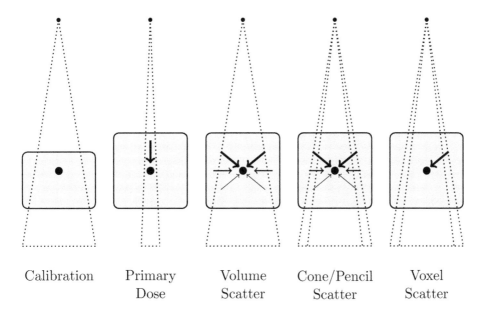

Figure 3.5: Illustration of differential forms of scatter functions, isolating the scatter influenced by conic shells, pencil beams, and voxels. For simplicity, only single-scattering pathways are illustrated .

The scattering effect from surrounding voxels to a destination point can be visualized as iso-influence functions by inverting energy deposition kernels. In this reciprocal view, irradiation with x-rays activates scattering voxels with an intensity proportional to the local primary TERMA. Figure 3.6 (top row) shows two kernels for large and small absorbing spheres of water with radii of 60 and 10 cm. The scattered dose contributions to a voxel depend on the overlying depth of water and field size. In Figure 3.6 (bottom row) the *fractional* energy (i.e. scatter/total) deposited by each type of interaction is segregated for *monoenergetic* 1.25 and 6 MeV photons.

The primary contributions for each energy are essentially the same for the two kernel sizes because each is large enough to encompass the range of primary charged particles. For a large field irradiation with 1.25 MeV photons representing cobalt radiation, the scatter fraction agrees well with values exhibited in Figure 3.4 for wide fields and deeper locations. This level of agreement shows the linkage that exists between theoretical energy deposition kernels and traditionally measured data sets for cobalt-60.

3.2.1.3 Scattered Dose Distribution

Figure 3.7 shows the total and scattered dose distributions for a cobalt beam (SSD = 100 cm, FS = 30×30 cm^2) obtained by a convolution procedure described later (Section 3.3.1). The scatter fraction (D_s/D) is also mapped in Figure 3.7C; it is

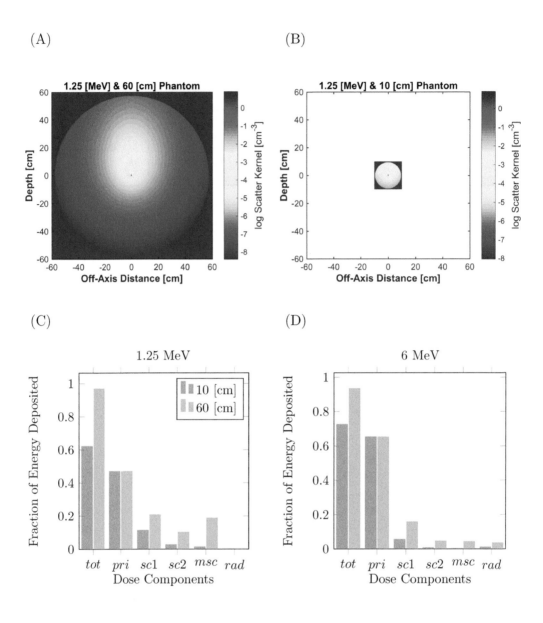

Figure 3.6: (A) Very large (120 cm width) and (B) Small clipped iso-influence kernels (20 cm width) for 1.25 MeV photons. (C) Fraction of energy deposited for monoenergetic photons of 1.25 MeV. D. Fraction of energy deposited for monoenergetic photons of 6 MeV. tot = total; pri = primary; sc1 = single Compton scattering; sc2 = double Compton scattering; msc = higher-order multiple Compton scattering; rad = bremsstrahlung and annihilation radiation. The total fraction is less than unity because some energy escapes the finite kernel space.

compatible with empirical SAR/TAR data shown in Figure 3.4 for a large field size and depth.

3.2.2 A Precursor Pencil Beam Algorithm

An algorithm that extended Clarkson's radial integration method for application to irregularly-shaped intensity modulated fields was proposed by Johns and Cunningham over 45 years ago (Johns and Cunningham 1974)! This method was well ahead of its time, before IMRT was even in the vocabulary of radiation oncology. The goal of this retrospective analysis is to demonstrate the conceptual link that exists between an SAR-based method and contemporary pencil beam methods. Consider a beamlet that passes through the mechanical compensator, depicted in Figure 3.8. The patient is assumed to be composed entirely of water-like tissue bounded by an irregular surface. The Cartesian origin is located at the central axis point (P_0). The calibration dose to a small mass of tissue in open air space, D_0, is known at a source distance $(SSD + d_m)$, with the filter removed. The filter function f(x,y) essentially describes the primary transmission through the filter, relative to an open uniform field. The goal is to calculate the dose at a generic point Q_1, by summing the primary dose along the trajectory $P_1 - Q_1$ and the scattering from all neighbouring pencil beams along rays such as $P_2 - Q_2$.

The attenuated *primary* dose in tissue at point Q_1, with the beam filter in position, is given by:

$$D_p(Q_1) = \left[D_0 \left(\frac{SSD + d_m}{SSD + d} \right)^2 f(x_1, y_1) \times TAR(d_1, FS \approx 0) \right] \tag{3.2}$$

where the inverse square term accounts for primary source divergence and d_1 is the overlying oblique depth at point Q_1 in water. The primary attenuation through tissue is embedded in the zero-field TAR value on the right of equation 3.2.

The computation of the scattered dose component requires a two-dimensional (2D) integration of contributions from pencil beams, each exposed to a different photon fluence transmitted through the beam filter, $f(x_2, y_2)$. If we consider $P_1 - Q_1$ as a pseudo-central ray, we obtain:

$$D_s(Q_1) = D_0 \left[\frac{SSD + d_m}{SSD + d} \right]^2 \times \sum_\theta \sum_r f(x_2, y_2)$$
$$\times \left[SAR(d_2, r + dr) - SAR(d_2, r) \right] \frac{d\theta}{2\pi} \tag{3.3}$$

where d_2 is the oblique depth to point Q_2 along the pencil ray line, and r is the radial distance from the off-axis point Q_1. The quantity to the left of the double-summation

(A) (B)

(C)

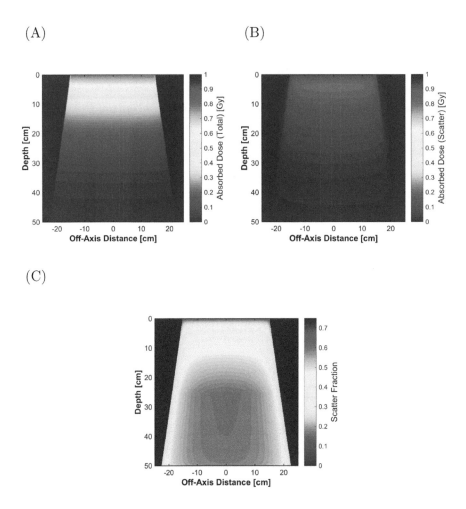

Figure 3.7: A. Total dose distribution. B. Scattered dose distribution C. Ratio of scatter-to-total dose (SAR/TAR) for a cobalt-60 beam (SSD = 100 cm, field size = 40×40 cm^2. Data obtained by convolution method, without tilting of kernels along divergent beam rays.

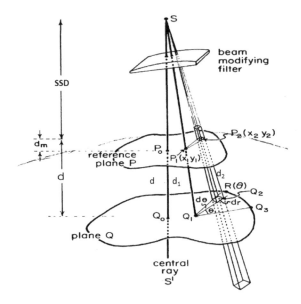

Figure 3.8: Pencil beam prototype based on differential scatter-air ratios (dSARs). Reproduced with permission (Johns and Cunningham 1974).

is the dose to Q_1 in air, without the filter in place. The double-summation produces a custom scatter-air ratio for the beam. The filter function inside the summation term tracks the variation in fluence incident on each pencil beam at the patient surface. The pencil beam scattering is derived by differentiating SAR values with respect to a radial increment $(r + dr)$ along $Q_1 - Q_2$, and normalizing for angular increment $\frac{d\theta}{2\pi}$. The summation *is* a 2D integration that will re-appear much later in the development of the convolution-superposition algorithm (Section 3.3.3.1). The total dose is given by the sum of primary and scattered dose for any point Q_1:

$$D(Q_1) = D_p(Q_1) + D_s(Q_1) \qquad (3.4)$$

This pencil-beam model demonstrated the power of decomposing primary and scatter components and re-assembling them in smart ways to predict dose distribution for a much wider range of field conditions. However, the calculation assumed a patient composed of homogeneous tissue. This simplified view of the patient would soon change with the introduction of x-ray CT scanning.

3.2.3 Second Giant Leap: CT Tissue Densitometry

Historically, dose distributions were first calculated for a homogeneous water-like patient and secondary correction factors were subsequently applied to modify the dose distribution for radiation passing through heterogeneous tissue. In traditional algorithms, secondary corrections are then applied point wise to account for internal tissue inhomogeneity. The matrix of inhomogeneity correction factors (ICF) is applied in 3D space ($\mathbf{r}$) per discrete voxel location (i,j,k) as follows:

$$D_{inhomo}(\mathbf{r}(i,j,k)) = D_{homo}(\mathbf{r}(i,j,k)) \times ICF(\mathbf{r}(i,j,k)) \tag{3.5}$$

Note that this is *not* a matrix multiplication procedure with the usual mathematical interpretation applied to vector analysis. It is rather an element-wise multiplication of paired matrix elements (i,j,k). Algorithms that use this 2-step approach are called *correction-based* (Type A as described in Section 3.6.2). The tissue inhomogeneity correction factor (ICF) most often relied on primary ray tracing through voxels in a slice of the patient's anatomy, assuming symmetry of adjacent slices. Before CT scanning became routinely available to radiotherapy departments, the internal shapes of lungs, for example, were inferred from anatomy atlases, radiographic images, or non-computed tomography. Interior tissue regions were outlined by a set of contours entered by a digitizer pen and enclosed areas were assigned bulk tissue densities. Primary fluence corrections were applied using a water-equivalent depth to each dose calculation point. Beam rays were traced through contoured structures with each having an assigned bulk density. The Batho power-law method introduced better accounting of variable scattering conditions by accounting for the proximity to an inhomogeneity border.

CT scanning rapidly developed in the mid-1970s and opened the possibility for 3D conformal therapy based on digital imaging and quantitative *in vivo* densitometry. In his Nobel Prize lecture, Dr. Allan Cormack emphasized that his personal motivation for developing computed tomography (CT) was to improve treatment planning of cancer patients. *Pixel-by-pixel* algorithms rapidly emerged and replaced vector-based dose calculation methods. CT numbers expressed in Hounsfield units (HU) and normalized to water (μ_w) as a universal substance were adopted by the diagnostic imaging community, as follows:

$$N_{CT}[HU] = 1000 \times \left(\frac{\mu_i - \mu_w}{\mu_w} \right) \tag{3.6}$$

where attenuation coefficients (μ_i, μ_w) are specific to the absorber and imaging energy – typically 120 kVp or effectively 70 keV for a diagnostic scanner. This renormalization assigns a value of 0 HU to water as a baseline, -1000 to vacuum, and 1000 to dense tissue that attenuates x-rays at double the rate of water (per cm). Muscle, fat, and respiratory-averaged lung tissue have typical values of $+30$ HU, -100 HU, and -700 HU. Bone normally spans a very wide range of CT number values because of its variable core density and content (e.g. bone marrow); values exceeding $+500$ HU are common for bone or for prosthetic materials. The ratio of linear attenuation coefficients can be re-expressed in terms of electronic interaction cross sections for megavoltage photon interactions, as follows:

$$\frac{\mu_i}{\mu_w} = \frac{\sigma_{e,i} \times \rho_{e,i}}{\sigma_{e,w} \times \rho_{e,w}} = 1.00 + 0.001\, N_{CT} \tag{3.7}$$

where σ_e is the total interaction cross section (per electron) and ρ_e is electron density (e per cm^3). The ratio of electron densities is often used in dose algorithms to re-scale radiation parameters such as photon and charged particle path lengths, relative to their values in water. The electron density, relative to liquid water, ρ'_e, is related to CT Numbers as follows:

$$\rho'_e = \frac{\rho_{e,i}}{\rho_{e,w}} = \left(\frac{\sigma_{e,w}}{\sigma_{e,i}}\right) \times (1.00 + 0.001\, N_{CT}) \tag{3.8}$$

The correlation of relative electron density with CT Number is therefore linear *for a given tissue type only*, with a slope dependent on tissue composition (Battista and Bronskill 1981). Fortunately, for water-like absorbers, the ratio of cross sections is close to unity for all soft tissues. For fatty tissue, this ratio has a slightly elevated value of 1.05 because of a richer carbon content. Conversely for bone, the leading term is approximately 0.8 for a CT effective energy of 70 keV. For example, a CT Number of 1000 corresponds to a relative electron density of 1.6 for bone while the same CT number corresponds to a relative density of 2.0 for soft tissue. The gradual transition in slope from soft-tissue to bone is seen in Figure 3.9. The plot was produced from images of common anatomical regions in the same patient, using kV and MV CT scans. Kilovoltage CT numbers were plotted against relative electron density measured independently using megavoltage (MV) CT imaging on a tomotherapy unit. For MVCT scan data, the leading term in Equation 3.8 is unity for a very wide range of tissues and elements, up to and including heavy metals (Rogers et al. 2005).

Note that CT numbers obtained by cone-beam scanning (CBCT) require correction for in-patient scattering that is *not* reflected in the above discussion and equations related to fan-beam scanning. The direct use of raw CBCT numbers can lead to erroneous and misleading dose results and is not encouraged (Disher et al. 2013; MacFarlane et al. 2018).

We have over-stated the need to determine electron density of tissue, because the Compton effect is clearly dominant for megavoltage x-rays incident on soft tissue (see Chapter 2). For higher atomic number (Z) materials such as bone and teeth, and prosthetics or implanted devices, pair production and Coulomb nuclear scattering are accentuated. If these effects are deemed important clinically, atomic composition can be determined using dual energy CT scanning (DECT) (McCollough et al. 2015) or segmentation of tissue types in MRI scans (Edmund and Nyholm 2017).

The calculation of primary dose involves ray tracing along beam rays from the radiation source into the patient space. Better accuracy is achieved in primary dose calculations if the linear attenuation coefficient for each medium encountered along a ray is used instead of density-scaled values (e.g. not approximating bone

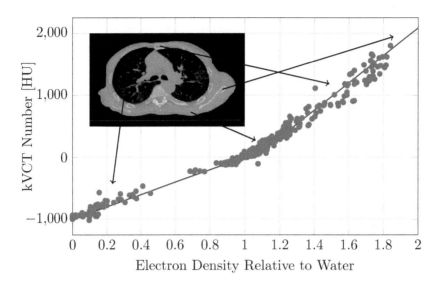

Figure 3.9: Correlation of kilovoltage (kV) CT number with relative electron density measured independently by megavoltage (MV) CT in homologous tissue in a patient who underwent both scans. Adapted from (Prasad et al. 2005).

as compressed water). The practice varies across and even *within* dose computation algorithms that may rely partially or wholly on density-scaling for primary and scatter dose calculation. For example, a convolution-superposition algorithm may use media-specific attenuation coefficients for the primary (i.e. TERMA) ray tracing calculation, but use density-scaling for the energy deposition kernels in a heterogeneous medium.

3.2.4 First Voxel-Based Algorithms

The equivalent-TAR (ETAR) method heralded the start of 3-dimensional dose computations with incorporation of patient-specific tissue density maps obtained by CT scanning (Sontag and Cunningham 1978). Each voxel was viewed as a scattering element with relative weight resembling the values in an iso-influence kernel later introduced for more advanced algorithms (Figure 3.6A). In addition to using a water-equivalent *depth* along primary beam ray line for primary dose calculation, the ETAR method introduced a novel scaling of *field size* based on the integration of weighted scatter contributions. The effective field size was intended to reflect the scattering environmental conditions for each dose calculation point. Because of computer limitations at the time, a 3D integration was not practical. Using a separation of variables technique, the three-dimensional (3D) scatter contribution was collapsed onto a single scattering slab placed at an effective distance from the dose calculation slice. Because of this simplification, the original version of the ETAR method was often described as being 2.5D in its mathematical nature. The delta-volume method (Wong and Henkelman 1983) implemented a 3D structure using analytically-derived kernels based on Compton scattering (review Chapter 2, Figure 2.20). A 3D version of ETAR was also eventually implemented, taking advantage of fast Fourier transform techniques (Yu and Wong 1993). In retrospect, these methods were precursors of superposition-convolution algorithms, using inverted scatter kernels (Cunningham and Battista 1995).

These calculation methods still relied on the assumption of *on-the-spot* absorption of charged particles. This became progressively untenable for higher megavoltage energy incident on low-density tissue. Higher energy beams interacting in lung released charged particles with stretched ranges. Comparisons of dose predicted by traditional algorithms with measured data revealed large errors in both central depth-dose and lateral beam profiles. With a large influx of megavoltage accelerators replacing cobalt-60 units, the weakening foundation of semi-emprical algorithms became apparent. The problem is defined in Figure 3.10. Figure 3.10A shows a depth profile of inhomogeneity correction factors along the central axis of a polystyrene-cork slab phantom for a 15 MV x-ray beam(Mackie et al. 1985). For a field size of 5×5 cm^2, the large depression in the central axis dose within the cork section was grossly over-predicted by first generation algorithms, by as much as 15%. This central effect was most pronounced in smaller fields, but ubiquitous near beam edges for *any* field shape and size (Figure 3.10B) (Young and Kornelsen 1983). With growing interest in radiosurgical and conformal irradiation techniques for small lung tumours, these observations spurred developments of the next generation of dose calculation algorithms. The initial focus of attention was on grafting charged particle transport to photon-only algorithms (Yu et al. 1995).

During the same era, Monte Carlo simulation was also introduced to medical physics. The credibility of virtual Monte Carlo results began to surpass that of real experimental data in some situations! However, the computation speed was still unacceptably slow for interactive treatment planning. The Monte Carlo method was relegated to the roles of commissioning and dose verification of faster algorithms.

3.3 SECOND GENERATION ALGORITHMS

The next two generations of algorithms are often described as *model-based* because of their reliance on the underlying fundamental physics of radiation interactions rather than empirically-derived quantities.

3.3.1 Convolution and Superposition Concepts

The concept of convolution was introduced in Chapter 2 to demonstrate the release of primary photon energy (TERMA) and associated spreading of scattered energy. In dose computations, the pattern of energy spread from a single interaction point was described as an *energy deposition kernel* (EDK) (Mackie et al. 1988). The ker-

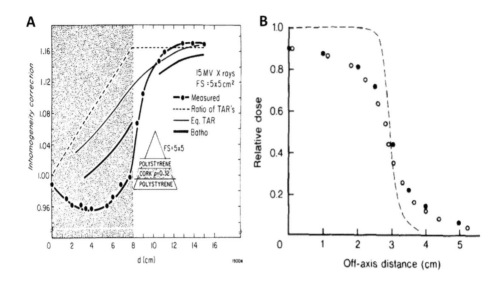

Figure 3.10: A. Hammock-shaped depression in the measured central axis dose for a low-density medium exposed to 15 MV x-rays. Reproduced with permission from (Mackie et al. 1985). B. Measurements of degraded penumbra (points) for a 10 MV x-ray beam incident on a low-density absorber. Reproduced with permission from (Young and Kornelsen 1983).

nel is closely related to the *point spread function* used to characterize the spatial resolution limits of any imaging system. In the context of a radiation treatment beam, energy deposition kernels cause blurring of the sharply-defined TERMA distribution. The size of the kernel ultimately limits how small a tumour can be treated!

To introduce the concept of convolving, a variety of analogies may appeal to different readers: (1) The observable universe was traced back 13.8 billion years to the *Big Bang* impulse (i.e. TERMA). This event released all the mass-energy that now constitutes all known galaxies. The scatter spread kernel contains an estimated total of 10^{23} stars (https://www.space.com/26078-how-many-stars-are-there.html); (2) The sharpness of a pointillist's painting is limited by the bristle pattern (i.e. the kernel) and stroke intensity (i.e. TERMA) of the artist's paint brush; (3) In a fireworks display, the burst pattern of each explosive element represents a kernel. If several elements overlap at nearly the same time, the observer's eye integrates the light to form a colourful collage (i.e. the dose distribution); (4) A perfume bottle with an atomizer (i.e. kernel) can be used to superimpose a desired pattern of aromatic molecules onto a skin surface (i.e. dose distribution); (5) Painting of a large surface (i.e. dose distribution) can be accomplished through intensity-modulated bursts from the nozzle of a spray paint can (i.e. kernel); (6) A lawn can be watered quite uniformly using a hose with variable pressure (i.e. TERMA) and nozzle spray pattern (i.e. kernel). This last analogy was explored in Chapter 2.

The development of convolution-superposition methods spawned new insights and raised awareness of tissue inhomogeneity issues for clinical cases that could potentially be influenced negatively by charged particle disequilibrium. The primary photon fluence was therefore clearly decoupled from the spread of energy by progeny charged particles. The convolution-superposition framework allowed consideration of the longer range of recoil charged particles. As computational power escalated the algorithm's predictive power was further improved by replacing the convolution integral with a more general superposition integral. The energy deposition kernels were modified locally to account for tissue inhomogeneity, x-ray spectra, and beam divergence. The kernels were no longer spatially invariant; this gave rise to concerns over the validity of using standard Fourier techniques (Wong et al. 1997).

We begin by refreshing the concept of convolution before presenting the mathematical basis for superposition. Consider the x-ray interaction sites as launch pads located at energy-sending positions ($\mathbf{s}$). The portion of energy arriving and deposited at surrounding *receptor* locations is defined for each type of interaction (i) occurring at the launch point. This is an energy-sensitive process so that it requires a summation over the x-ray spectrum, E, (Ahnesjö et al. 1987). The dose at any receiving location $\mathbf{r}$ is then given by

$$D_I(\mathbf{r}) = [1/\rho(\mathbf{r})] \int_E \iiint_V T_E(\mathbf{s})\, \rho(\mathbf{s})\, k_i(E, \mathbf{r} - \mathbf{s})\, \mathrm{d}^3 s\, \mathrm{d}E \qquad (3.9)$$

where T_E is the energy spectrum of the primary TERMA ($\frac{\mathrm{dTERMA}}{\mathrm{d}E}$), $k_i(E)$ designates a kernel for interactions of type i for a monoenergetic launch energy, E. The variable $\mathbf{r}$ specifies the location of any voxel in the dose calculation matrix. The kernels are normalized as fractions of the total energy released (i.e. TERMA × mass) at $\mathbf{s}$ that is deposited in a voxel at $\mathbf{r}$, *per unit volume* of this receiving voxel.The integral then sums the energy contributions from all sending voxels, in units of energy per unit volume of the receiving voxel. In the convolution integral, the exchange of energy between interaction-deposition voxel pairs is wholly defined by the vector that joins them, ($\mathbf{r} - \mathbf{s}$). The leading density term is necessary to obtain dose (i.e. energy per unit mass) at the receiving voxel. The expression for total dose from all interaction processes (i) is the sum:

$$D(\mathbf{r}) = \sum_i D_i(\mathbf{r}) \qquad (3.10)$$

where the energy spread pattern is assumed to be identical from all energy launch sites. However, realistic conditions (Field and Battista 1987) invalidate the assumption of *spatial invariance* of kernels at each location $\mathbf{s}$. In order of importance, the real-life complications are:

(1) **Tissue Inhomogeneity and Tissue-Air Boundary**: Kernels are determined for a large homogeneous water phantom. These can be severely perturbed by tissue inhomogeneities or proximity to the external surface of the patient. These can cause severe distortions of the in-water kernel shapes. Density-scaling along kernel rays between interaction-deposition pairs of voxels provides an approximate technique for remapping scattered energy, provided that variations in tissue atomic number are minimal (Mackie et al. 1984).

(2) **Polyenergetic Beams**: The x-ray spectrum changes at each interaction point in the patient due to beam hardening. The attenuation of the primary TERMA and the application of kernels for spectral energies are therefore dependent on each interaction location in the patient tissue. A convolution over each spectral component accommodates this effect, but comes at the expense of greater computation time. Practical approximations have been devised to reduce the number of convolutions (i.e. one per spectral component) to a single convolution using energy-effective composite kernels. The energy spectrum of the TERMA, normally chosen at a typical depth in tissue, is used to weight each monoenergetic kernel in the formation of a polyenergetic kernel (Sharpe et al. 1997). The clear separation of KERMA into collisional KERMA$_c$ and scatter SCERMA terms with associated normalized kernels (Chapter 2) allows inclusion of x-ray beam *hardening* at depth in tissue and beam *softening* at off-axis locations of a beam.

(3) **Beam Divergence**: Strictly speaking, kernels should be re-oriented along divergent rays emerging from the x-ray source. In the Cartesian grid most often used in dose computations, the kernel orientation must be tilted to align with off-axis rays (Sharpe and Battista 1993). Alternatively, a fan-beam coordinate system can be implemented to force parallel-beam geometry during convolution, followed by the inverse re-mapping of voxels back to Cartesian space.

Collectively, the above complications can have a significant impact on dose accuracy. To preserve accuracy, the convolution integral *must* be replaced by the more general superposition integral. The interaction-deposition voxel pairs are then described by their *absolute* positions $(\mathbf{r}; \mathbf{s})$ instead of *relative* locations $(\mathbf{r} - \mathbf{s})$. This allows consideration of the density environment in the intervening space in-between $\mathbf{s}$ and $\mathbf{r}$. The superposition integral thereby replaces equation 3.10 as follows:

$$D_I(\mathbf{r}) = [1/\rho(\mathbf{r})] \int_E \iiint_V T_E(\mathbf{s}) \, \rho(\mathbf{s}) \, k_i(E, \mathbf{r}; \mathbf{s}) \, \mathrm{d}^3 s \, \mathrm{d}E \qquad (3.11)$$

3.3.2 Tissue Inhomogeneity Revisited

3.3.2.1 *Effect on Primary* TERMA

Tissue inhomogeneity has posed the biggest challenge in the design of any dose calculation algorithm. This is also the case for convolution-superposition algorithms.

However, the explicit splitting of primary and secondary dose has made the solution more evident. The *primary* energy released at each point in the patient is subjected to non-uniform attenuation. The attenuated TERMA can be computed relatively easily by ray tracing through voxels along primary routes (review Equations 2.11 and 2.12). The mass and linear attenuation coefficients vary along these paths and are specific to the tissue or material traversed. Identification of the voxel contents and electron density is most often accomplished by using CT number ranges, with conversion to required coefficients for the photon energy under consideration (via Table 2.1).

3.3.2.2 Effect on Energy Deposition Kernels

The secondary impact of tissue inhomogeneity on the energy spread kernels, however, is more complex and profound. For extended regions of a different density but common atomic number, dose spread kernels can be simply re-scaled by the relative electron density, ρ'_e (Equation 3.8). This is justified by two theorems developed originally by O'Connor for scattered photon fluences and by Fano for recoil electron fluences (Bjarngard 1987). These results are the consequence of interaction probabilities *per unit mass* not being affected strongly by the physical state of the material (e.g. gas or solid). This condition generally holds true for atomic elements normally found in human tissue. There is a small exception caused by polarization effects on stopping powers for megavoltage electrons and positrons. According to these theorems, kernel values for *large* homogeneous media of non-unit density can be indexed in an all-water kernel using water-equivalent ray-line distance (r) from the kernel origin. This radiological distance in a medium (med) is $r_w = \rho'_e \times r_{med}$, where ρ'_e is the relative electron density of the uniform neighbourhood (Mackie et al. 1984). Note that this density scaling ignores atomic number effects. In contrast, consideration of attenuation coefficients in the primary TERMA calculation is sensitive to atomic number, provided the type of traversed tissue is identifiable by CT number, for example.

The presence of small inhomogeneities localized along individual kernel rays merits further discussion. We still use water-equivalent distances, i.e. $r_w = \int \rho'_e \, dr$ where dr is a ray-line increment along fan-line directions. The kernel shape is deformed by stretching or contraction *along each fan-line*. This is justified because the bulk of the energy is deposited by primary charged particles and first-scattered photons (recall Figures 3.6C and D). These particles travel *paraxially* along the *line of sight* between source and destination points – the way the crow flies. The charged particle trajectories deviate somewhat from pure straight line paths, but their range is much more confined.

A *primary-only* energy deposition kernel of 6 MeV photons for a layered phantom is highlighted in Figure 3.11. The distortion in kernel shape, due to a large cylindrical volume of air, was calculated by two methods (Woo and Cunningham

1990): (a) density-scaling along kernel ray lines between source-dose pairs of voxels, and (b) more accurate Monte Carlo simulation of charged particle trajectories. The kernel shapes are in general agreement, but there are localized discrepancies. Practically-speaking, these effects are less exaggerated for a *total energy* kernel defined for a polyenergetic beam incident on more typical anatomical structures with greater densities.

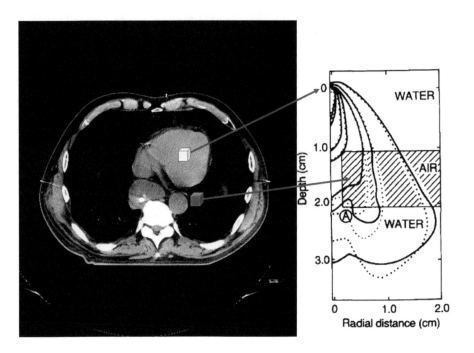

Figure 3.11: Distortion of *primary* water dose spread kernel using ray tracing and density-scaling from the interaction to dose deposition sites (dotted lines) or detailed Monte Carlo simulation (solid lines) for 6 MeV photons. Data on the right was adapted with permission (Woo and Cunningham 1990).

For multiple photon scattering, the diversity of photon pathways suggests that density averaging should take place over a much wider space, beyond direct flight paths. The original implementation used a global average of all tissue voxels (Mackie et al. 1984). Others have suggested an average restricted to the radiation-exposed sub-volume. Sharpe (Sharpe 1997) proposed a *weighted* average over surrounding exposed voxels, using the kernel values for each type of interaction as weights (as in Figure 3.6A) but this refined concept was not widely adopted in practice. This type of averaging process would then have been similar to that used in the ETAR method (Sontag and Cunningham 1978).

Figure 3.12 shows the result of a convolution calculation for an all-water phantom and superposition calculation for a lung-like slab phantom exposed to a mo-

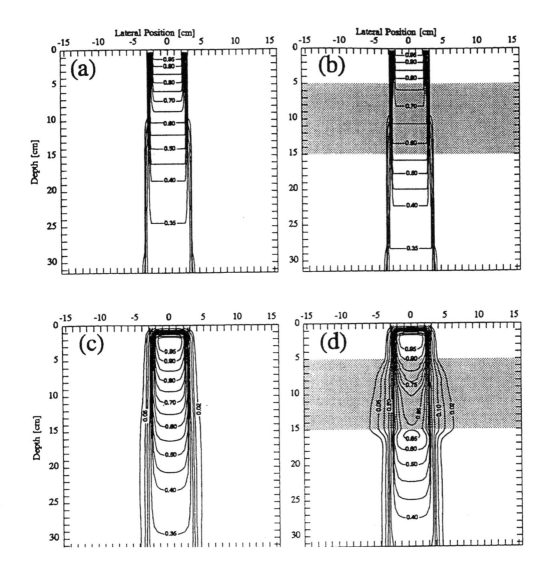

Figure 3.12: (a) Primary TERMA distribution in water. (b) Primary TERMA distribution in a lung slab phantom. (c) Convolution dose distribution in water. (d) Superposition dose distribution in a lung slab phantom. Results are for an ideal 5 MeV monoenergetic non-divergent photon beam (5×5 cm^2). Reproduced with permission (Battista and Sharpe 1992).

noenergetic 6 MeV photon beam (FS $= 5 \times 5$ cm^2) (Battista and Sharpe 1992). The difference between the TERMA (top row) and total dose (bottom row) is clearly visible. This demonstrates the decoupling that arises between primary TERMA and absorbed dose where charged particle equilibrium is disrupted.

To recapitulate, these algorithms can predict the local TERMA intensity using ray tracing through voxels and summing linear attenuation coefficients, ideally accounting for changes in atomic composition. However, the algorithms differ in the approach to modifying scattering patterns in heterogeneous tissue. Density-scaling is based on the assumption that most tissues have atomic constituents similar to that of water, and that primary and first-scatter kernels contribute the majority of dose. If there are regions of tissue or foreign elements with higher atomic number relative to water, inaccuracies will be introduced by simple density-scaling of the kernels. Most notably, localized dose perturbations at tissue interfaces will *not* be predicted by convolution-superposition algorithms, but can be predicted by Monte Carlo and Boltzmann methods.

3.3.3   Algorithm Acceleration Tactics

A number of fast techniques have been attempted to reduce the computation time for the convolution-superposition algorithm. Some are based on explicitly reducing the dimension of scatter integrals. Others rely on fast convolution techniques executed in Fourier space (Boyer 1984; Field and Battista 1987). Accelerating computer hardware, discussed in Chapter 1, can also be implemented with over 100-fold speed gains using a graphics processor (GPU) (Jacques et al. 2011).

3.3.3.1   *Down-Sizing the Scatter Integral*

Integration over the type of interaction (Equation 3.11), x-ray spectrum, and spatial dimensions (Equation 3.12) yields maximum accuracy and flexibility for the convolution-superposition algorithm. The most general superposition algorithms therefore requires a 5-dimensional integration of kernels. However, higher dimensionality increases the computational demand and lengthens the turnaround times for clinical treatment planning. Approximations have therefore been introduced to reduce the multi-dimensional integration and speed up the algorithm. For example, a polyenergetic kernel can be composed from a weighted set of monoenergetic kernels in order to reduce the integration by one dimension. Kernels of different types of interactions may also be fused together if the scattering trajectories follow similar ray-line pathways (e.g. primary and first-scattered photons). Under certain conditions of beam and tissue symmetry, portions of the integrand can be pre-integrated to form pencil beams or slabs. The kernel format (point, pencil beam, or slab) is paired with the dimension of scatter integration (3D, 2D, 1D). Figure 3.13 illustrates the geometric nature of pre-integrated kernels with similarity to Figure 3.5.

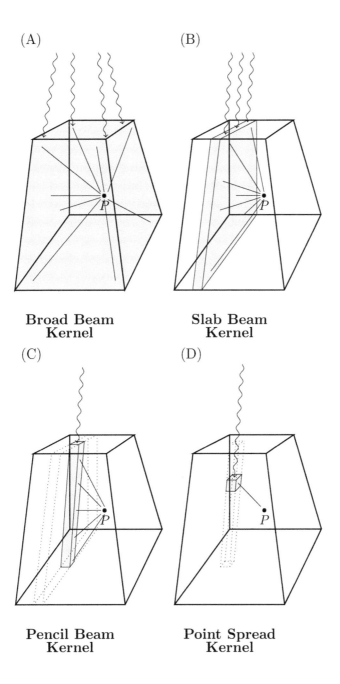

(A)

Broad Beam Kernel

(B)

Slab Beam Kernel

(C)

Pencil Beam Kernel

(D)

Point Spread Kernel

Figure 3.13: Scatter dose contribution to point P originating from: (A) irradiated volume. (B) slabs. (C) pencil beams. (D) voxels. Graphics design adapted from (Sharpe 1997).

Note that pre-integration is only permissible when symmetry in the beams and the patient anatomy permits mathematical separation of variables. Each reduction of dimension in the integral entails some loss of generality that places restrictions on the possible clinical applications. For example, in using only 1D integration of scattering slabs, the TERMA is restricted to change along a single direction (Figure 3.13B) only, as in a wedge-filtered beam. If pencil beams are used instead of a slab kernel (Figure 3.13C), a 2D integration allows intensity-modulation within the radiation field.

In the equations below, we have removed the indices for interaction kernels (i) and energy (E) variables to avoid mathematical clutter. We used the simple symbol k to represent a generic kernel. In Cartesian coordinates, the 3D dose distribution recast from equation 3.11 is as follows:

$$D(x,y,z) = \iiint T(x',y',z')\, k(x,y,z;x',y',z')\, \mathrm{d}x'\mathrm{d}y'\mathrm{d}z' \qquad (3.12)$$

For reduction to a 2D superposition, the separation of variables in TERMA is as follows:

$$T(x',y',z') = T(x',y',0)\frac{T(x',y',z')}{T(x',y',0)} = T_0(x',y')\, F(z') \qquad (3.13)$$

where $T_0(x',y')$ is the *entrance* TERMA for the beamlet at the patient's surface. With the assumption that an effective attenuation coefficient, μ', provides sufficient accuracy, the attenuating function $F(z')$ reduces to

$$F(z') = \exp[-\mu'\, d'(z')] \qquad (3.14)$$

where d' is the *oblique* depth to the interaction point at location (x',y',z'). $F(z')$ is the *relative* TERMA, normalized to the incident TERMA at the patient surface and central axis of the beam. Historically, this method was used in the example described previously in Section 3.2.2. Note that this 2D simplification still maintains the option for beam intensity modulation over the entrance surface (x,y). Using TERMA-based normalization the dose distribution is given by

$$D(x,y,z) = \int_y \int_x T_0(x',y') \left[\int_z \exp[-\mu'\, d'(z')]\, k(x,y,z;x',y',z')\mathrm{d}z' \right] \mathrm{d}x'\, \mathrm{d}y'$$

$$= \int_y \int_x T_0(x',y')\, k_{\mathrm{pencil,z}}(x,y;x',y')\, \mathrm{d}x'\, \mathrm{d}y' \qquad (3.15)$$

Note that some 2D implementations have a different normalization of the kernel. The kernel values can be defined as fractional energy deposited per unit mass at the destination voxel per unit energy fluence entering each pencil beam at its surface

(Ahnesjö et al. 1992). This is intuitive for application to intensity-modulated fields.

If the incident TERMA is uniform along beam strips (e.g. in the y-direction), then pencil-beam kernels can be recombined into slab kernels, each exposed to constant incident TERMA along the y'-direction, changing only along the x'-direction, $T_0(x')$. This uni-directional ramp of beam intensity modulation can be achieved with a mechanical or virtual wedge filter, for example. The dose distribution then collapses to a simpler 1D superposition of slab contributions. Historically, this simplification was used in Cunningham's CBEAM program (Cunningham 1972; Cunningham and Battista 1995). Note that the contribution from a pencil beam will depend on the depth of the dose calculation point because of the scattering angle subtended. The dose expression becomes

$$
\begin{aligned}
D(x, y, z) &= \int_x \int_y T_0(x', y') \, k_{\text{pencil,z}}(x, y; x', y') \, \mathrm{d}y' \, \mathrm{d}x' \\
&= \int_x T_0(x') \left[\int_y k_{\text{pencil,z}}(x, y; x', y') \, \mathrm{d}y' \right] \mathrm{d}x' \\
&= \int_x T_0(x') \, k_{\text{slab,y,z}}(x; x') \, \mathrm{d}x'
\end{aligned}
\tag{3.16}
$$

This section has outlined the main features of the convolution-superposition method in 3D, 2D, and 1D. For consistency, we maintained normalization of kernels to TERMA quantities across all dimensions. Three-dimensional integration of point kernels is performed on the Tomotherapy Planning system (Accuray) and the Pinnacle system (Philips Healthcare) using a collapsed kernel approximation that reduces the computational burden while maintaining a 3D integration (Ahnesjö 1989). Reduced integration techniques that are less rigorous and flexible were adopted on many of the early commercial treatment planning systems, such as the PC-12 (Artronix) and TP-11 (Atomic Energy of Canada). Pencil beam kernels were also applied on Plato (Nucletron) and CadPlan (Varian Oncology Systems) systems, re-introduced with improvements for handling tissue inhomogeneity in the Anisotropic Analytical (2.5D) Algorithm on the Eclipse system (Varian Oncology Systems) (Tillikainen et al. 2008). These methods are described in Chapter 4.

3.4   THIRD GENERATION ALGORITHMS

This most advanced generation of algorithms introduces further improvement in modelling dose in regions of electron disequilibrium, including those caused by variation in atomic composition of the absorbers. The third generation algorithms include Monte Carlo simulation and Boltzmann transport equation solvers.

3.4.1   Monte Carlo Simulation

The Monte Carlo method has been applied to a wide range of fields including mathematics, socio-economics, radiation sciences, and the design of games and lotteries.

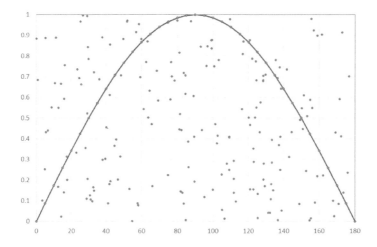

Figure 3.14: Monte Carlo evaluation of the integral of the semi-sine function, by counting the fraction of points landing under the curve.

The general applicability stems from the random sampling technique for evaluating complex integrals numerically. A simple example is illustrated in Figure 3.14. The function under consideration is a semi-sine curve. Random points are plotted using pairs of random numbers in the interval [0,1.00] and rescaled for the range of angles on the x-axis (180 degrees). The fraction of points landing under the sine curve determines the area under the curve and hence evaluates the integral. This is the essence of the Monte Carlo technique in the evaluation of a complex integral representing the dose at a voxel in a dose distribution.

The *random walk* nature of the method is particularly suited to modelling a sequence of multiple x-ray interactions by random selection (Andreo 1991; Raeside 1976). First consider a simple example of determining the probability of obtaining a specific hand in a card game of poker. For a *straight* hand with 5 cards of any suit in numerical succession (e.g. 5, 6, 7, 8, 9), the probability is approximately 1 in 250 dealt hands, assuming a 52-card deck and 5-cards per deal. If we simulate a large number of random draws, the fraction of *straight* hands, will converge to the deterministic probability known from combinatorial theory, subject to statistical fluctuations. Now if we consider a poker game with multiple hands dealt to many players, the probability that a player will eventually win with a superior hand becomes more difficult to express analytically in an equation. The outcome of possible succession of cards and grouping of hands in different combinations and permutations can be more easily modelled by a Monte Carlo simulation.

A sequence of random events is considered Markovian if the future state of a particle only depends on its immediate state, with no influence of prior history lead-

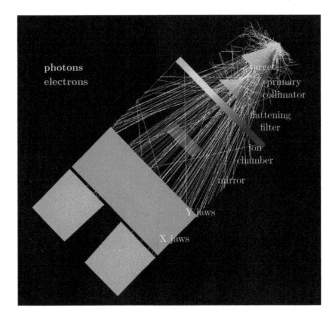

Figure 3.15: Simulation of radiation produced within the head of a Varian 2100 model accelerator operated in x-ray mode. X-ray trajectories are shown in yellow and rarer trajectories of secondary electrons are shown in violet.

ing up to the current state. For example, the chances of a photon interacting after traversing a free path is identical to the chances of a photon of identical energy just entering the absorber. Similarly, a 5 MeV photon that has experienced multiple scattering leading to a degraded energy of 1 MeV has the same odds of interacting as a naive 1 MeV primary photon that has yet to experience any interaction.

The Monte Carlo method mimics a radiation experiment in *slow motion*, modelling the individual radiation interactions with fewer simplifications. For troubleshooting or educational purposes, the resultant particle trajectories can be displayed individually in *slow motion* on a computer screen (e.g. using RadSim software, McGill University) (Verhaegen et al. 2006). Sketches of a group of sequential interactions resemble observations made with cloud chambers or nuclear emulsion plates. Figure 3.15 shows particles emerging from an accelerator operated in x-ray mode.

In addition to being intuitive, this method provides valuable radiation data that is almost impossible to gather experimentally. For example, energy deposition kernels reflect the spread of energy from a *single location* of forced x-ray interactions, a process that normally occurs randomly throughout an absorber. A futuristic dosimeter array with coincidence detection circuitry may permit such measurements. Until recently, clinical applications of Monte Carlo simulation have been hampered by two impediments (Brualla et al. 2017): (1) excessive computation times in achieving ac-

Figure 3.16: Factors that affect computational speed in Monte Carlo simulations. Cartoon courtesy of creativecommons.org.

ceptable dose accuracy and (2) the commissioning process that required special input beam data generated ironically by the Monte Carlo method.

The main parameters affecting overall computational times are listed in Figure 3.16. There is a clear interplay between *dose noise* and spatial resolution (i.e. voxel size) — a compromise often seen in the field of medical imaging. The most obvious strategy for gaining speed is to reduce the number of photon histories in the computer simulation. However, this can lead to imprecision which, in turn, limits the achievable dose accuracy. This uncertainty is subject to Poisson statistics in view of the binary nature of voxel hits leading to dose deposition. Improvement in dose precision therefore comes with a considerable penalty in computation burden, inefficiently proportional to the *square root* of the number of dose-depositing particles. These particles are related to the number of photons launched in the simulation within the patient space. Fortunately, the incident photons can be distributed across all treatment beams because of the confluence of secondary particles that produces the dose to a voxel. In other words, the computer run time is almost independent of the number of beams in a treatment plan, assuming that the fields have similar intensity and modulation. In addition, wise choices of simulation parameters and biasing towards more relevant radiation events can reduce simulation times; the artificial bias can be accounted for in the results. For example, *bremsstrahlung* occurring in x-ray targets, can be forced to happen more frequently in forward directions of interest, but emerging photons are given a reduced weight. Unimportant simulated particles, for example those with insufficient range to reach the borders of the scoring voxels, can undergo Russian roulette with the survivors receiving higher weight. Variance reduction techniques will be described in more detail in Chapter 5 on

Monte Carlo methods. Rapid advances in computer hardware and software are also alleviating the computational speed problems. The statistical noise in dose results is receding as computer workstations with fast multi-core processing units, both central (CPUs) and graphical (GPUs), become very cost-effective (Chapter 7). As a point of historical interest, these kernels were originally computed in 1987, running 1 million histories per kernel over 0.5 CPU-years of aggregated computational time on a cluster of VAX11 750/780 (Digital Equipment Corporation) installed across Canada. Today, the entire database can be regenerated in less than 1 hour using a personal desk-top computer!

As to commissioning issues (Chetty et al. 2007), first-generation algorithms only require measured data sets acquired in a water phantom. For more advanced algorithms, the fluence of each accelerator beam must be known in detail. Fortunately, Monte Carlo codes (Rogers et al. 2009, 1995; Sheikh-Bagheri and Rogers 2002) can model the radiation particles emerging from linear accelerators. Energy and directional spectra of particles crossing a reference plane can be stored in files and re-used by dose algorithms. These data sets are binned in a *phase space* for photons or secondary electrons, with specification of location, energy, and particle direction. The concept of phase space will be described more fully in Section 3.4.2.1. Data files are shared across the scientific community (www-nds.iaea.org/phsp/phsp.htmlx) or provided by manufacturers of medical linear accelerators. Generally, the reference plane is located *above* the final collimators, for each beam type and energy. This characterizes the x-ray source and primary collimator effects common to all downstream interaction events. The lower portion of the accelerator head needs to be re-simulated for patient-specific treatment conditions such as the configuration of the multi-leaf collimator (MLC) (Lobo and Popescu 2010). In summary, phase space data sets are used to *bootstrap* the modelling of radiation passing through final beam-shaping stages and into the patient space (Rogers et al. 2003).

The reader is reminded that accuracy in dose can only reflect the quality of the underlying physics infrastructure: *you cannot fool the physics!* Monte Carlo results can be misleading when using a substandard random number generator or insufficient particle histories. Incorrect input geometry and tissue parameters extracted from CT or MRI data can also produce faulty cascades of scattering events.

Tracking of photons is relatively straightforward as evidenced by many *home-grown* codes (Battista and Bronskill 1978). In contrast, charged particles have very short mean free paths, interacting quasi-continuously along their tracks. For example, consider an electron slowing down at a rate of 2 MeV per cm in soft tissue (i.e. linear stopping power). Assuming that the formation of an ion pair requires an injection of approximately 30 eV, an average of 6,500 soft interactions occurs per mm of electron travel! In practice, multiple scattering histories are usually *condensed* (Nelson et al. 1985) near the main particle track, punctuated occasionally by energetic bursts of bremsstrahlung x-rays and delta rays. Assuming that valid input data and appropriate run-time parameters are set, the accuracy of Monte Carlo results will be only limited by the Poisson uncertainty of scored results. This is ultimately affected by the initial number of particles launched in the simulation. The number required to obtain 1% precision in x-ray dose regions of interest is ideally greater than 10 million starting histories.

Consider a radiation problem where an analytical closed-form integral solution would be difficult to derive. We use an example of a pencil beam of 1.25 MeV photons incident along the central axis of a water phantom (see Figure 3.17) and follow multiple Compton scattering events (Battista and Bronskill 1978). We pose the question — *How many scattering events take place before the incoming photon is either absorbed or exits the phantom?*. This is an important question because it can help specify what scattering events can be dropped in a simulation, leading to a potential reduction in computation time. In our example, photons are launched sequentially until they reach a cut-off energy of 0.100 MeV *or* exit the phantom borders. All interactions are assumed to be Compton scattering, avoiding the triage of possible interaction processes. The Compton scattering angle was randomly chosen, based on the differential cross section $(d\sigma/d\theta)$. The azimuthal angle was distributed evenly over 360 degrees. The photon is redirected into 3D space, and its energy is downgraded, affecting subsequent free paths and scattering events. The result of this simulation is shown as a projected view in Figure 3.17B. For 10,000 incident photons, 2,810 are transmitted and fewer events occur with increasing order of scattering, as expected. This information would be difficult, if not impossible, to express mathematically in a closed-form expression or to measure experimentally. This highlights one of the unique advantages of Monte Carlo simulation. Particles can be tagged along their historical pathway, including their decaying energy profile, location of key inflection points and cumulated number of events, for example. The particle histories can be displayed graphically for software troubleshooting or educational purposes (Verhaegen et al. 2006).

Figure 3.18 shows energy deposition kernels used in convolution-superposition algorithms. Total energy kernels for monoenergetic photon energies of 0.4, 1.25, and 10 MeV were obtained by Monte Carlo simulation with at least 500,000 histories per energy (Ahnesjö et al. 1987). The dispersion of points reflects the statistical variation in energy deposits. The dash-dotted curves for single Compton scattering are noiseless because they were computed analytically. These point spread functions are similar to those presented in Chapter 2, but displayed as isometric contours with logarithmic spacing. The data are expressed as *fractions* of the total energy released at the origin, imparted to neighbouring voxels *per unit volume* of recipient voxels. Within a few cm's of an interaction site, the contribution is on the order of 0.1% per cm^3. In far zones, the contributions drop by orders of magnitude.

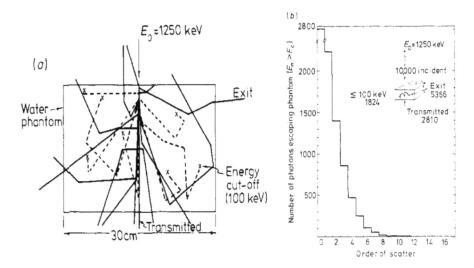

Figure 3.17: Simulation of a cobalt-60 pencil beam irradiation. (a) water phantom configuration with sample photon trajectories. (b) number of scattering events prior to escaping the phantom. Reproduced with permission (Battista and Bronskill 1978).

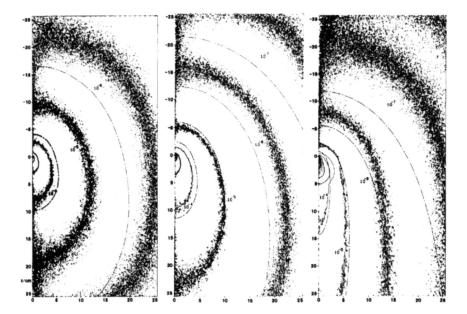

Figure 3.18: Monte Carlo total (points) and analytical first-scatter (dash-dotted isocurves) kernels for incident photon energies of 0.4, 1.25, and 10 MeV, from left to right. The displayed units are fractions of energy released at the origin deposited in neighbouring voxels per unit volume. Reproduced with permission (Ahnesjö et al. 1987).

3.4.2 Solving the Boltzmann Transport Equation

The most recent quest for a fast dose computation with accuracy comparable to Monte Carlo simulations has led to the renaissance of deterministic solutions to the Boltzmann transport equations (MacDermott 2016b; Gifford et al. 2006; Varian 2015b; Failla et al. 2015). This approach has been applied to study a vast range of natural phenomena including electrodynamics, fluid dynamics, aerodynamics, thermodynamics, meteorology, semi-classical quantum mechanics, and space radiation dosimetry. The core principle of the Boltzmann equation is conservation of mass-energy. In radiation oncology, the concept has been used successfully for electron beam dose computations, but only recently applied to x-ray beams with the added degree of complexity due to the 2-step process of energy transfer and deposition. Mutual exchanges of energy between photons and charged particles result in *coupled* equations (Kan et al. 2013; Han et al. 2013; Vassiliev et al. 2010; Andreo et al. 2017b). This approach requires efficient mathematical tools (www.vareximaging.com/products/attila-software) for solving a large set of simultaneous interleaved equations. In contrast to stochastic solutions, deterministic methods consider the flow of *cohorts* of particles rather than tracking individual particles. The solution to the Boltzmann equations leads directly to a spatial distribution of the *average* scalar fluence of charged particles, without statistical fluctuation.

3.4.2.1 Introduction to Phase Space

A significant shift in mentality is required to understand the Boltzmann method. The collective flow of particles is tracked in a *multi-dimensional virtual phase space*. At a moment in time t , particles of a certain type (e.g. photons or electrons) have a 3D location, $\vec{r}(x, y, z)$, instantaneous energy (E) and trajectory direction $(\hat{\Omega})$. This hyper-dimensionality with six coordinates per particle type may be difficult to visualize. However, if we focus on photons only and a voxel at a fixed location in real space and restrain attention to only the polar angle of their trajectory, then the phase space can be interpreted as a two-dimensional (2D) *histogram* with angle-energy coordinates for each cell or bin. This space resembles a checker game board with play pieces moving throughout the board and stacking up to count the particles per bin at any moment in time. For example, consider a 1 cm^3 water voxel at a particular location in 3D space. Its electron density is 3.34×10^{23} electrons per cm^3. A bi-energetic fluence consisting of 10^{27} per cm^2 of 1 MeV photons and 10^{26} per cm^2 of 0.5 MeV photons enters the voxel at polar angles of 0 and 60 degrees, respectively. As shown in Figure 3.19, before any interactions take place inside the voxel, incoming photons are counted in initial angle-energy bins at (0 degrees, 1.00 MeV) and (60 degrees, 0.5 MeV). After single-scattering occurs in the voxel, the counts will re-distribute to other bins, constrained by Klein-Nishina cross-sections (review Section 2.2.3). The counts after interaction form a quasi-diagonal line due to the strong angle-energy correlation of Compton events. For a polyenergetic and poly-directional incident fluence, scattering interactions would populate many more histogram bins. This is the basic concept used for Boltzmann accounting of *changes*

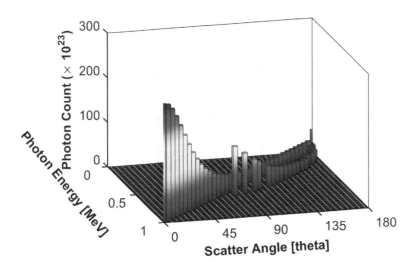

Figure 3.19: Phase space occupancy with initial angle-energy bins at (0, 1) and (60, 0.5), subjected to Compton scattering events. The photon bin count is normalized per unit solid angle at each of the scattering angles.

in particle fluence as the radiation propagates through patient tissue space. The redistribution rules are governed by quantum cross-sections of radiation interactions (Chapter 2) and are consistent with mass-energy conservation laws. The phase space movements can be viewed as in a game of chess, with each piece only allowed to move in a strict pattern. For a shower of photons and electrons, one can expand the game to a pair of boards with cross-talk allowed between them. Particle counts will *ping-pong* across two histogram spaces as the pieces transform between photonic and Coulomb states.

3.4.2.2 Radiometrics

Before we proceed to mathematical formulae, we define quantities of radiation transport theory (Figure 3.20) (Andreo et al. 2017a; Sinclair et al. 1991; Rossi and Roesch 1962). Quantities on the left specify the density and flow rates for particles travelling with direction $(\hat{\Omega})$ and energy (E) at location $(\vec{r})$ at a moment in time, t. The upper-case quantities in the middle column are the result of integrating left-side quantities over all particle directions and energies. Quantities on the right are integrated over an exposure time. Scalar quantities are used in radiation dosimetry.

The variable n represents the local density of particles (per cm^3) at a spatial location **r**. This density is differential with respect to energy (E) and trajectory direction $(\hat{\Omega})$, evaluated at an instant of time, t. When integrated over all energies

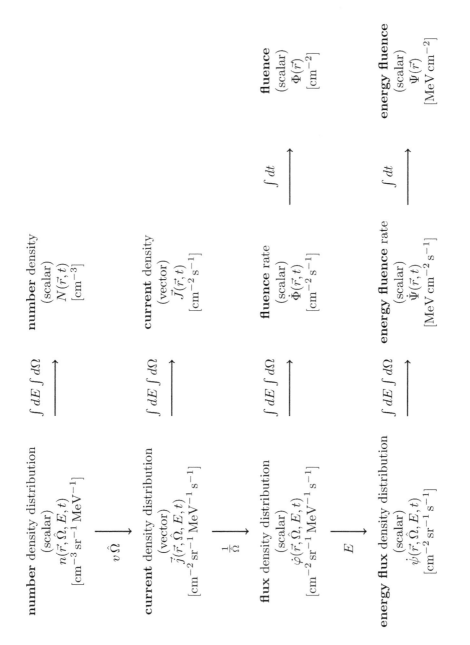

Figure 3.20: Radiometric quantities used in radiation transport theory.

and directions, one obtains N on the right side – the density of all rays (per cm^3).

The variables $\vec{j}$ and $\vec{J}$ are vector quantities associated with *currents* in the sense of electrical currents or hydrodynamic flow. $\vec{j}$ represents the flow rate of energy-selected particles with velocity $\vec{v}$ moving along a selected direction (Ω) orthogonal to a small surface element that is traversed. When integrated over all energies and directions, one obtains $\vec{J}$, the overall particle flow rate along an averaged direction. The surface elements are associated with a unit vector that points outward from the surface to specify the traffic direction. For surfaces that enclose a volume of interest V centred at position $\vec{r}$, overall surface integration can be used to determine the *net outflow* of particles using $\oiint \vec{J} \cdot d\vec{S}$.

The scalar quantities $\dot{\phi}$ and $\dot{\Phi}$ are the magnitude of the vectors $\vec{j}$ and $\vec{J}$, respectively. $\dot{\psi}$ considers the energy flow rate across a surface element. When integrated over all energies and directions, these quantities yield the familiar scalars of particle and energy fluence rates used in standard radiation dosimetry (Andreo et al. 2017a).

3.4.2.3 Particle Counting in Phase Space

The overarching Boltzmann principle is that energy must be conserved as particles transition through phase space. Changes in occupancy levels in phase space bins are cross-checked from two perspectives: (1) net flow across real dose voxel borders, and (2) accounting for particle interaction events occurring within the voxel. The Boltzmann equations essentially impose an equality of results obtained by these two approaches - tracking border crossings and tracking internal events that cause particles to be removed, created, or transformed. This is similar to the standard method of cross-checking sums of numbers in rows and columns in financial spreadsheets.

Three coordinates specify voxel locations in 3D space and two spherical coordinates (i.e. polar and azimuthal angles) specify particle directions. Adding the particle's instantaneous energy (E), the phase space becomes 6-dimensional for each particle type. Alternatively, this space could have been specified by coordinates used in molecular thermodynamics, i.e. (x, y, z, v_x, v_y, v_z) where v is velocity. However, $(x, y, z, \theta, \phi, E)$ is more intuitive for radiation physics applications because of the explicit energy dependence of interactions that produce particle deflections. Radiation interactions change the particle number density distribution, n, in phase space assigned to each of the particle species. The overall particle count changes as particles of interest cross into, start/stop in, transform, or leave a volume of interest. Particles that pass straight through a voxel without interacting will retain their incoming energy and orientation, and only change their 3D location. Particles are *removed* either by absorption or scattering events. Scattering also causes transposition of cell counts across energy bins ($E \rightarrow E'$) for the same particle type. In addition, interactions of other particles can launch particles of interest. For example, *bremsstrahlung* originates with charged particles but contributes to photon phase space counts. Particles may also be freshly supplied (i.e. births). These may emerge spontaneously from radioactive sources, including activity induced by external irradiation.

The Boltzmann transport equations *audit* the particle counts as they are transferred within and across phase space domains. For simple situations such as a monoenergetic collimated beam of radiation passing through a homogeneous absorber, the Boltzmann equations can be solved analytically for the photon fluence to obtain the familiar exponential attenuation expression. A similar example for an isotropic point radiation source yields the familiar inverse-square law (MacDermott 2016b). However, for intensity-modulated polyenergetic beams incident on heterogeneous media, the equations become too complex and they can only be solved by numerical methods. The final solution to these equations is the *scalar* fluence of electrons particles throughout the absorber; the local dose is then calculated using Equation 2.23 at every voxel of the dose distribution matrix. This abides by the dictum "All dose is deposited by charged particles!"

As an example of adapting a nuclear physics code to medical physics applications, the Atilla code (www.vareximaging.com/products/attila-software) originated at the Los Alamos National Laboratory. This method was then commercialized by Transpire Incorporated and implemented as the Acuros XB algorithm on the Eclipse treatment planning system (Varian 2015b). Simplifications and assumptions are required for practicality: (a) scattering cross sections are expanded into a finite series of terms (i.e. Legendre polynomials), (b) energy levels are quantized *via* the multigrouping method (c) particle directions are also quantized *via* discrete ordinates, and (d) the Cartesian voxel space of the patient anatomy is converted to a lattice of tetrahedrally-shaped elements. The details will be described in Chapter 6.

A simple analogy might instil the right frame of mind for understanding the Boltzmann book-keeping method. Consider the task of tracking a sub-population of individuals with a particular trait (e.g. red hair) within a country. The net gain or loss is the *divergence* measured at all international border crossings of the country. The immigration rate is variable, as is the emigration rate. The rate of increase in the red-headed sub-population will depend upon the difference between arrival and departure rates, plus the internal birth rate and hair tinting to a red colour. The rate of decrease will also depend on the attrition rate, including hair loss, colour transition to a non-red colour, and deaths of individuals.

A numerical example further explains the accounting method. Imagine a voxel located in real space at $\vec{r}(x, y, z)$ exposed to a bi-energetic bi-directional fluence of photons. Each arrow in Figure 3.21 represents the flow of 10 photons over the irradiation period. The two starting populations initially occupy different bins in phase space. Sixty of the incoming photons have initial conditions $(E_1, \hat{\Omega}_1)$, shown in green. Another 20 photons have a greater initial energy and different initial orientation $(E_2 > E_1, \hat{\Omega}_2 \neq \hat{\Omega}_1)$, shown in blue. On passing through the voxel, 20

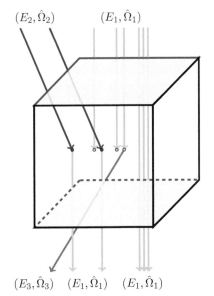

Phase Space	Incoming	Outgoing	Bin Count $(E_1, \hat{\Omega}_1)$
$(E_2, \hat{\Omega}_2)$	20	0	+20 In Scatter
$(E_3, \hat{\Omega}_3)$	0	10	−10 Out Scatter
Absorb	0	20	−20 Loss
Net Change $(E_1, \hat{\Omega}_1)$	60	50	−10

Figure 3.21: A voxel is centred at a coordinate location $\vec{r}$. Each arrow represents a cohort of 10 photons. Green denotes the energy and direction of interest. Interactions in the voxel (dots) will remove or redistribute particles in phase space bins defined by photon energy (E) and vector direction $(\hat{\Omega})$, as shown on the right side tally.

photons with incoming energy E_1 undergo total absorption inside the voxel and are eliminated. Another 10 similar particles are out-scattered to a lower energy E_3 and redirected $(\hat{\Omega}_3 \neq \hat{\Omega}_1)$, shown as a colour change from green to red. These photon counts are transferred to another phase space bin $(E_3, \hat{\Omega}_3)$. This process was described as an *out-scattering* event because counts were moved out of the phase space bin of interest $(E_1, \hat{\Omega}_1)$. The remaining 30 photons are transmitted through the voxel, and their count is therefore retained in their original phase space cell. Twenty photons with a greater energy of E_2 are *in-scattered*, producing energy and directional coordinates $(E_1, \hat{\Omega}_1)$, incrementing the bin count of interest by 20. In the book-keeping table shown as an inset of Figure 3.21, the total loss in phase space bin $(E_1, \hat{\Omega}_1)$ due to removal by absorption and out-scattering is -30, while the gain from higher energy in-scattering is $+20$. The *net change* in the count at phase space bin $(E_1, \hat{\Omega}_1)$ is therefore -10. This change is equally predictable by the divergence of photons through the voxel:

$$\texttt{Divergence} = \texttt{Particles}_{\texttt{exiting}} - \texttt{Particles}_{\texttt{entering}} = 50 - 60 = -10 \quad (3.17)$$

which *must* match:

$$\texttt{Particles}_{\texttt{gained}} - \texttt{Particles}_{\texttt{lost}} = 20 - 30 = -10 \quad (3.18)$$

The formal expression of the equality of Equations 3.17 and 3.18 *is* the time-independent or steady-state Boltzmann equation for constant exposure over a fixed period of irradiation. It is shown below as equation 3.19 and will be explained further in Chapter 6 (Equation 6.53). For the moment, we will just accept the following result without an explicit derivation:

$$\hat{\Omega} \cdot \vec{\nabla} \varphi(\vec{r}, \hat{\Omega}, E) = \int_E^\infty dE' \int d\hat{\Omega}' \, \Sigma_s(\vec{r}, \hat{\Omega}' \to \hat{\Omega}, E' \to E) \, \varphi(\vec{r}, \hat{\Omega}, E')$$
$$- \Sigma(\vec{r}, E) \, \varphi(\vec{r}, \hat{\Omega}, E) \tag{3.19}$$

Re-arranging of the terms leads to a more common form of the Boltzmann equation

$$\left[\hat{\Omega} \cdot \vec{\nabla} + \Sigma(\vec{r}, E) \right] \varphi(\vec{r}, \hat{\Omega}, E) = \int_E^\infty dE' \int d\hat{\Omega}' \, \Sigma_s(\vec{r}, \hat{\Omega}' \to \hat{\Omega}, E' \to E)$$
$$\times \varphi(\vec{r}, \hat{\Omega}, E') \tag{3.20}$$

where $\varphi(\vec{r}, \hat{\Omega}, E)$ is the flux density distribution integrated over time. The Σ term refers to the *total macroscopic* cross section for photon interactions that entails a loss of photons through absorption and scattering. In our numerical example, this term corresponded to a total of 30 photons removed from the phase space bin of interest. The Σ_s term refers to the scattering (only) macroscopic cross section that can produce photons with the designated energy and direction of interest by "in-scattering". This term contributed 20 photons to the phase space bin of interest.

Note that we have ignored the creation of fresh photons originating from other types of sources. For example, charged particles entering the voxel could create bremsstrahlung or annihilation photons of the right energy and direction. Internal radioactive sources, real or induced, could also emit photons of interest and have been ignored. Note that energy loss of positrons released in pair production is often treated using electron stopping data. This simplification is permissible if gamma rays from subsequent pair annihilation are expected to produce negligible dose (see Figure 3.6C and D).

3.5 DOSE TO A VOXEL

The absorbed dose in a voxel *per se* can be determined by one of the following expressions from Chapter 2. We assume monoenergetic particles.

$$D(\mathbf{r}) \overset{\text{CPE}}{=} (\mu_{en}/\rho) \, \Psi \tag{3.21}$$

$$= [1/\rho(\mathbf{r})] \iiint T(\mathbf{s}) \, \rho(\mathbf{s}) \, k(E, \mathbf{r}; \mathbf{s}) \, d^3 s \tag{3.22}$$

$$= \frac{\Phi^{(cp)} \, (S/\rho)_c \, \rho \, dA \, dl}{dm} \tag{3.23}$$

$$\overset{\delta \text{CPE}}{=} \Phi^{(cp)} \, (S/\rho)_c \tag{3.24}$$

The dose reported by an algorithm depends on the type of quantities scored during program execution. If pure charged particle equilibrium (CPE) is assured, absorbed dose becomes synonymous with collision $KERMA_c$ as in Equation 3.21. While the effects of variations in electron density on primary and scattered radiation are considered to a variable extent by most algorithms, voxels are often implicitly assumed to be water-like in their atomic composition. For example, bone can be modelled as artificially compressed water. In other words, the change in mass energy absorption coefficient due to atomic composition is sometimes ignored in equation 3.21. For a given megavoltage x-ray beam photon fluence under equilibrium conditions, the dose to water, represented by D_w, is less than a few percent different for most soft tissues but approximately 15% different for bone compositions (Fernández-Varea et al. 2007; Siebers et al. 2000).

Where charged particle equilibrium cannot be guaranteed, convolution-superposition algorithms can be invoked to model charged particle transport. The primary TERMA term (i.e. $T(\mathbf{s})$ in Equation 3.22), can incorporate atomic number and density variations along beam rays, provided that the tissues crossed can be identified by CT numbers. The kernels (k) *per se*, on the other hand, are generally pre-computed for an all-water medium and only modified by density scaling.

The Monte Carlo method scores the energy deposited individually by charged particles passing through, stopping, or starting in a voxel using equation 3.23 where dl is the path length traversed inside the voxel. In the Boltzmann method, the fluence of charged particles, $\Phi^{(cp)}$ reaching a voxel is determined and a form of equation 3.24 is then applied *a posteriori*. These methods generally account for density and atomic number variations in the patient space. In summary, Equations 3.21 to 3.24 should all yield the same dose result for a given absorber and for conditions that satisfy the intrinsic assumptions of each equation.

3.5.1 Dose to What Medium?

Absorbed dose is the concentration of energy deposited *in a specific absorber*. The flippant use of the word *dose* without specifying the type of absorbing medium in a voxel leads to ambiguity across the results of different dose calculation algorithms (Ma and Li 2011). Some produce dose distributions expressed in terms of dose-to-water universally for all voxels. The justification is that beams are calibrated in terms of dose-to-water, D_w, under reference conditions, and clinical dose prescriptions and experience are historically based on soft tissue absorption. Other algorithms report dose-to-medium, D_{med}, where the medium is variable across voxels.

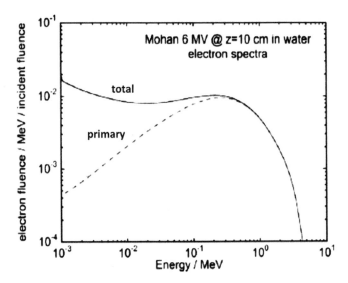

Figure 3.22: Spectra of electrons reaching a voxel at a depth of 10 cm in a water phantom exposed to a 6 MV x-ray beam. Primary designates inclusion of only electrons set in motion by photon interactions. Total also includes secondary δ-rays set in motion by charged particles. Reproduced with permission (Andreo 2015).

When the differential particle fluence arriving at a voxel, $\mathrm{d}\Phi^{(cp)}/\mathrm{d}E$, is known, the ratio of these doses is often derived from polyenergetic expansions of Equation 3.24:

$$\frac{D_w}{D_{med}} = \frac{\int \dfrac{\mathrm{d}\Phi_w^{(cp)}}{\mathrm{d}E} \left(\dfrac{S}{\rho}\right)_w \mathrm{d}E}{\int \dfrac{\mathrm{d}\Phi_{med}^{(cp)}}{\mathrm{d}E} \left(\dfrac{S}{\rho}\right)_{med} \mathrm{d}E} \tag{3.25}$$

where the mass stopping powers, $\left(\frac{S}{\rho}\right)$, are averaged over a spectrum of charged particles at the voxel in question. This equation is intentionally ambiguous. The spectrum includes or excludes δ-rays in averaging stopping powers that are restricted or unrestricted (Andreo 2015; Chetty et al. 2007). The choice depends on the relative size of voxels compared with the range of charged particles set in motion, in accordance with cavity theory. The reader is referred to a standard textbook on radiation dosimetry for a full description of this theory (Andreo et al. 2017a). Figure 3.22 shows electron spectra reaching voxels located at a depth of 10 cm inside a water phantom exposed to 6 MV x-rays.

3.5.2 Microscopic Awareness

Dose computations determine the average macroscopic dose to voxels that are much larger (i.e. mm's) than microscopic biological entities (i.e. microns) contained therein. For example, the energy deposited in soft tissue cells residing within bone niches can be of greater interest than the dose to the bone matrix *per se*. Examples include tumour cells, bone marrow, and osteoblast/osteoclast cells that are responsible for continually remodelling bone structure. Monte Carlo simulations performed at ultra-high spatial resolution simulate the microscopic and stochastic nature of energy deposition by charged particle tracks passing through biological entities. The specific energy (i.e. energy deposited per unit mass) exhibits large statistical deviations about the mean macroscopic dose and varies with cellular distribution in tissue (Oliver and Thomson 2017; Zhang et al. 2017). Futhermore, indirect radiochemical damage of DNA also plays a significant role (Moiseenko et al. 2001). It is therefore important for the reader remain aware of the microscopic nature of cellular response to radiation. There is a large spatio-temporal communication gap lying between the disciplines of macrodosimetry described in this book and microdosimetry.

3.6 CLASSIFICATION OF ALGORITHMS

The introduction of CT in the 1970's caused a sudden paradigm shift in visualization of multi-slice anatomy and access to *in vivo* tissue densitometry. Following the development of pixel-based methods, a classification scheme was needed to distinguish algorithmic structures and their assumptions. Many commercial algorithms were branded as *convolution* methods for competitive marketing purposes. All algorithms indeed had access to the CT image matrix for display, but only made partial use of the density information for dose computation purposes. Scatter integration was often performed over limited dimensions with under-utilization of the 3D image information (recall Figure 3.13). The purpose of a classification scheme was to overcome the confusion across nomenclature of algorithms and to help uncover pitfalls and blind-spots in their clinical implementation.

3.6.1 Categories 1 through 4

Numerical categories (Wong and Purdy 1990) were defined in a matrix according to the extent of tissue density sampling and considerations of electron transport (Table 3.1). Simpler methods of the first generation were adapted to the new wealth of geometric and density CT information but their physics infrastructure was weak. They were classified as Category 1 because they only performed ray tracing along beam rays, and intrinsically assumed charged particle equilibrium (CPE). The scattered dose contributions were also assumed to be correlated with the primary radiological depth along these rays. Category 2 algorithms separated primary and scatter dose components, with expanded coverage of the tissue density environment surrounding each dose calculation point. The ETAR algorithm (Sontag and Cunningham 1978) was the first *commercialized* algorithm to account for lateral photon scattering from

Table 3.1: Categories of dose algorithms in the 1990s based on anatomical sampling and electron transport (Papanikolaou et al. 2004)

CT Voxel Sampling	Charged Particle Equilibrium (on-the-spot absorption)	Electron Transport Included
	Category 1	**Category 3**
1D Ray Tracing ONLY	Ratio of Tissue-air ratios (TAR) Pencil Scatter-air ratios (dSAR) Batho Power Law Simple Pencil Beam	Pencil-beam Convolution (2D) Fourier Convolution
	Category 2	**Category 4**
3D CT Image Matrix	Voxel SARs (d^2SAR) Equivalent TAR (ETAR) (Sontag) Delta Volume (Wong)	Equivalent TAR (ETAR+) (Woo) Delta Volume+ (Yu) 3D Convolution-Superposition (with point kernels)

anatomic voxels in 2.5D. A more rigorous 3D integration of scattered photon dose was implemented later in the delta-volume method (Wong and Henkelman 1983). In time, the energy of knock-on charged particles was appended in Category 4 algorithms, leading to the 3D convolution-superposition method. The Monte Carlo and Boltzmann methods were not in use at the time when these categories were defined; they certainly would have merited a Category 4 designation.

3.6.1.1 Dimensionality of Scatter Integral

Most of the confusion surrounding algorithms relates to pre-integration of point kernels. The Wong classification scheme was therefore further clarified by focusing on the nature of the scatter integral (Battista and Sharpe 1992). Faster, simpler algorithms with reduced dimensionality trigger implicit assumptions that could limit clinical applicability. Early pencil-beam convolution models were classified as Category 3 because tissue density was only considered longitudinally along the pencil beam axis. With later improvements, the scattering effects of tissue outside of the central core of each pencil beam was also considered. The scatter function was decomposed into depth and radial components. Examples include the singular value decomposition (SVD) (Bortfeld et al. 1993) and anisotropic analytical (AAA) algorithms (Varian 2015a; Tillikainen et al. 2008). Since scattered radiation actually travels along many possible pathways surrounding each pencil beam column, these methods only partially account for lateral disequilibrium. Algorithms of Category 4 overcome this limitation with a more rigorous 3-dimensional treatment of secondary radiation.

Principle #5: "The versatility and accuracy of an x-ray dose algorithm is driven by the dimension of the scatter integral"

3.6.2 Types A-B-C

As megavoltage photon energy increased beyond that of cobalt-60 radiation, the assumption of charged particle equilibrium in heterogeneous absorbers became the usual suspect. Tracking of charged particles therefore came to the forefront, especially in lung where particle ranges are stretched. Accurate modelling, under conditions of lateral disequilibrium has now become *the* distinguishing hallmark for an algorithm suitable for application to contemporary clinical techniques. If an algorithm *cannot* predict the enlarged beam penumbra and depression of central axis dose for small fields incident on a low-density absorber, it is triaged to Type-A. Type-C algorithms (Zhou et al. 2017) are beyond Type-B algorithms with richer physics content and greater consideration of atomic composition.

Principle #6: "An algorithm's performance under conditions of lateral charged particle disequilibrium is a strong indicator of its trustworthiness over a wide range of clinical applications"

In this chapter, three generations of algorithms were essentially aligned with A-B-C designations, as summarized in Figure 3.23. First generation algorithms automatically fall into Type-A with a limited scope of clinical applicability beyond cobalt-60 energy. Pencil beam algorithms that only sample anatomy longitudinally along beamlet rays, also cannot predict accurate doses in zones of lateral disequilibrium. The *3D* superposition method, on the other hand, is considered a Type-B algorithm. The most comprehensive methods can further predict dose under the more extreme conditions of charged particle disequilibrium conditions, including media interface effects. Monte Carlo and Boltzmann transport methods have a Type-C designation. Results from these advanced models of radiation propagation can serve as benchmarks to test algorithms of Types A-B.

3.7 SUMMARY

In summary, accurate modelling of x-ray beam interactions in human tissue has steadily improved, driven by conceptual and technological advances. The accuracy of algorithms has been ranked in descending order (Lu 2013): Monte Carlo and Acuros XB, Superposition, Analytical Anisotropic Algorithm (AAA), pencil beam, and correction-based first generation algorithms. This ranking was judged on the basis of a set of comparisons with measured data in homogeneous and inhomogeneous test phantoms. Given the same fundamental database and run conditions, dose

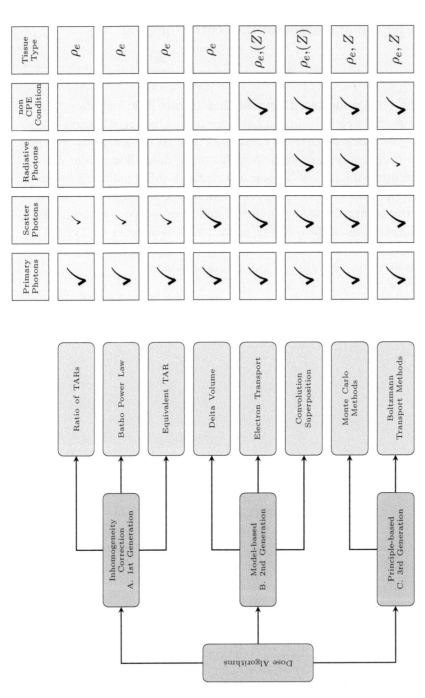

Figure 3.23: Re-categorization of dose algorithms. ρ_e is electron density and Z is atomic number of tissue. Type-A algorithms are based on inhomogeneity corrections of in-water dose distributions, and normally ignore Z-effects. Type-B methods are based on a physical model that normally can account for charged particle disequilibrium. They partially account for Z-effects (Z) used in the TERMA calculation. Type-C algorithms are based on tracking particle flow and energy losses *via* using fundamental interaction probabilities. They have intrinsic potential to account for Z-effects not only in primary but also in secondary dose contributions.

results from Monte Carlo and Boltzmann transport algorithms were convergent to within statistical uncertainty. These competitive methods (MacDermott 2016b; Ojala 2014) will attract commercial interest and together with fast computers will accelerate deployment into clinical radiation oncology.

BIBLIOGRAPHY

Ahnesjö, A. (1989). Collapsed cone convolution of radiant energy for photon dose calculation in heterogeneous media. *Medical Physics 16*(4), 577–592.

Ahnesjö, A. and M. M. A. (1999). Dose calculations for external photon beams in radiotherapy. *Physics in Medicine and Biology 44*(11), R99–R155.

Ahnesjö, A., P. Andreo, and A. Brahme (1987). Calculation and application of point spread functions for treatment planning with high energy photon beams. *Acta Oncologica 26*(1), 49–56.

Ahnesjö, A., M. Saxner, and A. Trepp (1992). A pencil beam model for photon dose calculation. *Medical Physics 19*(2), 263–273.

Andreo, P. (1991). Monte Carlo techniques in medical radiation physics. *Physics in Medicine and Biology 36*(7), 861–920.

Andreo, P. (2015). Dose to water-like media or dose to tissue in MV photons radiotherapy treatment planning: still a matter of debate. *Physics in Medicine and Biology 60*(1), 309 – 337, 2619.

Andreo, P. (2017). The Monte Carlo Simulation of the Transport of Radiation Through Matter. In *Fundamentals of Ionizing Radiation Dosimetry* (Second ed.)., Chapter 8, pp. 349–396. Weinheim, Germany: Wiley-VCH.

Andreo, P., D. Burns, A. Nahum, J. Seuntjens, and F. Attix (2017a). *Fundamentals of Ionizing Radiation Dosimetry.* Weinheim, Germany: Wiley-VCH.

Andreo, P., D. Burns, A. Nahum, J. Seuntjens, and F. Attix (2017b). Macroscopic aspects of the transport of radiation through matter. In P. Andreo, D. Burns, A. Nahum, J. Seuntjens, and F. Attix (Eds.), *Fundamentals of Ionizing Radiation Dosimetry*, Chapter 6, pp. 279–313. Weinheim, Germany: Wiley-VCH.

Battista, J. and M. Bronskill (1978). Compton-scatter tissue densitometry: calculation of single and multiple scatter photon fluences. *Physics in Medicine and Biology 23*(1), 1–23.

Battista, J. and M. Bronskill (1981). Compton scatter imaging of transverse sections: An overall appraisal and evaluation for radiotherapy planning. *Physics in Medicine and Biology 26*(1), 81–99.

Battista, J. and M. Sharpe (1992). True three-dimensional dose computations for megavoltage x-ray therapy: a role for the superposition principle. *Australasian Physical and Engineering Sciences in Medicine 15*(4), 159–178.

Bielajew, A. (2016). History of Monte Carlo. In J. Seco and F. Verhaegen (Eds.), *Monte Carlo Techniques in Radiation Therapy*, Chapter 1, pp. 3–16. CRC Press, Taylor & Francis Group.

Bjarngard, B. (1987). On Fano's and O'Connor's Theorems. *Radiation Research 109*(2), 184–189.

Borgers, C. (1998). Complexity of Monte Carlo and deterministic dose-calculation methods. *Physics in Medicine and Biology 43*(3), 517–528.

Bortfeld, T., W. Schlegel, and B. Rhein (1993). Decomposition of pencil beam kernels for fast dose calculations in threedimensional treatment planning. *Medical Physics 20*(2), 311–318.

Boyer, A. (1984). Shortening the calculation time of photon dose distributions in an inhomogeneous medium. *Medical Physics 11*(4), 552–4.

Brualla, L., M. Rodriguez, and A. Lallena (2017). Monte Carlo systems used for treatment planning and dose verification. *Strahlentherapie und Onkologie 193*(4), 243–259.

Chetty, I., B. Curran, J. Cygler, J. DeMarco, G. Ezzell, B. Faddegon, I. Kawrakow, P. Keall, H. Liu, C. Ma, D. Rogers, J. Seuntjens, D. Sheikh-Bagheri, and J. Siebers (2007). Report of the AAPM Task Group No. 105: Issues associated with clinical implementation of Monte Carlo-based photon and electron external beam treatment planning. *Medical Physics 34*(12), 4818–4853.

Cunningham, J. (1972). Scatter-air ratios. *Physics in Medicine and Biology 17*(1), 42–51.

Cunningham, J. (1989). Keynote address: development of computer algorithms for radiation treatment planning. *International Journal of Radiation Oncology, Biology, and Physics 16*(6), 1367–1376.

Cunningham, J. and J. Battista (1995). Calculation of dose distributions for x ray therapy. *Physics in Canada 51*(4), 190–195.

Cunningham, J. and L. Beaudoin (1974). Calculations for tissue inhomogeneities with experimental verification. In *International Congress of the XIIIth International Congress of Radiology. Vol. #2.*, pp. 653–657. Excerpta Medica, Amsterdam.

Disher, B., G. Hajdok, A. Wang, J. Craig, S. Gaede, and J. Battista (2013). Correction for artificial electron disequilibrium due to cone-beam CT density errors:

Implications for on-line adaptive stereotactic body radiation therapy of lung. *Physics in Medicine and Biology 58*(12), 4157–4174.

Edmund, J. and T. Nyholm (2017). A review of substitute CT generation for MRI-only radiation therapy. *Radiation Oncology 12*(1), 28.

Failla, G., T. Wareing, Y. Archambault, and S. Thompson (2015). Acuros XB Advanced Dose Calculation For the Eclipse Treatment Planning System. Report RAD 10156A, Varian Medical Systems, Palo Alto, California.

Fernández-Varea, J., P. Carrasco, V. Panettieri, and L. Brualla (2007). Monte Carlo based water/medium stopping-power ratios for various ICRP and ICRU tissues. *Physics in Medicine and Biology 52*(21), 6475–6483.

Field, C. (1988). Dose Calculations for Megavoltage Photon Beams using Convolution. Master's thesis, University of Alberta.

Field, G. and J. Battista (1987). Photon dose calculations using convolution in real and Fourier space: assumptions and time estimates. In *Proceedings of the IXth International Conference on the Use of Computers in Radiation Therapy*, pp. 103–100.

Fleckenstein, J., L. Jahnke, F. Lohr, F. Wenz, and J. Hesser (2013). Development of a Geant4-based Monte Carlo Algorithm to evaluate the MONACO VMAT treatment accuracy. *Zeitschrift fur Medizinische Physik 23*(1), 33–45.

Gibbons, J. (2016). Treatment Planning Algorithms: Photon Dose Calculations. In F. Khan, J. Gibbons, and P. Sperduto (Eds.), *Khan's Treatment Planning in Radiation Oncology* (Fourth ed.)., Chapter 4, pp. 47–60. Lippincott Williams & Williams.

Gifford, K., J. Horton, T. Wareing, G. Failla, and F. Mourtada (2006). Comparison of a finite-element multigroup discrete-ordinates code with Monte Carlo for radiotherapy calculations. *Physics in Medicine and Biology 51*(9), 2253–2265.

Han, T., D. Followill, J. Mikell, R. Repchak, A. Molineu, R. Howell, M. Salehpour, and F. Mourtada (2013). Dosimetric impact of Acuros XB deterministic radiation transport algorithm for heterogeneous dose calculation in lung cancer. *Medical Physics 40*(5), 051710–051711.

Hissoiny, S., B. Ozell, H. Bouchard, and P. Després (2011). GPUMCD: A new GPU-oriented Monte Carlo dose calculation platform. *Medical Physics 38*(2), 754–764.

Jacques, R., J. Wong, R. Taylor, and T. McNutt (2011). Real time dose computation: GPU accelerated source modeling and superposition/convolution. *Medical Physics 38*(1), 294–305.

Jahnke, L., J. Fleckenstein, F. Wenz, and J. Hesser (2012). GMC: a GPU implementation of a Monte Carlo dose calculation based on Geant4. *Physics in Medicine and Biology 57*(5), 1217–1229.

Jenkins, T., W. Nelson, and A. Rindi (2012). *Monte Carlo Transport of Electrons and Photons.* Heidelberg, Germany: Springer.

Johns, H. and J. Cunningham (1974). *The Physics of Radiology* (3rd ed.). Charles C. Thomas.

Johns, H., M. Morrison, and G. Whitmore (1956). Dosage calculations for rotation therapy with special reference to cobalt-60. *The American Journal of Roentgenology, Radium Therapy, and Nuclear Medicine 75*(6), 1105.

Kan, M., P. Yu, and L. Leung (2013). A review on the use of grid-based Boltzmann equation solvers for dose calculation in external photon beam treatment planning. *BioMed Research International ID 692874*, 1–10.

Kancharla, S. (2013). Mastering model risk: assessment, regulation and best practices. Report, Numerix Research.

Knöös, T. (2017). 3D dose computation algorithms. *Journal of Physics: Conference Series 847*(1), 021037.

Lobo, J. and I. Popescu (2010). Two new DOSXYZnrc sources for 4D Monte Carlo simulations of continuously variable beam configurations, with applications to RapidArc, VMAT, TomoTherapy and CyberKnife. *Physics in Medicine and Biology 55*(16), 4431–4443.

Lu, L. (2013). Dose calculation algorithms in external beam photon radiation therapy. *International Journal of Cancer Therapy and Oncology 1*(2), 01025–01028.

Ma, C. and J. Li (2011). Dose specification for radiation therapy: dose to water or dose to medium? *Physics in Medicine and Biology 56*(10), 3073–3089.

MacDermott, P. (2016a). Convolution/Superposition Dose Computation Algorithms. In *Tutorials in Radiotherapy Physics* (First ed.)., Chapter 3, pp. 120–169. CRC Press, Taylor & Francis Group.

MacDermott, P. (2016b). Deterministic Radiation Transport: a Rival to Monte Carlo Methods. In *Tutorials in Radiotherapy Physics* (First ed.)., Chapter 4, pp. 170–216. CRC Press, Taylor & Francis Group.

MacFarlane, M., D. Wong, D. Hoover, E. Wong, C. Johnson, J. Battista, and J. Chen (2018). Patient specific calibration of cone beam computed tomography data sets for radiotherapy dose calculations and treatment plan assessment. *Journal of Applied Clinical Medical Physics 19*(2), 249–257.

Mackie, T., A. Bielajew, D. Rogers, and J. Battista (1988). Generation of photon energy deposition kernels using the EGS Monte Carlo code. *Physics in Medicine and Biology 33*(1), 1–20.

Mackie, T., E. El-Khatib, J. Battista, J. Scrimger, J. Van Dyk, and J. Cunningham (1985). Lung dose corrections for 6 and 15 MV x-rays. *Medical Physics 12*(3), 327–332.

Mackie, T., J. Scrimger, and J. Battista (1984). A convolution method of calculating dose for 15 MV x rays. *Medical Physics 12*(2), 188–96.

Mayles, P., A. Nahum, and J. Rosenwald (2007). *Handbook of radiotherapy physics: theory and practice.* CRC Press, Taylor & Francis Group.

McCollough, C., S. Leng, L. Yu, and J. Fletcher (2015). Dual- and multi-energy ct: Principles, technical approaches, and clinical applications. *Radiology 276*(3), 637–653.

Metcalfe, P., T. Kron, and P. Hoban (2007a). Beam models: Part I. In P. Metcalfe, T. Kron, and P. Hoban (Eds.), *The physics of radiotherapy x-rays and electrons*, Chapter 9, pp. 573–616. Madison, Wisconsin: Medical Physics Publishing.

Metcalfe, P., T. Kron, and P. Hoban (2007b). Beam models: Part II. In *The physics of radiotherapy x-rays and electrons*, Chapter 10, pp. 619–670. Madison, WI, USA: Medical Physics Publishing.

Moiseenko, V., R. Hamm, A. Waker, and W. Prestwich (2001). Calculation of radiation-induced DNA damage from photons and tritium beta-particles. *Radiation and Environmental Biophysics 40*(1), 23–31.

Nelson, W., D. Rogers, and H. Hirayama (1985). The EGS4 Code System. Manual SLAC-0265, Stanford University, Stanford, California.

Oelkfe, U. and C. Scholz (2006). Dose calculation algorithms. In *New Technologies in Radiation Oncology*, pp. 187–196. Springer.

Ojala, J. (2014). The accuracy of the Acuros XB algorithm in external beam radiotherapy: a comprehensive review. *International Journal of Cancer Therapy and Oncology 2*(4), 1–12.

Oliver, P. and R. Thomson (2017). A Monte Carlo study of macroscopic and microscopic dose descriptors for kilovoltage cellular dosimetry. *Physics in Medicine and Biology 62*(4), 1417–1436.

Papanikolaou, N., J. Battista, A. Boyer, C. Kappas, E. Klein, and T. Mackie (2004). Tissue Inhomogeneity Corrections for Megavoltage Photon Beams: Report of the AAPM Radiation Therapy Committee Task Group 65. Technical Report AAPM Report 85, American Association of Physicists in Medicine.

Papanikolaou, N. and S. Stathakis (2009). Dose-calculation algorithms in the context of inhomogeneity corrections for high energy photon beams. *Medical Physics 36*(10), 4765–4775.

Paudel, M., A. Kim, A. Sarfehnia, S. Ahmad, D. Beachey, A. Sahgal, and B. Keller (2016). Experimental evaluation of a GPU-based Monte Carlo dose calculation algorithm in the Monaco treatment planning system. *Journal of Applied Clinical Medical Physics 17*(6), 230–241.

Prasad, S., T. Kron, J. Battista, and J. Kempe (2005). Conversion of kVCT Numbers to electron density of human tissues in vivo: Validation using megavoltage CT scanning on a tomotherapy machine. *Medical Physics 32*(7), 2412.

Raeside, D. (1976). Monte Carlo principles and applications. *Physics in Medicine and Biology 21*(2), 181–197.

Reynaert, N., S. van der Marck, D. Schaart, W. Van der Zee, C. Van Vliet-Vroegindeweij, M. Tomsej, J. Jansen, B. Heijmen, M. Coghe, and C. De Wagter (2007). Monte Carlo treatment planning for photon and electron beams. *Radiation Physics and Chemistry 76*(4), 643–686.

Rodriguez, M. and L. Brualla (2018). Many-integrated core (MIC) technology for accelerating Monte Carlo simulation of radiation transport: A study based on the code DPM. *Computer Physics Communications 225*, 28–35.

Rogers, D. (2006). Fifty years of Monte Carlo simulations for medical physics. *Physics in Medicine and Biology 51*(13), R287–R301.

Rogers, D., B. Faddegon, G. Ding, C. Ma, J. We, and T. Mackie (1995). BEAM: a Monte Carlo code to simulate radiotherapy treatment units. *Medical Physics 22*(5), 503–524.

Rogers, D., I. Kawrakow, J. Seuntjens, B. Walters, and E. Mainegra-Hing (2003). NRC User Codes for EGSnrc. Manual PIRS-702 (Rev. B), National Research Council of Canada, Ottawa, Canada.

Rogers, D., B. Walters, and I. Kawrakow (2009). BEAMnrc User's Manual. Report PIRS-509A, National Research Council of Canada, Ottawa, Canada.

Rogers, M., A. Kerr, G. Salomons, and L. Schreiner (2005). Quantitative investigations of megavoltage computed tomography. In *Medical Imaging 2005: Physics of Medical Imaging*, Volume 5745, pp. 685–695. International Society for Optics and Photonics.

Rossi, H. and W. Roesch (1962). Field equations in dosimetry. *Radiation Research 16*(6), 783–795.

Schey, H. (1997). *Div Grad Curl and all that.* New York: WW Norton & Company.

Seco, J. and F. Verhaegen (2016). *Monte Carlo Techniques in Radiation Therapy.* Boca Raton, Florida, USA: CRC press, Taylor & Francis Group.

Seuntjens, J., E. Lartigau, S. Cora, G. Ding, S. Goetsch, and J. Nuyttens (2014). Prescribing, recording, and reporting of stereotactic treatments with small photon beams. *Journal of the ICRU 14*, 1–160.

Sharpe, M. (1997). *A unified method of calculating the dose rate and dose distribution for therapeutic x-ray beams.* Ph. D. thesis, Department of Medical Biophysics, University of Western Ontario.

Sharpe, M. and J. Battista (1993). Dose calculations using convolution and superposition principles: the orientation of dose spread kernels in divergent x-ray beams. *Medical Physics 20*(6), 1685–1694.

Sharpe, M., E. Wong, J. Van Dyk, and J. Battista (1997). Accounting for the x-ray energy spectrum in superposition dose calculations. In *Proceedings of the XIIth International Conference on the Use of Computers in Radiation Therapy*, pp. 103–107.

Sheikh-Bagheri, D. and D. Rogers (2002). Monte Carlo calculation of nine megavoltage photon beam spectra using the BEAM code. *Medical Physics 29*(3), 391–402.

Siebers, J., P. Keall, A. Nahum, and R. Mohan (2000). Converting absorbed dose to medium to absorbed dose to water for Monte Carlo based photon beam dose calculations. *Physics in Medicine and Biology 45*(4), 983–995.

Sinclair, W., H. Rossi, R. Alsmiller, M. Berger, A. Kellerer, W. Roesch, L. Spencer, and M. Zaider (1991). *Conceptual Basis for Calculations of Absorbed-Dose Distributions.* Bethesda, Maryland: National Council on Radiation Protection and Measurements.

Sontag, M. and J. Cunningham (1978). The equivalent tissue-air ratio method for making absorbed dose calculations in a heterogeneous medium. *Radiology 129*(3), 787–794.

Spiegel, M. (1959). *Theory and Problems of Vector Analysis.* New York: Schaum Publishing Company.

Thomas, S. (2016a). Treatment Planning Algorithms: Part 1. *SCOPE, Institute of Physics and Engineering in Medicine, September 2016*, 30–36.

Thomas, S. (2016b). Treatment Planning Algorithms: Part 2. *SCOPE, Institute of Physics and Engineering in Medicine, December 2016*, 28–33.

Tillikainen, L., H. Helminen, T. Torsti, S. Siljamäki, J. Alakuijala, J. Pyyry, and W. Ulmer (2008). A 3D pencil-beam-based superposition algorithm for photon dose calculation in heterogeneous media. *Physics in Medicine and Biology 53*(14), 3821–3839.

Varian (2015a). Aaa algorithm in eclipse 15. Manual, Varian Medical Systems, Palo Alto, California.

Varian (2015b). Eclipse photon and electron reference guide. Manual P1008611-003-C, Varian Medical Systems, Palo Alto, California.

Vassiliev, O., T. Wareing, J. McGhee, G. Failla, M. Salehpour, and F. Mourtada (2010). Validation of a new grid-based Boltzmann equation solver for dose calculation in radiotherapy with photon beams. *Physics in Medicine and Biology 55*(3), 581–598.

Verhaegen, F., S. Palefsky, and F. DeBlois (2006). RadSim: a program to simulate individual particle interactions for educational purposes. *Physics in Medicine and Biology 51*(8), N157–N161.

Verhaegen, F. and J. Seuntjens (2003). Monte Carlo modelling of external radiotherapy photon beams. *Physics in Medicine and Biology 48*(21), R107–R164.

Wong, E., J. Van Dyk, and Y. Zhu (1997). Lateral electron transport in FFT photon dose calculations. *Medical Physics 24*(12), 1992–2000.

Wong, J. and R. Henkelman (1983). A new approach to CT pixel-based photon dose calculations in heterogeneous media. *Medical Physics 10*(2), 199–208.

Wong, J. W. and J. A. Purdy (1990). On methods of inhomogeneity corrections for photon transport. *Medical Physics 17*(5), 807–814.

Woo, M. and J. Cunningham (1990). The validity of the density scaling method in primary electron transport for photon and electron beams. *Medical Physics 17*(2), 187–194.

Woo, M., J. Cunningham, and J. Jezioranski (1990). Extending the concept of primary and scatter separation to the condition of electronic disequilibrium. *Medical Physics 17*(4), 588–595.

Young, M. and R. Kornelsen (1983). Dose corrections for low density tissue inhomogeneities and air channels for 10 MV x rays. *Medical Physics 10*(4), 450–455.

Yu, C., T. Mackie, and J. Wong (1995). Photon dose calculation incorporating explicit electron transport. *Medical Physics 22*(7), 1157–1165.

Yu, C. and J. Wong (1993). Implementation of the ETAR method for 3D inhomogeneity correction using FFT. *Medical Physics 20*(3), 627–632.

Zhang, Y., Y. Feng, W. Wang, C. Yang, and P. Wang (2017). An expanded multiscale monte carlo simulation method for personalized radiobiological effect estimation in radiotherapy: a feasibility study. *Scientific Reports 7*(ID45019), 1–10.

Zhou, C., N. Bennion, R. Ma, X. Liang, S. Wang, K. Zvolanek, M. Hyun, X. Li, S. Zhou, W. Zhen, C. Lin, A. Wahl, and D. Zheng (2017). A comprehensive dosimetric study on switching from a Type-B to a Type-C dose algorithm for modern lung SBRT. *Radiation Oncology 12*(80), 1–11.

Pioneers of 3D convolution-superposition methods of x-ray dose computation.
Thomas Rockwell Mackie (left) and Anders Ahnesjö (right) formulated the
concepts, published extensively, and translated their ideas into commercial
software products across many treatment planning platforms.
Photos courtesy of University of Wisconsin-Madison and Uppsala University.

CHAPTER 4

Convolution and Superposition Methods

Jeff Z. Chen

London Health Sciences Centre and University of Western Ontario

CONTENTS

4.1 INTRODUCTION

S UPERPOSITION and convolution methods for radiation dose calculation were first introduced during a very productive period of algorithm developments from 1983-1985 (Mackie 1983; Mackie et al. 1985; Ahnesjö 1984; Boyer and Mok 1985; Mohan and Chui 1985b). Many of the ideas seemed to converge at the VIIIth *International Conference on the Use of Computers in Radiotherapy* (ICCR) held in 1984 in Toronto, Canada. Fast implementations of the three-dimensional (3D) convolution method evolved, including collapsed cone convolution (CCC) (Ahnesjö 1989; Ahnesjö and Aspradakis 1999) and fast Fourier transform convolution (FFTC) (Mohan and Chui 1987; Boyer et al. 1989; Wong et al. 1996). With some loss of generality and accuracy, two-dimensional (2D) models were also developed to achieve speed gains, including variations of pencil beams (Mohan and Chui 1986; Ahnesjö et al. 1992), singular value decomposition (SVD) (Bortfeld et al. 1993), and the analytic anisotropic algorithm (AAA) (Ulmer et al. 2005; Tillikainen et al. 2008). These adaptations were particularly effective in accelerating dose optimization procedures. All these methods had a historical speed advantage over Monte Carlo and Boltzmann methods and were therefore widely adopted in commercial radiation treatment planning systems [e.g. CCC in Pinnacle (Philips Healthcare, Fitchburg, WI), RayStation (RaySearch Laboratories AB, Stockholm, Sweden), Monaco (Elekta CMS, Maryland Heights, MO, USA), and AAA in Eclipse (Varian Medical Systems, Palo Alto, CA) (Tillikainen et al. 2008; Varian 2016)].

In principle, if we know the fluence spectrum $\Phi_{\mathbf{r}}^{\mathrm{cp}}(E_k)$ of charged particles with kinetic energy E_k at any position $\mathbf{r}$, similar to Equation 2.23, we can then calculate the dose using

$$D(\mathbf{r}) = \int_0^{E_{k,\max}} \frac{\mathrm{d}\Phi_{\mathbf{r}}^{\mathrm{cp}}(E_k)}{\mathrm{d}E_k} (S/\rho)_c \, \mathrm{d}E_k \qquad (4.1)$$

where $(S/\rho)_c$ is the mass collisional stopping power for kinetic energy E_k. However, it is difficult to calculate the fluence spectrum of charged particles unless we use Monte Carlo methods presented in Chapter 5 or the analytical semi-classical approach using Boltzmann transport equations presented in Chapter 6.

As discussed in Chapter 2, radiation energy deposition in a patient from a photon beam involves a two-step process. First, primary photons (those that have not yet had any interactions in the patient) interact with matter, transferring their energy to charged particles (electrons or positrons) and scattered photons, by Compton scattering, photoelectric effect, or pair production. Second, the energy is deposited by charged particles close to the interaction site or transferred further away by scattered photons or radiated photons generated by bremsstrahlung or positron-electron annihilation.

The dose calculation is simplified greatly when we separate the energy transfer of primary photons from more complicated energy propagation by secondary charged particles and scattered photons. This was described in Chapter 3 as the "divide and conquer" strategy. The first step is to calculate the total energy transferred from primary photons to charged particles and scattered photons. The second step is to calculate the spatial spreading of energy by charged particles and scattered photons. The first step is relatively straightforward, since the total energy released by the primary photons can be calculated using total attenuation coefficients defined in Chapter 2. For the second step, the dose spread per unit energy released by the primary photons can be pre-calculated by the Monte Carlo method for a large homogeneous water absorber. These baseline kernels can then be modified for tissue inhomogeneity under certain approximations. Recycling of pre-established kernels avoids a full re-computation of individual dose spread kernels per interaction point, producing a more efficient algorithm. This concept leads to the superposition and convolution dose calculation methods.

4.2 GENERAL CONCEPTS

Superposition methods are widely used in science, such as in Fourier analysis, wave superposition, and superposition principle in quantum mechanics. Each application relies on an impulse-response concept for summing up contributions from multiple sources. If each photon interaction point is considered a source that launches energy (i.e. an impulse), the dose to a surrounding location can be calculated by summing up the dose depositions contributed from all such photon interaction sites (i.e. combined response). The dose pattern from photons interacting at one spot in an absorber is called a "dose spread array" (Mackie et al. 1985), "energy deposition kernel" (Mackie et al. 1988), or "point spread function" (Ahnesjö 1989) as shown previously in Figure 2.18. The energy spread pattern is calculated by the Monte Carlo method for monoenergetic photons forced to interact at a fixed point in a large all-water absorber (Mackie et al. 1988). By way of analogy, we can consider each interaction site as a virtual radioactive source emitting gamma and beta rays anisotropically in the forward direction. The total energy transferred from primary photons to charged particles and scattered photons mimics source strength, calculated by using the local TERMA defined in Chapter 2 (Ahnesjö et al. 1987).

Principle #7: "The dose computation is greatly simplified when the energy *transfer* from primary photons is separated from the energy *propagation* by charged particles and scattered photons"

If the dose spread kernel is the same at every launch point, depending only on the relative coordinates of the interaction point source at location $\mathbf{s}$ and dose calculation point "target" at location $\mathbf{r}$, i.e. $(\mathbf{r} - \mathbf{s})$, the superposition is simplified as a convolution. A pure convolution only holds true when the photon beam is monoenergetic, non-divergent, and the dose region is within a large homogeneous absorber, sufficiently far away from any external boundary. However, the basic concept of spreading and integrating energy launched from interaction points can be generalized to account for x-ray spectra and tissue inhomogeneity. The relationship between a restricted convolution and more general superposition is described in the following sections.

4.2.1   General Superposition

We start with a general superposition method for dose calculation of a polyenergetic photon beam. As shown in Figure 4.1, the total energy *released* by primary photons at a source voxel with a volume $\mathrm{d}^3 s$ and density $\rho(\mathbf{s})$ is

$$\Delta E_T(\mathbf{s}) = T_E(\mathbf{s})\rho(\mathbf{s})\mathrm{d}^3 s \mathrm{d}E \tag{4.2}$$

where $T_E(\mathbf{s})$ symbolizes the energy spectrum of TERMA (i.e. $T_E = \mathrm{d}T/\mathrm{d}E$). The dose spread kernel $k(E, \mathbf{s}, \mathbf{r})$ is defined as the fraction of the total energy released by primary photons with energy E at location $\mathbf{s}$, which is absorbed per unit volume at a surrounding location $\mathbf{r}$, as follows:

$$k(E, \mathbf{s}, \mathbf{r}) \equiv \frac{\Delta\epsilon(\mathbf{r})/\Delta E_T(\mathbf{s})}{\Delta V} \tag{4.3}$$

where $\Delta\epsilon(\mathbf{r})$ is the energy absorbed in the volume ΔV at destination location $\mathbf{r}$. If we sum up the dose contributions from all voxels, the total dose is given by the general superposition integral

$$D(\mathbf{r}) = [1/\rho(\mathbf{r})] \int_E \iiint_V T_E(\mathbf{s})\rho(\mathbf{s})k(E, \mathbf{s}, \mathbf{r})\mathrm{d}^3 s \mathrm{d}E \tag{4.4}$$

where the factor $[1/\rho(\mathbf{r})]$ converts the energy deposited per unit volume to the energy deposited per unit mass (i.e. dose). Please note that the dose given by Equation 4.4 is the dose to the *medium* with mass density $\rho(\mathbf{r})$, provided the kernel gives the fractional energy absorbed per unit volume in the *medium* at location $\mathbf{r}$. We will get back to this important point when we discuss the dose calculation in a heterogeneous medium.

The above superposition integral is very general, but very time consuming to calculate, especially for polyenergetic beams incident upon a heterogeneous medium.

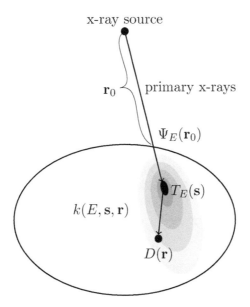

Figure 4.1: Schematic for dose spread kernels.

It requires the calculation of the local dose spread kernel for each combination of source-target locations, $\mathbf{r}$ and $\mathbf{s}$, and for each energy bin. This could be done using the Monte Carlo method, but this would negate any time savings gained by the convolution-superposition dose calculation algorithm. In the following sections, we will discuss various approximations to improve the efficiency of superposition dose calculations.

4.2.2 Polyenergetic Kernels

The integral with the energy spectrum of TERMA in Equation 4.4 can be simplified if we first define an average polyenergetic kernel by pre-integrating over energy, as follows:

$$\bar{k}(\mathbf{s}, \mathbf{r}) = \frac{\int T_E(\mathbf{s}) k(E, \mathbf{s}, \mathbf{r}) \mathrm{d}E}{T(\mathbf{s})} \tag{4.5}$$

where $T(\mathbf{s})$ is the overall TERMA from primary photons of all energies

$$T(\mathbf{s}) = \int T_E(\mathbf{s}) \mathrm{d}E \tag{4.6}$$

The polyenergetic kernel is thus obtained as a TERMA-weighted sum of monoenergetic kernels. Using the polyenergetic kernel, the 3D superposition integral (Equation 4.4) then simplifies to

$$D(\mathbf{r}) = [1/\rho(\mathbf{r})] \iiint T(\mathbf{s}) \rho(\mathbf{s}) \bar{k}(\mathbf{s}, \mathbf{r}) \mathrm{d}^3 s \tag{4.7}$$

One challenge that arises with using Equation 4.7 is that the beam spectrum changes with depth in an absorber due to the beam hardening effect (see Chapter 2). This affects (1) the TERMA which relies on energy-dependent attenuation coefficients, and (2) the polyenergetic kernel $\bar{k}(\mathbf{s}, \mathbf{r})$ at every launch location, $\mathbf{s}$. As an approximation, the average polyenergetic kernel can be calculated at a reference location $\mathbf{r_0}$ such as the *entrance* surface location. Since TERMA is proportional to energy fluence and average mass attenuation coefficient (see Chapter 2), the average polyenergetic kernel can then be approximated as a spectrum-weighted sum of monoenergetic kernels with discrete energy bins (Ahnesjö 1989), as follows:

$$\bar{k}(\mathbf{s}, \mathbf{r}) \approx \sum_i w_i k(E_i, \mathbf{s}, \mathbf{r}) \tag{4.8}$$

where $w_i = \Psi_i(\mathbf{r_0})/\Psi(\mathbf{r_0})$ is the spectrum weight, $\Psi(\mathbf{r_0})$ is the primary photon energy fluence at the reference location, and $\Psi_i(\mathbf{r_0})$ is the energy fluence for photons with only energy E_i. The construction of the average polyenergetic kernel is illustrated in Figure 4.2. In this approximation, the polyenergetic kernel is calculated only once for the incident beam spectrum at the reference location. It can be pre-calculated numerically with Equation 4.8 using Monte Carlo calculated monoenergetic kernels and stored as a table or approximated by analytical functions with curve-fitted parameters to be discussed in the following sections. The beam hardening along depth and the beam softening off axis can be accounted for approximately with some correction methods (Ahnesjö 1989; Ahnesjö and Aspradakis 1999; Tillikainen et al. 2007; Varian 2016).

As introduced in Chapters 2 and 3, there are clear advantages to separating the dose calculation into primary and scattered contributions. In the convolution-superposition algorithm, this is performed as two key steps: (1) primary photon energy release and (2) dispersal of this released energy. First, the total energy released by primary photons (TERMA) is divided into short-range collisional kinetic energy of charged particles, KERMA$_\text{c}$, and longer-range scattered photon energy per unit mass, SCERMA = TERMA − KERMA$_\text{c}$. Secondly, the scatter kernels are divided further according to type of scattering interaction (Chapter 3). This dual strategy simplifies modeling of the x-ray source and makes the convolution-superposition method efficient and accurate (Ahnesjö and Aspradakis 1999; Ahnesjö et al. 2005, 2017).

4.2.3 Convolution

To make the two-step dose calculation efficient using the superposition method, we need to make repeated use of pre-calculated in-water kernels for the energy range of interest. This can be achieved if the kernels depend *only* on the relative coordinates $(\mathbf{r} - \mathbf{s})$ between the source and the target. This condition is called *spatial invariance*. Under this condition, the superposition integral in Equation 4.7 becomes

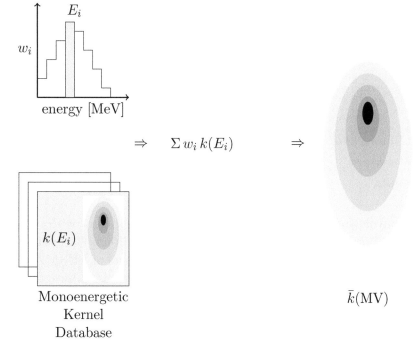

Figure 4.2: A polyenergetic dose spread kernel $\bar{k}(\text{MV})$ can be calculated for an x-ray spectrum as a weighted sum of monoenergetic kernels $k(E_i)$ with relative weight w_i for photons with energy E_i.

the convolution integral with kernel values depending only on relative positions of the source-destination pair of voxels, as follows:

$$D(\mathbf{r}) = [1/\rho(\mathbf{r})] \iiint T(\mathbf{s})\rho(\mathbf{s})\bar{k}(\mathbf{r} - \mathbf{s})\mathrm{d}^3 s \qquad (4.9)$$

This reduces the computational burden significantly, since the average polyenergetic kernel can be pre-calculated.

In homogeneous media, $\rho(\mathbf{r})$ is a constant, the convolution integral is simplified further to

$$D(\mathbf{r}) = \iiint T(\mathbf{s})\bar{k}(\mathbf{r} - \mathbf{s})\mathrm{d}^3 s \qquad (4.10)$$

For a monoenergetic beam in a uniform water medium, the convolution integral has the following simplest form:

$$D(\mathbf{r}) = \iiint T(\mathbf{s})k(\mathbf{r} - \mathbf{s})\mathrm{d}^3 s \qquad (4.11)$$

As a simplified example, the convolution dose calculation for a Co-60 beam with field size 30 x 30 cm^2 is shown in Figure 4.3. The beam is divergent, but this was

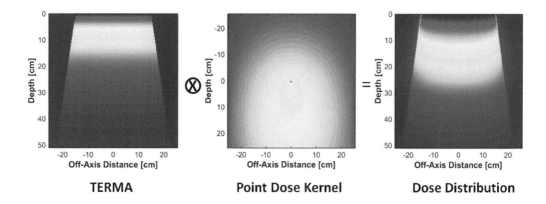

TERMA   **Point Dose Kernel**   **Dose Distribution**

Figure 4.3: Convolution dose calculation for a Co-60 beam with field size 30×30 cm^2. The point dose kernel was alculated by the Monte Carlo simulation. The dose distribution was obtained by the convolution integral of TERMA with the point dose kernel (Equation 4.11). The calculation was simplified by ignoring the re-orientation of kernels along beam divergent rays.

ignored in the orientation of dose spread kernels. For divergent beams, the kernels should be slightly tilted along primary beam rays, especially for large fields or short source-to-surface-distance (SSD) beam setups (Sharpe and Battista 1993).

To use a convolution-style algorithm for a heterogeneous medium, we need to find efficient ways to modify the pre-calculated in-water kernels based on densities of the voxels lying between the primary photon interaction point and the dose receiving point. This allows the kernel to change its spreading pattern at each interaction point. With a spatially-varying kernel shape, *superposition* becomes the more appropriate mathematical description.

4.2.4 Analytical Representations of Kernels

There are some advantages in representing the kernels with analytical expressions. For example, this facilitates ad-hoc access to kernel data and simplifies kernel modification for tissue inhomogeneity. It also provides physical insight into the dependence of the kernel values on the density distribution in the medium.

The splitting of both the primary photon energy release and kernels into primary and scattering components results in clear segregation of tasks for enhanced computational clarity and efficiency. X-ray beam hardening effects and tissue inhomogeneity effects can be handled more explicitly in this way. For example, the kernel can be divided into primary and secondary components as follows:

$$\bar{k}(\mathbf{r}) = \bar{k}_p(\mathbf{r}) + \bar{k}_s(\mathbf{r}) \tag{4.12}$$

For the primary kernel $\bar{k}_p(\mathbf{r})$, the dose is deposited by charged particles set into motion locally by primary photons at the interaction site, while for the secondary kernel $\bar{k}_s(\mathbf{r})$, the dose is deposited by charged particles set into motion more distantly by scattered photons, *bremsstrahlung*, or pair production-annihilation γ-rays. Each component of the dose deposition kernel can be calculated by the Monte Carlo method as introduced in Chapters 2 and 3.

Ahnesjö (Ahnesjö 1989) proposed that *polyenergetic* kernels for water can be well approximated by an analytic expression in spherical coordinates (r, θ)

$$\bar{k}(r, \theta) = (A_\theta e^{-a_\theta r} + B_\theta e^{-b_\theta r})/r^2 \tag{4.13}$$

where θ is the polar angle coordinate of the dose calculation point with respect to the direction of the impinging primary photon and r is the distance from the primary photon interaction point (at the origin). The first term represents the primary dose kernel with parameters A_θ and a_θ, which depend on the energy spectrum of the beam and the polar angle, while the second term presents the secondary dose kernel with associated parameters B_θ and b_θ.

The analytical approximation of point dose kernels by Equation 4.13 has similar dependence on distance as the dose distribution of a fictitious anisotropic brachytherapy source that emits both beta and gamma rays. The primary kernel resembles a β-particle source with the attenuation parameter a_θ proportional to the stopping power, while the secondary kernel resembles a γ-ray source with the photon attenuation coefficient b_θ.

4.2.5 Accounting for Tissue Inhomogeneities

The dose in homogeneous medium can be calculated fairly easily using various empirical methods derived from water phantom dosimetry (Johns and Cunningham 1983). The real challenge arises with a heterogeneous medium. Despite the existence of various semi-empirical methods to correct for inhomogeneity (Johns and Cunningham 1983), their accuracy and scope of application were very limited, particularly for small-field beams or beamlets of higher megavoltage energy incident on low density tissues, as discussed in Chapter 3. Compared with Monte Carlo and Boltzmann transport methods, convolution-superposition methods provide a fast and sufficiently accurate way to incorporate inhomogeneities in the dose calculation for most clinical situations.

Using the convolution of Equation 4.9, tissue inhomogeneity can be easily accounted for in the TERMA by ray-tracing from the entrance surface location $\mathbf{r}_0$ to each energy release site $\mathbf{s}$ and applying the inverse square law to account for primary beam divergence, as follows:

$$T(\mathbf{s}) = \left(\frac{r_0}{s}\right)^2 \frac{\bar{\mu}(\mathbf{s})}{\rho(\mathbf{s})} \Psi(\mathbf{r}_0) \exp\left(-\int_{\mathbf{r}_0}^{\mathbf{s}} \bar{\mu}(l)\mathrm{d}l\right) \tag{4.14}$$

where $\bar{\mu}(\mathbf{s})$ is the mean attenuation coefficient calculated by spectrum-weighted averaging (similarly to Equation 4.8)

$$\bar{\mu}(\mathbf{s}) = \sum_i w_i \mu(E_i, \mathbf{s}) \tag{4.15}$$

$\Psi(\mathbf{r}_0)$ is the energy fluence at a reference location $\mathbf{r}_0$ as shown in the Figure 4.1, and l is the path from $\mathbf{r}_0$ to $\mathbf{s}$. The greater difficulty is how to account for tissue inhomogeneities in the kernels.

As noted previously, recalculating the actual kernel for the heterogeneous surrounding of each interaction site loses the advantage of an efficient dose calculation using a convolution-superposition strategy. Alternatively, the dependence of kernels on the density of the medium can be exploited using the analytical expression given by Equation 4.13. The analytical kernel has similar dependence on distance as the dose distribution of a β and γ source at the origin with attenuation coefficient a_θ for the primary kernel and b_θ for the secondary kernel, both falling off with distance exponentially and by inverse square law. We may assume that the "attenuation coefficient" a_θ for charged particles is approximately proportional to the collisional stopping power of the medium, while the attenuation coefficient b_θ for scattered photons is approximately proportional to the Compton-scattering cross section as discussed in Chapter 2. Both the collisional stopping power for charged particles and the Compton scattering cross section are mainly proportional to the relative electron density of the absorber, if we ignore atomic number effects. Therefore, in a heterogeneous medium, the kernel can be scaled approximately by the water-equivalent distance along rays between the source and dose-receiving points. The in-water kernel values are then "looked up" along rays at this radiological distance instead of physical distance, as follows:

$$\bar{k}(\mathbf{s}, \mathbf{r}) = \rho'_e(\mathbf{r})(\bar{\rho}_e')^2 \bar{k}_\mathrm{w}[\bar{\rho}_e'(\mathbf{r} - \mathbf{s})] \tag{4.16}$$

where $\rho'_e(\mathbf{r}) = \rho_e(\mathbf{r})/\rho_e^\mathrm{w}(\mathbf{r})$ is the local relative electron density at $\mathbf{r}$, $\bar{\rho}_e'$ is the average relative electron density along the path from point $\mathbf{s}$ to $\mathbf{r}$

$$\bar{\rho}_e' = \frac{\int_\mathbf{s}^\mathbf{r} \rho'_e(l)\mathrm{d}l}{|\mathbf{r} - \mathbf{s}|} \tag{4.17}$$

and $\bar{k}_\mathrm{w}$ is the kernel for a uniform water medium.

Note that the kernel scaled by Equation 4.16 gives the fractional energy absorbed per unit volume in the *medium* at location $\mathbf{r}$. The multiplication of the local relative electron density $\rho'_e(\mathbf{r})$ converts the energy deposited per unit volume in water to that in the medium, and $(\bar{\rho}_e')^2$ factor is used to offset the inverse-square fall off with physical distance that is embedded in the water kernel values. The dose-receiving point experiences a different density environment but actually stays at the same physical location relative to the source. The numerator in Equation

4.17 is the radiological distance, while the denominator is the geometric distance.

Depending on the resolution of the dose calculation grid and density sampling, some dose calculation algorithms calculate the dose to a water voxel inside a heterogeneous medium. In this case, the kernel is calculated as the fractional energy absorbed per unit volume in a *water* voxel embedded inside *non-water* medium. Thus the local relative electron density $\rho'_e(\mathbf{r})$ should *not* be included in Equation 4.16. The ambiguity of whether dose is calculated for a water or local tissue medium was discussed in Section 3.5 of Chapter 3. On a practical level, this potential difference in the implementation of dose algorithms complicates the comparison of dose results in Section 4.5. On a fundamental level, microscopic entities (i.e. micron scale) are much smaller than voxel sizes (i.e. mm scale). Radiobiologically, it is the energy absorbed in cells, cell nuclei, and important macromolecules (e.g. DNA) that determines cell survival (Oliver and Thomson 2017). This is considered within the domain of microdosimetry, beyond reach of macroscopic dose computations described here.

The kernel-scaling method is illustrated in Figure 4.4. The dose to point $\mathbf{r}$ in the shadow of a low density inhomogeneity from a point source at $\mathbf{s}$ is approximately equal to the dose given by a water dose kernel to point $\mathbf{r}'$ with a shorter radiological distance from the source, provided the inverse-square fall off with physical distance is maintained. The inhomogeneity makes dose kernels *spatially variant* with the density-scaling approximation, since the average relative electron density depends on both source and dose receiving locations. This spoils the universal symmetry of the kernels, and the dose thus needs to be calculated with the superposition integral in Equation 4.7 instead of the convolution integral in Equation 4.9.

The density scaling applied only along the direct path between the source $\mathbf{s}$ and the dose-receiving point $\mathbf{r}$ is justified by the fact that the dominant dose contributions are from the primary and 1st scatter kernels as was illustrated in Figure 3.6. Most of the primary charged particles reaching the dose receiving point $\mathbf{r}$ travel closely along the direct-flight path between $\mathbf{s}$ and $\mathbf{r}$, while the singly-scattered photons reaching point $\mathbf{r}$ have a scattering angle corresponding exactly to this straight-line path.

For higher-order scattering, density-scaling between source and destination points is more approximate because the photon paths are not direct. For a large homogeneous structure with effective atomic number similar to water, it is a very good approximation, because it is compliant with O'Connor's (O'Connor 1984) and Fano's theorems (Bjarngard 1987) for photons and charged particles, respectively. The effect of a general inhomogeneity depends on its composition, structure and *surrounding* density environment. Simple linear density-scaling can introduce large errors near the interface of such an inhomogeneity (Woo and Cunningham 1990).

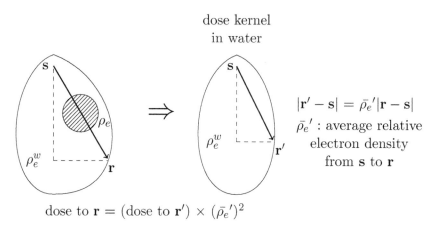

dose kernel
in water

$$|\mathbf{r}' - \mathbf{s}| = \bar{\rho}_e'|\mathbf{r} - \mathbf{s}|$$

$\bar{\rho}_e'$: average relative electron density from $\mathbf{s}$ to $\mathbf{r}$

$$\text{dose to } \mathbf{r} = (\text{dose to } \mathbf{r}') \times (\bar{\rho}_e')^2$$

Figure 4.4: Illustration of density scaling of dose spread kernels along a sample ray-tracing path from $\mathbf{s}$ to $\mathbf{r}$. The dose to point $\mathbf{r}$ in the shadow of a low density circular inhomogeneity with relative electron density ρ_e is approximately equal to the dose given by a water dose kernel to point $\mathbf{r}'$ with a shortened radiological distance from the interaction point $\mathbf{s}$, provided the inverse-square fall off with physical distance is restored. Adapted from reference (Woo and Cunningham 1990).

As mentioned in Chapter 3 earlier, we need to be clear about the interpretation of *dose* calculated in a heterogeneous medium, whether it is the dose to a *water* voxel in the medium or the dose to the *medium* in each voxel. For photon beams, O'Connor proposed that the doses in two media with different electron densities are the same provided they have the same effective atomic number and all dimensions in the media are scaled inversely with the density of the media (O'Connor 1957, 1984). Similar to O'Connor's theorem, the kernel with density scaling given by Equation 4.16 is approximately the fractional energy deposited to the *medium* located at $\mathbf{r}$ per unit volume. When it is used *strictly in accordance* with the superposition integral defined in Equation 4.7, it results in the dose to the *medium*. In commercial software, there may be implementation variations and the reader should consult the vendor's software manual for clarification of the absorbing medium.

4.2.6 Reciprocity Theorem

The concept of reciprocity was introduced in Chapters 2 and 3. We now extend this concept to three dimensions. The attenuation of energy between a launch and dose deposition site does not depend on the direction of travel of the radiation particles. For a monoenergetic beam in an infinite homogeneous medium, the kernel shape remains identical when we invert the kernel orientation and exchange the roles of the interaction source $\mathbf{s}$ and the destination dose-receiving point $\mathbf{r}$, as illustrated in Figure 4.5. Mathematically, the inverted kernel is defined as

$$k_{\text{inv}}(\mathbf{r}) \equiv k(-\mathbf{r}) \tag{4.18}$$

so that the inverted kernel has the same value as the original kernel $k_{\text{inv}}(\mathbf{s} - \mathbf{r}) = k(\mathbf{r} - \mathbf{s})$ when the source and target positions are interchanged. This means that the dose contribution from $\mathbf{s}$ to $\mathbf{r}$ remains the same if we use the inverted kernel as shown in Figure 4.5B, where the origin of the inverted kernel is now at the dose-receiving location $\mathbf{r}$. When we apply this relation to multiple interaction source points, such as the two sources in Figure 4.5, we obtain the reciprocity theorem for the dose calculation of a monoenergetic beam in an infinite homogeneous medium

$$D(\mathbf{r}) \;=\; \iiint T(\mathbf{s})k(\mathbf{r} - \mathbf{s})\mathrm{d}^3 s \tag{4.19a}$$

$$\;=\; \iiint T(\mathbf{s})k_{\text{inv}}(\mathbf{s} - \mathbf{r})\mathrm{d}^3 s \tag{4.19b}$$

The reciprocity theorem gives us two equivalent views for the dose calculation with the convolution integral. One view is the forward "pitcher's" point of view as given by Equation 4.19a and Figure 4.5A, in which TERMA from multiple source points contributes fractional dose to each of the dose-receiving voxels. The other view is the inverted "catcher's" point of view as given by Equation 4.19b and Figure 4.5B, in which the dose-receiving voxel receives the fractional dose from all surrounding sources within the inverted kernel space that is irradiated. If we just need to calculate dose to a few voxels, it is more efficient to use the inverted catcher's view, since fewer density-scaled kernels are required for dose calculations.

The reciprocity theorem can be generalized to the dose calculation for a heterogeneous medium using Equation 4.7 if the beam spectrum is assumed to be the same everywhere (i.e. ignoring beam hardening or softening effects) and density-scaling approximation is used for kernels. Since the kernel is scaled by the radiological distance between the source and the dose-receiving locations (Equation 4.16), we can define the inverted kernel as

$$\bar{k}_{\text{inv}}(\mathbf{s}, \mathbf{r}) \equiv \bar{k}(\mathbf{r}, \mathbf{s}) \tag{4.20}$$

so that the dose contribution from $\mathbf{s}$ to $\mathbf{r}$ remains the same if we use the inverted kernel as shown in Figure 4.6B. The reciprocity with heterogeneity is illustrated in Figure 4.6. The reversed kernel generally has a different composite shape compared to individual kernels in the forward pitcher's view, since inhomogeneities appear differently from different points. However, the dose calculated to a point in the forward pitcher's view is the same as that calculated from the inverted catcher's view, since the radiological distances between the interaction source and dose-receiving point pairs are preserved in both viewpoints.

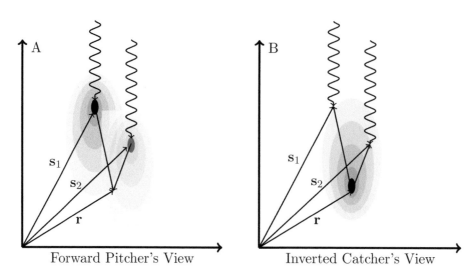

Figure 4.5: Illustration of reciprocity for a monoenergetic non-divergent beam in an infinite homogeneous medium.

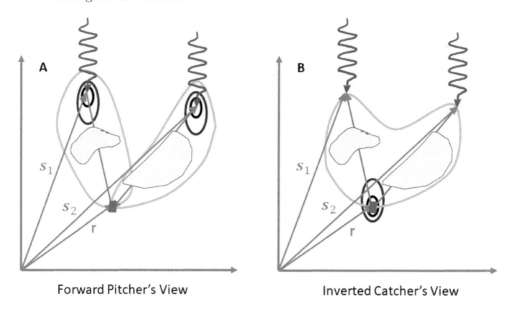

Figure 4.6: Illustration of reciprocity in a heterogeneous medium for a dose-receiving point $\mathbf{r}$. In the forward pitcher's viewpoint, the kernels for interaction source points 1 and 2 are shown separately. In the inverted catcher's viewpoint, a composite kernel shape is shown and radiological paths are preserved between energy launch sites $\mathbf{s}$ to the dose-receiving point at $\mathbf{r}$ shown as a red voxel.

4.3 FAST IMPLEMENTATIONS

The superposition method is inherently faster than the Monte Carlo or Boltzmann methods because kernels are pre-computed and reused. However, it is still very time consuming to perform the superposition operation directly for a large volume. For example, if we want to calculate the dose distribution for a 3D volume with N voxels along each dimension, we need to calculate the dose for N^3 voxels. For each dose voxel, we need to calculate potential kernel contributions from neighboring N^3 voxels and associated ray-tracing from each of these voxels to surrounding dose receiving voxels with order-N computations, thus bringing the total number of computations to the order of N^7 (Ahnesjö and Aspradakis 1999). This has led to the development of various faster implementations of convolution-superposition methods.

For didactic reasons, we will reverse the order of presentation *from 2D to 3D* convolution-superposition, compared with the original order of presentation in Chapter 3. In other words, these algorithms are now resequenced in order of complexity and progressively improved accuracy. We begin with fast Fourier transform convolution that can be used for both 2D and 3D algorithms if kernels are assumed to be spatially invariant. Next we will describe a series of 2D implementations, which include pencil-beam algorithms and singular-value decomposition (SVD). Then we will describe a more accurate generalized pencil-beam algorithm called anisotropic analytical algorithm (AAA), which can be categorized as a 2.5D algorithm. Finally we will build up to the most accurate and widely used contemporary 3D superposition algorithm called collapsed cone convolution (CCC).

4.3.1 Fast Fourier Transform Convolution

One way to speed up the convolution dose calculation is to use fast Fourier transforms (FFT) (Boyer and Mok 1986; Mohan and Chui 1987; Boyer et al. 1989; Wong et al. 1996). We can use the convolution theorem to convert a convolution integral to a product in the Fourier frequency domain (Boyer and Mok 1986; Wong et al. 1996), as follows:

$$\mathcal{F}(T * k) = \mathcal{F}(T) \times \mathcal{F}(k) \qquad (4.21)$$

where $\mathcal{F}$ is the Fourier transform operator and $T * k$ is the convolution integral, both of which can be calculated in a continuous or discrete form. In order to use the FFT in a discrete form, we also need to calculate the convolution integral in the discrete form, as follows:

$$(T * k)(\mathbf{r}_i) = \sum_{j=1}^{N_t} T(\mathbf{s}_j) k(\mathbf{r}_i - \mathbf{s}_j) \qquad (4.22)$$

where $N_t = N^3$ is the total number of dose voxels. Even though the convolution theorem converts the convolution integral to a simple product in the Fourier domain, the Fourier transform itself still requires a number of calculations on the order of N_t^2. Therefore, we need to look for a faster implementation of the Fourier transform, which is called *fast* Fourier transform (FFT). This method was first discovered by Gauss in 1805 and became popular after Cooley and Tukey published their method in 1965 (Cooley and Tukey 1965). FFT reduces the number of calculations from N_t^2 to $N_t \log_2 N_t$. For example, for one beam with 100 dose voxels in each dimension, $N_t = 10^6$, $N_t^2 = 10^{12}$, while $N_t \log_2 N_t \approx 2 \times 10^7$. If each computation takes 1 ns (e.g. using a 1 GFLOP/s processor), then 10^{12} computations take about 17 minutes, while 2×10^7 computations take only about 20 ms per beam. A clinical volume-modulated arc therapy (VMAT) plan with 100 beams would take about 28 hours without the FFT, while taking only 2 s with the FFT.

The FFT uses periodicity properties of the Fourier transform to achieve the remarkable speed gain of the transformation. The Fourier transform (X_k) of a discrete variable x_n (e.g. discrete TERMA or kernel values on a 3D dose grid) can be written as

$$X_k = \sum_{n=0}^{N_t - 1} x_n e^{-i \frac{2\pi k}{N_t} n} \qquad k = 0, ..., N_t - 1 \qquad (4.23)$$

Since for each of the N_t terms in the Fourier transform, we need N_t calculations in the summation, the total number of calculations is proportional to N_t^2. If we can reduce the dimension N_t of the Fourier transform, we can then reduce the number of calculations. One can make use of the periodicity property in the Fourier transform to divide the Fourier transform into even and odd series to reduce the dimension of the Fourier transform. This division can be continued down to very small dimensions for each series. Assuming N_t is an even number $N_t = 2^m$ with m an integer, we can then split the Fourier transform m times, which is equal to $\log_2 N_t$. It can be shown mathematically that the number of computations using FFT is proportional to the product of the dimension N_t and the number of splits. This results in the reduction of the number of computations from N_t^2 to $N_t \log_2 N_t$.

In order to use the FFT approach, we need to use the Fourier convolution theorem given by Equation 4.21 that requires *spatial invariance of the kernel*, namely, that the kernel depends only on the relative coordinate $(\mathbf{r} - \mathbf{s})$ as in Equation 4.22. This makes it very challenging to incorporate inhomogeneity corrections as shown in Figure 4.7, where the inhomogeneity is different between the path from $\mathbf{s_1}$ to $\mathbf{r_1}$ and the path from $\mathbf{s_2}$ to $\mathbf{r_2}$, even though the geometric distances are the same, $\mathbf{r_1} - \mathbf{s_1} = \mathbf{r_2} - \mathbf{s_2}$.

Despite these challenges, there were a few studies that incorporated tissue inhomogeneity using the FFT convolution, such as the studies by Boyer *et al.* (Boyer and Mok 1986), Ahnesjö (Ahnesjö 1987), and Wong *et al.* (Wong et al. 1996). In addition to accounting for the inhomogeneity in the TERMA term through ray-tracing, a correction term for the first-scattering kernel was also calculated based

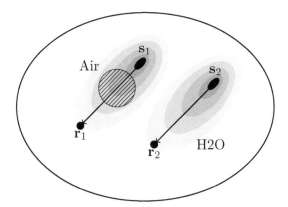

Figure 4.7: Illustration of the limitation of using spatially-invariant dose spread kernels in heterogeneous media. Geometrically, $\mathbf{r_1} - \mathbf{s_1} = \mathbf{r_2} - \mathbf{s_2}$, but the dose to $\mathbf{r_1}$ from $\mathbf{s_1}$ will be different compared to the dose to $\mathbf{r_2}$ from $\mathbf{s_2}$ because of different radiological paths. This difference cannot be accounted for by using a spatially-invariant dose kernel shown in the figure.

on a second-order perturbation theory for Compton scattering (Wong et al. 1996). These studies demonstrated some improvements over semi-empirical methods, but also showed the limitations (Field and Battista 1987) of using spatially invariant kernels, especially near the interface of an inhomogeneity.

4.3.2 Pencil Beam Algorithms (2D)

As explained in Chapter 3, a 3D dose distribution can be simplified to a 2D convolution-superposition under certain symmetry assumptions. Any clinical broad beam can be divided into many small beamlets. If we calculate the dose distribution from each beamlet, we can obtain the total dose distribution by simply summing up all their contributions. This is the basis for the 2D pencil-beam algorithm shown in Figure 4.8 for a homogeneous absorber. This approach has given rise to a different class of convolution-superposition algorithms that are *not* as accurate or versatile as 3D implementations. However, there is an advantage to using pencil beams – faster execution for dose optimization of intensity-modulated (IMRT) beams.

Generally it takes N^7 calculations for a 3D convolution-superposition algorithm, where N is the number of dose calculation voxels along one dimension. Using 2D convolution, for each of N^3 dose calculation voxels, we reduce the number of calculations from N^3 to N^2, resulting in the total number of computations on the order of N^5. If we assume spatially-invariant pencil-beam kernels, we can use the fast Fourier transform (FFT) for the 2D convolution integral. For each depth, the number of 2D calculations using the FFT is about $N^2 \log_2 N^2$. This results in a significant overall reduction in number of calculations from N^7 to $2N^3 \log_2 N$.

Incident Energy Fluence   **Pencil Beam Kernel**   **Dose**

Figure 4.8: Illustration of 2D pencil beam convolution. The primary energy fluence on the surface is convolved with pre-calculated pencil-beam kernel to calculate the total dose. Adapted from reference (McDermott 2016).

Pre-calculated finite-size pencil-beam dose kernels (also called beamlet dose distributions) are often used to optimize IMRT beam intensity pattern quickly. The beamlet dose distributions can be pre-calculated based on the actual patient anatomy using any one of the dose calculation algorithms described in this book. For example, in helical tomotherapy, these beamlet dose distributions are pre-calculated with the 3D superposition method using a graphical processing unit (GPU). The final dose distribution is the superposition of these beamlet dose distributions with optimized incident fluence, which determines the leaf opening times. In this way, the inhomogeneity is accounted for relatively accurately using the 2D superposition method because each finite-size pencil beam individually accounts for tissue inhomogeneities all along its path using 3D point kernels. However, the computation of patient-specific finite-size pencil beam dose kernels is time-consuming, depending on computational power of the treatment planning system (e.g. GPU acceleration).

The pencil-beam convolution algorithm introduced in Chapter 3 was reformulated intuitively as follows. For simplicity, let's consider a parallel beam incident on a water medium with a flat surface at z_s as shown in Figure 4.9. The polyenergetic pencil-beam kernel $k_z(x - x', y - y')$ is conventionally defined as the fraction of the incident energy at the surface location (x', y', z_s) from a pencil beam, which is absorbed per unit mass at a location (x, y, z). Based on this definition, the total dose distribution is simply the summation of the dose contributions from all pencil beams. If we assume that the pencil-beam kernel is the same everywhere, i.e., it is spatially invariant, the total dose is the 2D *convolution* of the pencil-beam kernel with the energy fluence given by

$$D(x, y, z) = \iint \Psi(x', y', z_s) k_z(x - x', y - y') \mathrm{d}x' \mathrm{d}y' \qquad (4.24)$$

where $\Psi(x', y', z_s)$ is the primary photon energy fluence at the surface that is closely related to the surface TERMA as described in Chapter 3. The pencil-beam kernel is

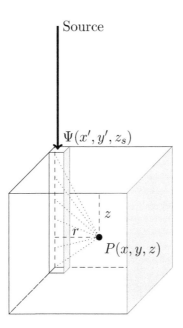

Figure 4.9: Illustration of a pencil beam incident on a water phantom surface at coordinate (x', y', z_s) with energy fluence $\Psi(x', y', z_s)$. The dose to the point P at location (x, y, z) is the sum of the dose contributions from all the points along the pencil-beam. This sum will depend on the position of the point $P(x, y, z)$, including its depth (z) and lateral distance $(r = \sqrt{x^2 + y^2})$ away from the pencil beam axis.

usually pre-calculated with the Monte Carlo method for a uniform water medium. Note that the pencil-beam kernel conventionally defined is related to the relative energy deposited per unit *mass* instead of per unit *volume*, which is usually defined for a point dose kernel (Equation 4.3). The 2D pencil-beam convolution (Equation 4.24) can also be derived from the 3D convolution integral given by Equation 4.10. The idea is to pre-integrate the point dose kernels with attenuated TERMA along the pencil-beam axis as described in Chapter 3.

In a homogeneous water medium, the pencil-beam kernel has cylindrical symmetry, depending only on the depth z and radius r from the beam axis. The depth dependence accounts for the variation in scattering angles and solid angle subtended along the length of the pencil beam for the dose calculation point P as shown in Figure 4.9. The kernel can be pre-calculated by the Monte Carlo method, tabulated in numerical tables, or approximated by an analytical expression of the following form (Ahnesjö et al. 1992):

$$k_z(r) = \frac{A_z \mathrm{e}^{-a_z r}}{r} + \frac{B_z \mathrm{e}^{-b_z r}}{r} \tag{4.25}$$

where A_z, a_z, B_z, and b_z are fitting parameters dependent on the dose calculation depth z and the energy spectrum. Comparing with point kernels in Equation 4.13,

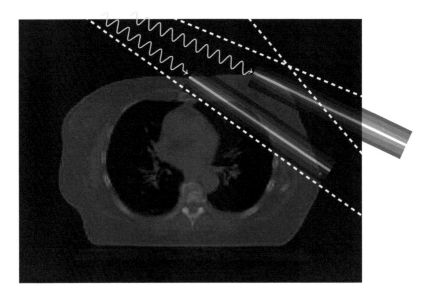

Figure 4.10: Illustration of limitations using pre-computed pencil-beam kernels with fixed shape to account for lung inhomogeneity. For example, the actual dose distribution of the pencil beam near low-density lung region should expand more laterally than the water pencil-beam kernel predicts.

the kernel now has a similar dependence on radial distance as the dose distribution of a linear source (e.g. brachytherapy line source) with $1/r$ geometric fall off, and attenuation parameters a_z for primary dose and b_z for scattered dose.

Since we pre-integrated along the depth z' for a uniform water medium, the pencil-beam algorithm using pre-calculated in-water pencil-beam kernels will lose some capabilities to account for 3D inhomogeneities. It can only account for inhomogeneities encountered along the beam axis or lateral position *separately*. One approximation that can be made is to account for inhomogeneity along the depth direction by a depth-dependent kernel with radiological depth and lateral inhomogeneity by a lateral radiological distance. The details and validity restrictions of this decomposition (or separation of variables) will be discussed further in the following Sections 4.3.3 and 4.3.4.

In general, the pencil-beam algorithm can account for beam shape and beam modifiers, such as blocks, attenuators, physical or dynamic wedges, but has difficulty accounting for tissue heterogeneity and missing tissue, as shown in Figure 4.10.

4.3.3 Singular-Value Decomposition (2D)

With the introduction of IMRT, there was a renewed pressure to speed up 3D dose calculations for the dose optimization process. As we can see from the previous section, we need to repeat a 2D convolution for each depth with a different pencil-beam kernel that depends on the depth of the dose calculation point. If we can decompose pencil-beam kernels into a sum of a few terms, each as a product of a depth dependent function and a radial dependent function, we can then speed up the 2D convolution. This is possible by the theory of the singular value decomposition (SVD) in mathematics first discovered independently by Eugenio Beltrami and Camille Jordan during 1873-1874.

The SVD theorem states that any matrix A can be decomposed into a product of three matrices $A = USV^T$, where U and V are orthonormal, V^T is the transpose of V, and S is a diagonal matrix with non-negative real numbers along the diagonal. If we write the decomposition in terms of matrix elements, we have

$$A_{ij} = \sum_{k=1}^{n} U_{ik} s_k V_{jk} \tag{4.26}$$

where s_k are called singular values which are the non-negative real numbers of diagonal elements of matrix S, and n is the number of singular values, also known as the number of principal components.

For a 2D pencil beam kernel with discrete coordinates along depth d_i and cylindrical radius r_j, we can write the kernel as a 2D matrix with elements

$$k_{ij} = k(d_i, r_j) \tag{4.27}$$

To apply SVD to the pencil-beam kernel, we can consider the 2D kernel as an arbitrary two-dimensional matrix A with index i for depth dependence and index j for radial dependence. In the SVD Equation 4.26, if we replace matrix element U_{ik} by a depth-dependent function $D'_k(d_i)$, and $s_k V_{jk}$ by a radially-dependent function $w_k(r_j)$, we can decompose the 2D kernel as follows:

$$k(d_i, r_j) = \sum_{k=1}^{n} D'_k(d_i) w_k(r_j) \tag{4.28}$$

with summing over the number of principal components n.

Bortfeld et al (Bortfeld et al. 1993) showed that, practically, the pencil-beam kernel can be decomposed into just three separate terms, each as a product of a radially-dependent function and a depth-dependent function, as follows:

$$k(r, d) \approx \sum_{k=1}^{3} w_k(r) D'_k(d) \tag{4.29}$$

where $w_k(r)$ are weighting factors as functions of cylindrical radial distance r from the pencil beam axis, and $D'_k(d)$ are depth-dose functions. Physically, these three depth-dose components result from small-field primary dose, small-field scatter dose, and large-field scatter dose. The small-field primary dose corresponds to the dose deposited by charged particles set into motion by primary photons within a small circular field just wide enough to achieve lateral electronic equilibrium. The small-field scatter dose corresponds to the dose due to scattered photons, bremsstrahlung and annihilation launched within the small field, while the large-field scatter dose corresponds to the dose due to the scattered photons, bremsstrahlung and annihilation originating within a large field beyond the small field region. The weighting factors $w_k(r)$ and depth dose function $D'_k(d)$ can be pre-calculated for a water medium. The decomposition replaces the multiple pencil-beam superposition for many depths (Equation 4.24) to a single superposition with only three terms, significantly speeding up the pencil-beam dose calculations (Bortfeld et al. 1993). The algorithm has been widely used in commercial treatment planning systems, such as Pinnacle and RayStation for IMRT or VMAT optimizations.

The effects of inhomogeneity can be approximately taken into account by using radiological (or water-equivalent) depth in the depth-dose function $D'_k(d)$. However, the validity of the approximation is very limiting. The density scaling along depth works only for slab-shaped layered inhomogeneities. Inhomogeneities lying away from the pencil-beam axis can be missed.

The advantage of the SVD method is its fast dose calculation. When combined with the fast Fourier transform for the 2D convolution, the number of calculations are further reduced from $2N^3\log_2 N$ to $6N^2\log_2 N$, since the 2D convolution is performed for only three decomposed terms instead of N dose voxels along the depth.

The SVD method has the same limitations to account for heterogeneity in a medium as the general pencil beam algorithm described in Section 4.3.2. However, the idea of decomposing pencil-beam kernels into depth and radial dependences opened up the possibility for more sophisticated methods to account for heterogeneity along the depth and the radial directions independently. One of these methods is the anisotropic analytical algorithm (AAA) described in the next section.

4.3.4 Anisotropic Analytical Algorithm (2.5D)

4.3.4.1 Overview of AAA

One of the generalized pencil-beam algorithms is called the anisotropic analytical algorithm (AAA), which is currently implemented in the Eclipse treatment planning system (Varian Medical Systems, Palo Alto, CA, USA) (Sievinen et al. 2005; Varian 2016). The model was initially developed by Ulmer and Kaissl (Ulmer and Harder 1995; Ulmer et al. 2005) and later improved by (Tillikainen et al. 2008).

The AAA consists of two major components: the source modeling for a treatment beam and the generalized pencil-beam dose calculation algorithm. Three sources are modeled for primary photons, scattered extra-focal photons, and contaminant electrons from linear accelerators. The Monte Carlo program (Kawrakow and Rogers 2003) is used to pre-compute energy spread kernels for monoenergetic *pencil* beams in water. A polyenergetic energy spread kernel is constructed as a spectrum-weighted sum of the monoenergetic kernels as shown previously in Figure 4.2 and described by Equation 4.8.

The AAA also assumes that the pencil-beam kernels can be separated into depth and lateral dependent components. The Monte-Carlo calculated pencil-beam kernel is factored into a product of a depth-dependent energy deposition function and a lateral scattering kernel. The lateral scattering is modeled by an anisotropic analytical kernel (hence the algorithm name) with six exponential functions, so that lateral heterogeneity can be taken into account anisotropically with density scaling along radial "spokes" radiating laterally from the pencil beam axis. The factorization and the analytical scattering kernels are accurate for a uniform water medium, since they are force-fitted to the kernels calculated by the Monte Carlo method. The approximation comes from the assumption that the density scaling can be performed along the depth and lateral directions *independently*. This is obviously an approximation, but an improvement over an algorithm that uses isotropic pencil-beam kernels.

In this section, we will focus on the AAA algorithm for primary photons only. Similar methods can be applied to extra-focal radiation (Sharpe and Battista 1993). For details of the AAA algorithm with all three radiation sources, the reader is referred to related publications and Eclipse user manuals (Tillikainen et al. 2008; Varian 2016). In the following sections, we will present the basic mathematical formulations of the AAA algorithm. First, we will define the pencil-beam kernels and associated coordinate systems. Second, we will express the kernel as a product of the energy deposition function and lateral scattering kernels, unique to the AAA method. Third, we will introduce the AAA approximations to account for heterogeneity in the medium. Finally, we will sum the pencil-beam contributions to calculate the total dose at a point of interest.

4.3.4.2 Pencil Beam Kernel and Coordinate Systems

A clinical broad beam is divided into a set of finite-size pencil beams with their edges aligned with the dose calculation grid as shown in Figure 4.11. To account for beam divergence and associated kernel tilting (Sharpe and Battista 1993), two coordinate systems are used in the AAA algorithm, as shown in Figure 4.12. One is a non-Cartesian beamlet coordinate system (x_p, y_p, z_p) attached to each finite-size pencil beam in a fan geometry and the other is a Cartesian coordinate system (x, y, z) associated with the patient. The lateral calculation surface (x_p, y_p) at every

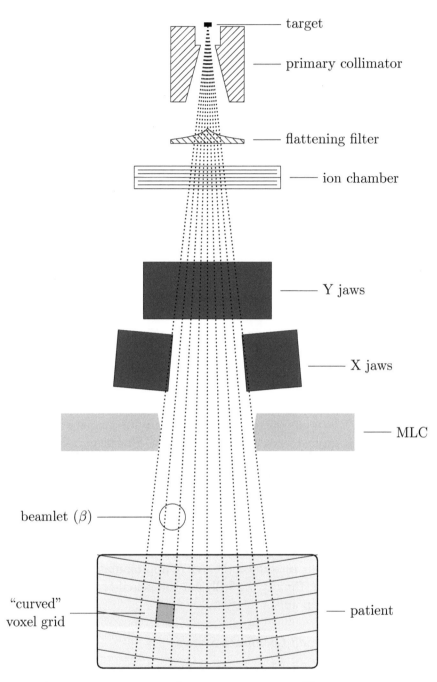

Figure 4.11: Major components of a treatment unit. Three sources are modeled in the AAA algorithm for primary photons, scattered extra-focal photons (mainly from flattening filter), and contaminant electrons. A clinical beam is divided into finite-size pencil beams with edges aligned with the dose calculation grid. Adapted from reference (Varian 2016).

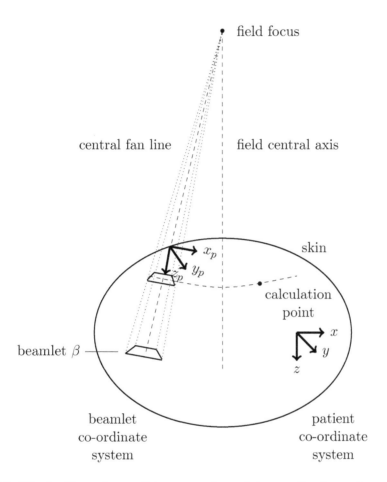

Figure 4.12: Illustration of pencil-beam or beamlet coordinate system (x_p, y_p, z_p) and patient coordinate system (x, y, z) used in the AAA algorithm. The origin of the beamlet coordinate system is at the intersection of the pencil-beam axis with the surface of a patient. In the beamlet coordinate system, the calculation point is on a curved surface with variable (x_p, y_p) coordinates and same distance from the focal spot of the x-ray beam. Adapted from reference (Varian 2016).

depth z_p along the pencil-beam axis is spherical in shape, with the center of the sphere located at the focal spot of the x-ray beam to account for beam divergence.

As shown in Figure 4.13, the finite-size pencil-beam (β) energy spread kernel $k_{z_p}(x_p, y_p)$ is defined as the energy deposited per unit volume (with units of J/cm^3) per incident primary photon in a semi-infinite water medium at location (x_p, y_p, z_p)

$$k_{z_p}(x_p, y_p) = \frac{\Delta E_\beta(x_p, y_p, z_p)/\Delta V}{\Phi_\beta \Delta A} \tag{4.30}$$

where Φ_β is the fluence of the pencil beam with the area ΔA on the surface of the medium. In a uniform water medium, the kernel has cylindrical symmetry, so we can write the kernel as $k_{z_p}(r_p)$ with r_p as the radius and z_p as the depth in the pencil-beam coordinate system. If we have two pencil beams, as shown in Figure 4.14, the total energy deposited per unit volume to the voxel at (x, y, z) in the patient coordinate system is then given by

$$\frac{\Delta E(x, y, z)}{\Delta V} = \Phi_1 k_{z_{p1}}(r_{p1})\Delta A_1 + \Phi_2 k_{z_{p2}}(r_{p2})\Delta A_2 \tag{4.31}$$

For multiple pencil beams, we have energy superposition of all the beamlets

$$\frac{\Delta E(x, y, z)}{\Delta V} = \iint_\beta \Phi(x', y')\tilde{k}_z(x', y'; x, y)dx'dy' \tag{4.32}$$

where $\tilde{k}_z(x', y'; x, y)$ is the kernel in the patient coordinate system, obtained by coordinate transformation from the beamlet coordinate system. Because of beam divergence and kernel tilting (Sharpe and Battista 1993), it is more convenient to perform the superposition in another intermediate patient coordinate system called the *fan-line system* or *divergent coordinate system*, where a divergent beam temporarily becomes a parallel beam and the lateral curved surface at every depth becomes a flat surface. For more details on the coordinate transform, the curious reader is referred to previous publications (Sharpe and Battista 1993; Bortfeld et al. 1993; Tillikainen et al. 2008). Since the heterogeneity is accounted for every finite-size pencil beam as discussed later, we lose the spatial invariance of the kernel. The sum of the pencil-beam contributions to the energy deposition is therefore given by the superposition integral of Equation 4.32 instead of a convolution integral.

4.3.4.3 Energy Deposition Function and Lateral Scattering Kernels

The AAA separates depth and lateral dependence of the kernel by multiplying and dividing the kernel by a depth-dependent function called the energy deposition function $I_\beta(z_p)$

$$\begin{aligned} k_{z_p}(x_p, y_p) &= I_\beta(z_p)\left[k_{z_p}(x_p, y_p)/I_\beta(z_p)\right] \\ &= I_\beta(z_p)K_{z_p}(x_p, y_p) \end{aligned} \tag{4.33}$$

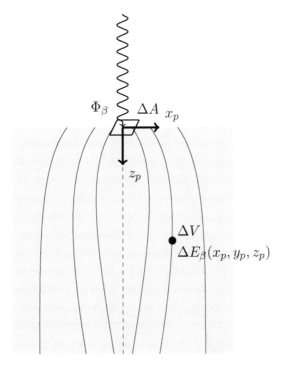

Figure 4.13: Dose spread kernel of a pencil beam.

where $K_{z_p}(x_p, y_p)$ is known as the lateral scattering kernel. The energy deposition function is calculated by integrating the pencil-beam kernel over the lateral curved surface (as shown in Figure 4.12) at depth z_p along the pencil-beam axis, as follows:

$$I_\beta(z_p) = \oiint_{\text{curved surface at } z_p} k_{z_p}(x_p, y_p)\, \mathrm{d}A \qquad (4.34)$$

The surface integration is equivalent to collapsing the lateral energy distribution onto the beam axis as illustrated in Figure 4.15B. The energy-deposition function represents the total energy deposited in the spherical shell (per unit thickness) at depth z_p by a single photon of the pencil beam, and has units of J/cm. As discussed in the previous section, to account for beam divergence, the lateral surface with fixed z_p of a pencil beam is spherical in shape. In a divergent patient coordinate system, all the pencil beams become co-aligned with beam's central axis and the same pencil-beam kernel can be used for all pencil beams in a uniform water medium. The lateral-scattering kernel $K_{z_p}(x_p, y_p)$ is the pencil-beam kernel normalized by the energy-deposition function value at each depth as shown in Figure 4.15C. It represents the fractional energy deposited per unit area at (x_p, y_p) on the curved surface at z_p with units of $1/\text{cm}^2$. In a uniform water medium, the lateral scattering kernel is normalized to unity when integrated over the whole curved surface at any depth z_p:

$$\oiint_{\text{curved surface at } z_p} K_{z_p}(x_p, y_p)\, \mathrm{d}A = 1 \qquad (4.35)$$

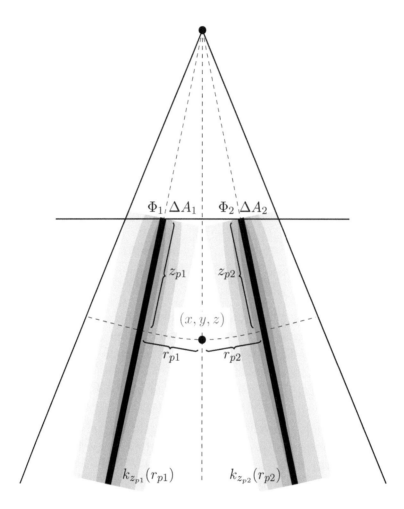

Figure 4.14: Dose summation of two pencil beams with incident fluence Φ_1 and Φ_2, pencil-beam kernels $k_{z_{p1}}(r_{p1})$ and $k_{z_{p2}}(r_{p2})$, respectively. The coordinates (r_{p1}, z_{p1}) and (r_{p2}, z_{p2}) are in a pencil-beam coordinate system, while the coordinate (x, y, z) is in a patient coordinate system.

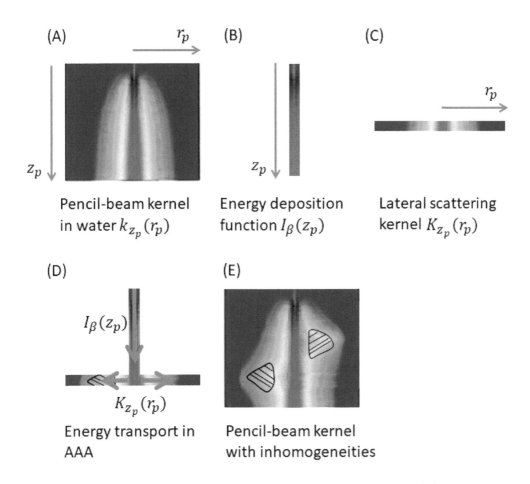

(A) r_p

z_p

Pencil-beam kernel in water $k_{z_p}(r_p)$

(B)

z_p

Energy deposition function $I_\beta(z_p)$

(C) r_p

Lateral scattering kernel $K_{z_p}(r_p)$

(D)

$I_\beta(z_p)$

$K_{z_p}(r_p)$

Energy transport in AAA

(E)

Pencil-beam kernel with inhomogeneities

Figure 4.15: Illustration of the in-water pencil-beam kernel (A), the energy-deposition function (B), and the lateral-scattering kernel (C) in the AAA algorithm. The energy-deposition function is obtained by integrating the pencil-beam kernel laterally or collapsing the lateral energy distribution onto the beam axis. The lateral-scattering kernel is obtained by normalizing the pencil-beam kernel by the energy-deposition function value at each depth z_p. The energy transport is assumed to travel along the beam axis first and then spread laterally (D). The inhomogeneity is accounted for by using radiological depth and radiological cylindrical radius, producing the anisotropic kernel shape (E). For simplicity, the projection of the lateral curved surface was drawn as a straight line.

We can see that there is no mathematical approximation made to the Monte Carlo calculated kernel in Equation 4.33; we simply multiply and divide the kernel by the energy deposition function. Therefore, for a uniform water medium, the AAA does not make any approximations except for the parameterization for the lateral scattering kernels. The approximation starts with inhomogeneity corrections to be discussed later.

4.3.4.4 Analytical Approximation

In order to account for inhomogeneities efficiently, the AAA models the lateral scattering kernel $K_{z_p}(r_p)$ by an analytical function that is the weighted sum of six exponential functions (Tillikainen et al. 2008), as follows:

$$K_{z_p}(r_p) = \sum_{k=1}^{6} c_k(z_p) \mathrm{e}^{-\mu_k r_p} / r_p \tag{4.36}$$

where μ_k are effective attenuation coefficients that were chosen by varying effective ranges $1/\mu_k$ from 1 to 200 mm with equal logarithmic intervals, c_k are the weights of the exponential functions at depth z_p, and r_p is the cylindrical radius of the dose receiving point (x_p, y_p, z_p). The weighting factors c_k are determined by a least-squares-fit of Monte Carlo calculated pencil-beam dose spread kernel for a water medium through Equation 4.33. Note that the lateral scattering kernel has similar decomposition as the SVD algorithm and spatial dependence as a radiation line source with $(1/r_p)$ geometric fall off and exponential attenuation terms.

4.3.4.5 Accounting for Tissue Inhomogeneities

We now generalize the AAA algorithm for a heterogeneous medium with mass density $\rho(x_p, y_p, z_p)$ and relative electron density $\rho'_e(x_p, y_p, z_p)$ distributions. Based on the definition of a pencil-beam energy spread kernel given by Equation 4.30 and the kernel factorization of Equation 4.33, the energy deposited per unit volume at location (x_p, y_p, z_p) from a finite-size pencil beam can be written as a product

$$\frac{\Delta E_\beta(x_p, y_p, z_p)}{\Delta V} = \Phi_\beta I_\beta(z_p; \rho) K_{z_p}(x_p, y_p; \rho) \Delta A \tag{4.37}$$

where Φ_β is the photon fluence assumed to be uniform over the small cross section ΔA of the finite-size pencil beam.

To account for tissue heterogeneity, the AAA assumes that particles first transport energy down along the central axis of the pencil beam with the energy deposition function $I_\beta(z_p; \rho)$, and then spread this energy laterally to destination voxels along the spherical shell with the lateral scattering kernel $K_{z_p}(x_p, y_p; \rho)$. Based on these assumptions and the density-scaling approximation, the energy deposition function in a heterogeneous medium with relative electron density $\rho'_e(x_p, y_p, z_p)$ is scaled by the radiological depth z_{eff} as

$$I_\beta(z_p; \rho) = \rho'_e(0, 0, z_p) I_\beta(z_{\text{eff}}) \tag{4.38}$$

with

$$z_{\text{eff}} = \int_0^{z_p} \rho'_e(0, 0, z'_p) dz'_p \tag{4.39}$$

Similarly, the lateral scattering kernel is scaled according to both the radiological depth z_{eff} and radius r_{eff}

$$K_{z_p}(x_p, y_p; \rho) = \rho'_e(x_p, y_p, z_p) \sum_{k=1}^{6} c_k(z_{\text{eff}}) e^{-\mu_k r_{z_{\text{eff}}}} / r \tag{4.40}$$

where:

$$r_{\text{eff}} = \int_0^{r_p} \rho'_e(r'_p) dr'_p \tag{4.41}$$

$$r_{z_{\text{eff}}} = r_{\text{eff}} \frac{(\text{SSD} + z_{\text{eff}})}{(\text{SSD} + z_p)}$$

and SSD is the source to surface distance for the pencil beam. Similar to the scaling of the field size for a divergent beam, the radiological radius is further scaled by the ratio $(\text{SSD} + z_{\text{eff}})/(\text{SSD} + z_p)$ to account for the finite-size pencil-beam divergence. The multiplication of the local relative electron density ρ'_e in Equations 4.38 and 4.40 converts the energy deposition in water to that in the medium.

The process of inhomogeneity correction by the AAA algorithm is illustrated in Figure 4.15. The pencil-beam kernel is first stretched or compressed along the pencil-beam axis (depth z_p direction) by using radiological depth in the energy-deposition function, then the kernel is expanded laterally by using radiological radii in the lateral-scattering kernels. For a general heterogeneous medium, the effective radiological distance r_{eff} is anisotropic around the pencil-beam axis; therefore the lateral scattering kernel becomes anisotropic, and hence the name of AAA algorithm. In a heterogeneous medium, the lateral scattering kernel changes its shape along the pencil-beam axis, as shown by the bulges in Figure 4.15E.

4.3.4.6 Total Dose from Primary Photons

Once we have calculated the absorbed energy distribution for each finite-size pencil beam based on the actual patient geometry as given by Equation 4.37, the total energy deposited per unit volume is then simply the sum of contributions from all pencil beams

$$\frac{\Delta E(x, y, z)}{\Delta V} = \sum_\beta \left[\frac{\Delta E_\beta(x, y, z)}{\Delta V} \right] \tag{4.42}$$

The total dose from primary photons is then given by

$$D(x, y, z) = \frac{1}{\rho(x, y, z)} \sum_{\beta} \left[\frac{\Delta E_{\beta}(x, y, z)}{\Delta V} \right] \qquad (4.43)$$

where $\rho(x, y, z)$ is the mass density at the dose receiving point.

The AAA improves inhomogeneity correction over a traditional pencil beam algorithm by introducing anisotropic lateral scattering kernels. This, however, leads to a more time-consuming superposition instead of simpler convolution with a spatially-invariant pencil-beam kernel. However, it gains computation efficiency by using an analytical expression for the kernels and separation of energy transport into radial and depth components. Overall, the AAA computation speed is slower than the traditional pencil beam convolution algorithm, but faster than the full 3D convolution method. It is a compromise between dose accuracy and computational speed.

Based on radiation physics, the energy transfer of primary photons is considered along the beam direction, while the energy transfer of charged particles and scattered photons can be in any directions with higher probability in the forward direction. Therefore, strictly speaking, the AAA assumption that the energy is transported along the depth first is valid for primary photons only, while the assumption that energy is subsequently dispersed laterally is partially valid for a portion of charged particles or scattered photons. Even though the AAA energy transfer assumption does not match the actual physical processes and scattering pathways, it does provide an approximate mathematical model. Based on phantom and clinical tests (Van Esch et al. 2006; Gagné and Zavgorodni 2007; Fogliata et al. 2006; Hasenbalg et al. 2007; Breitman et al. 2007), the dose calculation accuracy is acceptable for destination voxels that are not too close to the interface of a heterogeneity, but insufficient near the interface of severe heterogeneity, such as small tumours in lung.

4.3.5 Collapsed Cone Convolution (3D)

4.3.5.1 Overview

We now discuss an efficient approach called the collapsed cone convolution (CCC) method that simplifies *3D* convolution to make it faster while maintaining accuracy in heterogeneous media. The CCC was first introduced by Ahnesjö in 1989 (Ahnesjö 1989). It has become the most widely used algorithm in commercial treatment planning software such as Pinnacle (Philips), RayStation (RaySearch Laboratories), Monaco (Elekta CMS), Oncentra (Nucletron), and Mobius3D (Mobius Medical Systems).

As we described earlier, one of the major challenges of fast implementations of the superposition algorithm is how to handle inhomogeneity corrections. One common approach is density-scaling of the kernels according to the radiological distance between the source voxel and the dose-receiving voxel as given by Equation 4.16. However, the scaling made the kernels spatially variable, limiting the fast implementations discussed above. Instead of improving calculation speed by reducing 3D superposition to 2D superposition, such as in the pencil-beam algorithm discussed in the previous section, Ahnesjö et al. greatly reduced the number of voxels required to perform 3D superposition by quantizing the directions along which energy is propagated.

In this section, we will use the term *radiant energy* to describe the energy released by the primary photons, and then transferred, attenuated, and deposited in a medium. As we can see from Figure 4.16 and the analytical approximation of the kernel by Equation 4.13, the dose kernel values fall off rapidly from the point of interaction, approximately following an inverse square law with further exponential loss along ray-lines. It is natural to use spherical coordinates with origin at the point of interaction to describe the radiant energy distribution, as most of the energy deposition occurs near the interaction site and spherical volume elements become much smaller close to the origin. Another advantage is to incorporate the azimuthal symmetry of the point kernel conveniently. Because of a similar continuous drop-off pattern in all directions away from the interaction site, we may sample the energy transport along a few major cardinal directions with minimal loss of accuracy. This leads to the idea of propagating the energy transport into a set of adjacent cones and then *collapsing* the energy transport onto the cone axes. One can imagine that the collapsing operation is similar to closing an umbrella. If we assume that energy is added or removed only along points along cone axes, the computation becomes greatly simplified and much more efficient. It involves much fewer voxels than a full 3D superposition.

An analogy might help the reader. The idea of collapsed cone convolution is similar to a public transportation system. It is much more efficient to move people along major streets (i.e. bus routes) rather than to pick up and drop off individuals house-by-house. Similarly, instead of tracking radiant energy everywhere in the neighborhood of an interaction site, all the lateral energy (Equation 4.13) is artificially congregated along a cone axis at incremental radial distances (i.e. bus stops along the bus route).

As discussed earlier, dose deposition is a two-step process, the first step being the primary photon interaction, transferring energy to charged particles and secondary photons, and the second step being the spreading and deposition of energy. We need to track both steps to determine the energy transfer, propagation, and deposition. We will first define the cone geometry in a spherical coordinate system. Then we will use energy conservation and the analytical representation of the dose spread kernel to calculate the total energy fluence as it propagates in cone geometry. We will then discuss how to collapse the lateral energy onto cone axes and develop a very efficient recursive relation for energy propagation along cone axes, including

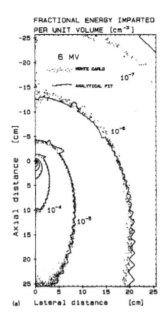 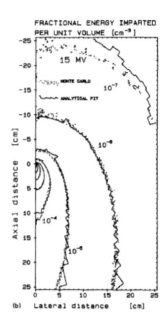

Figure 4.16: The point dose spread kernels in water for (a) 6 MV and (b) 15 MV x-rays, calculated by the Monte Carlo method (thin hatched lines) and the corresponding results from analytical kernels given by Equation 4.13 (thick solid lines). Reproduced with permission (Ahnesjö 1989).

inhomogeneity corrections. Finally, we will discuss the arguments to support and justify this approximation.

4.3.5.2 Cone Element Definition

From the interaction point of view, the dose spreading pattern can be divided into multiple adjacent cones with a common origin at the photon interaction sites. As shown in Figure 4.17, we can further subdivide a cone into *elements* or *segments* subtended by a solid angle Ω_m around the cone axis with direction (θ_m, ϕ_m) and of thickness Δr specified by

$$
\begin{aligned}
r_{i-1} \ &\leq \ r < r_i \\
\theta, \phi \ &\subset \ \Omega_m
\end{aligned}
\tag{4.44}
$$

where the primary photon is incident from the bottom, traveling along the positive z-direction. Note that our definition of a segment coincides with that of a volume element in spherical coordinates. Cones can be defined based on the geometry of the medium in which we need to calculate the dose distribution. For example, for a slab geometry with cylindrical symmetry, 48 ring-shaped cones can be defined (Ahnesjö 1989) for each point of dose calculation. For more general geometry, about 100 cones are typically used in treatment planning software (Mackie et al. 1996).

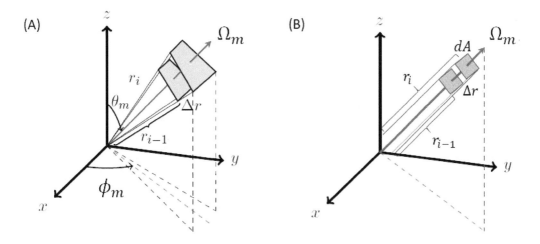

Figure 4.17: General definition of a cone element (bounded by the two shaded surfaces) (A), and the corresponding element on the cone axis (B) with infinitesimal area dA in spherical coordinates. The primary photon is incident from the bottom, traveling along the positive z-axis. Note that we flipped the usual beam direction from down to up for consistency with conventional spherical coordinates.

4.3.5.3 Energy Fluence of a Point Source

To calculate dose deposition from a single interaction point source, we consider the radiant energy released into the cone volume by primary photons interacting at the origin and its subsequent propagation along the cone in the form of "wavefronts" (as in Figure 4.16). As introduced in Chapter 2 (Equation 2.17) based on the energy conservation principle, the energy deposited in a cone voxel per unit volume ΔV can be calculated as the *divergence* of the total vectorial energy fluence $\boldsymbol{\Psi}$ incident at any location (r, θ), as follows:

$$\frac{\Delta \epsilon(r, \theta)}{\Delta V} = -\nabla \cdot \boldsymbol{\Psi} \tag{4.45}$$

As an approximation, by assuming the vectorial energy fluence is changing along the radial direction only, the divergence can then be calculated simply as

$$\nabla \cdot \boldsymbol{\Psi} = \frac{1}{r^2} \frac{\partial}{\partial r} (r^2 \Psi_r) \tag{4.46}$$

Based on the kernel definition given by Equation 4.3 and the analytical approximation of the kernel given by Equation 4.13, we can see that the energy conservation Equation 4.45 can be satisfied if the total energy fluence (wavefront) is as follows:

$$\Psi_r = \Delta E_T \left(\frac{A_\theta}{a_\theta} e^{-a_\theta r} + \frac{B_\theta}{b_\theta} e^{-b_\theta r} \right) / r^2 \tag{4.47}$$

where $\Delta E_T = T(\mathbf{0})\rho(\mathbf{0})\mathrm{d}V$ is the total energy released from a volume element $\mathrm{d}V$ at the origin by primary photons producing the TERMA, $T(\mathbf{0})$. Equation 4.47 essentially describes how the total energy fluence is dispersed throughout point kernels.

4.3.5.4 Accounting for Tissue Inhomogeneities

To account for inhomogeneities, it will be instructive to look at the dependence of the energy fluence produced by a point kernel as given by Equation 4.47. These methods can be viewed as the energy fluence contributed by primary charged particles and scattered photons with attenuation parameters a_θ and b_θ, and initial fractional energy fluence A_θ/a_θ and B_θ/b_θ, respectively. As discussed earlier, we assume that both a_θ and b_θ are scaled by relative electron density or equivalently that the physical distance in the exponential terms is substituted by radiological path length, as follows:

$$\Psi_r = \Delta E_T \left(\frac{A_\theta}{a_\theta} e^{-a_\theta r_{\text{eff}}} + \frac{B_\theta}{b_\theta} e^{-b_\theta r_{\text{eff}}} \right) / r^2 \tag{4.48}$$

where r_{eff} is the radiological path length from the origin to a cone voxel of interest

$$r_{\text{eff}} = \int_0^r \rho_e'(r', \theta, \phi)\mathrm{d}r' \tag{4.49}$$

and ρ_e' is the relative electron density of the medium along a ray-line path increment, $\mathrm{d}r'$.

We can now calculate the point dose kernel with inhomogeneity consideration using Equations 4.3, 4.45, 4.46, and 4.48 as follows:

$$\begin{aligned} \bar{k}(r, \theta) &= -\frac{1}{r^2} \frac{\partial}{\partial r} \left[r^2 \left(\frac{A_\theta}{a_\theta} e^{-a_\theta r_{\text{eff}}} + \frac{B_\theta}{b_\theta} e^{-b_\theta r_{\text{eff}}} \right) / r^2 \right] \\ &= \rho_e'(r, \theta, \phi) \left(A_\theta e^{-a_\theta r_{\text{eff}}} + B_\theta e^{-b_\theta r_{\text{eff}}} \right) / r^2 \end{aligned} \tag{4.50}$$

Comparing this equation with the point dose kernel for water (Equation 4.13), we see that it agrees with the general density scaling relation given by Equation 4.16.

4.3.5.5 Radiant Energy Released into a Cone from a Point Source

We can calculate the radiant primary energy (mainly electrons) released into a cone with solid angle Ω_m and its entrance spherical surface located at radial position r_i (shaded surface in Figure 4.17) by integrating the energy fluence over the cone entrance surface as follows:

$$\mathcal{R}_m^p(r_i) = T(\mathbf{0})\rho(\mathbf{0})\mathrm{d}V \iint_{\Omega_m} \left(\frac{A_\theta}{a_\theta} e^{-a_\theta r_{i,\text{eff}}} / r_i^2 \right) r_i^2 \sin\theta \mathrm{d}\theta \mathrm{d}\phi \tag{4.51}$$

where:

$$r_{i,\text{eff}} = \int_0^{r_i} \rho'_e(r', \theta, \phi) \mathrm{d}r' \tag{4.52}$$

In practice, when we work with a discrete array of adjacent cones along quantized directions (m) in spherical space, Equation 4.51 becomes

$$\mathcal{R}_m^p(r_i) = T(\mathbf{0})\rho(\mathbf{0})\mathrm{d}V\Omega_m \frac{A_m}{a_m} \mathrm{e}^{-a_m r_{i,\text{eff}}} \tag{4.53}$$

where $\mathrm{d}V$ is the infinitesimal volume located at the cone origin, A_m and a_m are discretized parameters of A_θ and a_θ respectively, and

$$r_{i,\text{eff}} = \sum_{k=1}^{i} \rho'_e(r_k, \Omega_m)\Delta r \tag{4.54}$$

Similarly for scattered photons, we obtain

$$\mathcal{R}_m^s(r_i) = T(\mathbf{0})\rho(\mathbf{0})\mathrm{d}V\Omega_m \frac{B_m}{b_m} \mathrm{e}^{-b_m r_{i,\text{eff}}} \tag{4.55}$$

where B_m and b_m are discretized parameters of B_θ and b_θ, respectively.

The above equations can be understood intuitively. The radiant energy released at the origin of a cone that reaches a cone segment at r_i is proportional to the total energy released by primary photons $T(\mathbf{0})\rho(\mathbf{0})dV$, initial fractional energy fluence A_m/a_m or B_m/b_m, the solid angle Ω_m encompassing the cone, and the exponential attenuation from the source to the entrance of the cone element.

4.3.5.6 Collapsing Radiant Energy onto the Cone Axis

In the CCC approximation, the radiant energy released into a cone is projected onto the cone axis (Ahnesjö 1989). The projected (or collapsed) energy is linearly transported, attenuated, and deposited only along the cone axis as shown in Figure 4.18. This allows us to establish a recursive relation for the propagation of the radiant energy along the cone axis.

For this purpose, we will first calculate the radiant energy released from an individual element on the cone axis and reaching a subsequent element downstream by integrating all the point-source contributions (Equations 4.53 and 4.55) along the element. Therefore, the incremental primary radiant energy released from an element (r varies from r_{i-1} to r_i with infinitesimal cross sectional area dA) and arriving at location r_i is given by

$$\begin{aligned}
\Delta\mathcal{R}_m^p(r_i) &= T_i\rho_i\mathrm{d}A\Omega_m \int_{r_{i-1}}^{r_i} \frac{A_m}{a_m} \mathrm{e}^{-a_m\rho'_{e,i}(r_i-s)}\mathrm{d}s \\
&= T_i\rho_i\mathrm{d}A\Omega_m \frac{A_m}{\rho'_{e,i}a_m{}^2}(1 - \mathrm{e}^{-a_m\rho'_{e,i}\Delta r})
\end{aligned} \tag{4.56}$$

where $\Delta r = r_i - r_{i-1}$, T_i, ρ_i, and $\rho'_{e,i}$ are respectively the local TERMA, mass density, and relative electron density.

For scattered photons, similar to Equation 4.56, we have

$$\Delta R_m^s(r_i) = T_i \rho_i \mathrm{d}A\Omega_m \frac{B_m}{\rho'_{e,i}b_m^2}(1 - \mathrm{e}^{-b_m\rho'_{e,i}\Delta r}) \tag{4.57}$$

In the CCC approximation, the radiant energies released into a cone element by Equations 4.56 and 4.57 are collapsed onto the corresponding element with infinitesimal cross-sectional area dA on the cone-axis as shown in Figure 4.17. Since the radiant energy released attenuates exponentially along radial distance (Equations 4.53 and 4.55), once the energy is collapsed onto a cone axis, the released energy is assumed to be attenuated only by medium along the cone axis instead.

When we apply the collapsed cone approximation to a 3D dose matrix defined in discrete Cartesian coordinates, the 3D volume is covered with a lattice of cone axes as shown in Figure 4.18. The radiant energy released by primary photons is transported along these cone axes only. Each line is divided into steps of Δr such that the densities and the TERMA can be approximated with constant values within each line increment.

4.3.5.7 Recursive Relations

Intuitively, if we follow along a cone axis such as the one shown in Figure 4.18, an element on the axis will inherit the total upstream primary radiant energy accumulated and attenuated along the cone axis, but will also pick up freshly released energy ΔR_m^p from primary photons interacting in an adjacent upstream element given by Equation 4.56, resulting in the following recursion expression for overall radiant primary energy

$$R_m^p(r_i) = R_m^p(r_{i-1})\mathrm{e}^{-a_m\rho'_{e,i}\Delta r} + \Delta R_m^p(r_i) \tag{4.58}$$

Note that the starting position $r_{i=0}$ for each cone axis should be outside the primary photon beam with $R_m^p(r_{i=0}) = 0$. The recursion can be rigorously derived using Equation 4.53 and Equation 4.56. Similarly, for scattered photons, we have the recursion relation

$$R_m^s(r_i) = R_m^s(r_{i-1})\mathrm{e}^{-b_m\rho'_{e,i}\Delta r} + \Delta R_m^s(r_i) \tag{4.59}$$

Since the attenuation coefficients b_m for scattered photons are much smaller than those for electrons, we can approximate the exponential functions by the first two terms of a Taylor expansion such that

$$R_m^s(r_i) = R_m^s(r_{i-1})(1 - b_m\rho'_{e,i}\Delta r) + T_i\rho_i \mathrm{d}A\Omega_m \frac{B_m}{b_m}\Delta r \tag{4.60}$$

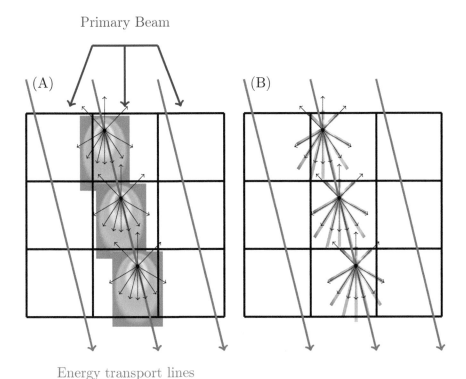

Figure 4.18: (A) Illustration of 2D dose matrix with 10 cone axes (black arrowed lines) and point dose kernels (aligned with the primary beam axis) at interaction points along one of the cone axes. Fresh TERMA is released at the intersection point of the cone axes. (B) Point dose kernels are collapsed onto cone axes. The radiant energy is released, transported, and attenuated along the transport lines shown in red. For clarity, energy release and propagation are shown only along one energy transport line direction. In practice, multiple transport lines will be crossing each dose voxel from different cone-axis directions. Inhomogeneities are accounted for in the calculations of TERMA and kernels by using radiological path lengths.

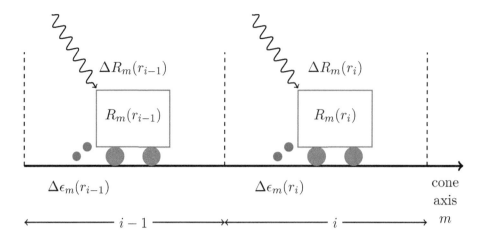

Figure 4.19: Illustration of the recursive relation by a train carrying the total radiant energy R_m along a cone axis m (a track). When the train moves from station $(i-1)$ to i, the accumulated radiant energy $R_m(r_{i-1})$ is exponentially attenuated (by friction of the track) and a fresh radiant energy $\Delta R_m(r_i)$ is received from primary photons. The losses of radiant energy, including self-absorption of the fresh radiant energy, becomes the energy $\Delta \epsilon_m$ shown as red circles deposited onto the track.

The nature of the recursion given by Equation 4.58 or 4.59 is illustrated by Figure 4.19 with an analogy of a train carrying the total radiant energy $R_m(r_i)$ (sum of both primary and scattered radiant energies) along a track (a cone axis). The total radiant energy carried by the train at the station i is the sum of the remaining radiant energy that had been accumulated up to the previous station $(i - 1)$ after exponential attenuation (caused by friction of the track) and a fresh supply of radiant energy $\Delta R_m(r_i)$. The attenuation of the radiant energy, including the self-absorption of the fresh supply of radiant energy (to be discussed in the following section), becomes the dose deposited along the track, which is proportional to the radiological distance between stations.

Mathematically, the recursive relation (Equations 4.58 or 4.59) represents a one-dimensional convolution (described in Chapter 2) along a cone axis. Starting from the beginning of a cone axis with $R_m^p(r_{i=0}) = 0$, the accumulated radiant energy has a build-up along the cone axis similar to that of a depth dose curve shown in Figure 2.22.

4.3.5.8 Dose Calculation

As described in Section 4.3.5.6, in the CCC approximation, the collapsed radiant energy is linearly transported, attenuated, and deposited along the cone axis. This simplifies the energy propagation and deposition to a one-dimensional problem along each of the cone axes. We can calculate the average dose deposited into a finite-sized volume element at r on the cone axis with cross-sectional area dA and length Δr

based on the absorption of the radiant energy by the element. Similar to Equations 4.58 and 4.56, the primary radiant energy going through the cross-sectional area dA on the cone axis m at r is given by

$$R_m^p(r) = R_m^p(r - \Delta r)e^{-a_m \rho_e'(r)\Delta r} + \Delta R_m^p(r) \tag{4.61}$$

where

$$\Delta R_m^p(r) = T(r)\rho(r)\mathrm{d}A\Omega_m \frac{A_m}{\rho_e'(r)a_m^2}\left(1 - e^{-a_m \rho_e'(r)\Delta r}\right) \tag{4.62}$$

If there was no energy absorption from the element, the primary radiant energy arriving at r would be the sum of the radiant energy arriving at $r - \Delta r$ and the freshly added radiant energy from primary photons interacting in the element, as follows:

$$\tilde{R}_m^p(r) = R_m^p(r - \Delta r) + T(r)\rho(r)\mathrm{d}A\Delta r\Omega_m \frac{A_m}{a_m} \tag{4.63}$$

The *average* primary radiant energy deposited into the volume element is then given by the energy difference between Equation 4.63 and Equation 4.61, as follows:

$$\begin{aligned}
\Delta \bar{\epsilon}_m^p(r, \Delta r) &= \tilde{R}_m^p(r) - R_m^p(r) \\
&= R_m^p(r - \Delta r)\left(1 - e^{-a_m \rho_e' \Delta r}\right) \\
&\quad + \left(T(r)\rho(r)\mathrm{d}A\Delta r\Omega_m \frac{A_m}{a_m} - \Delta R_m^p(r)\right)
\end{aligned} \tag{4.64}$$

The first term in the above equation is the attenuation of the primary radiant energy upstream, while the second term is the absorption of the newly released primary radiant energy from primary photons interacting in the element (Ahnesjö et al. 2017). It is the radiant energy that is released and then absorbed in the same element, and is described as *self-absorption*.

Similar to Equations 4.64 and 4.62, the average scattered radiant energy deposited into the finite element is

$$\begin{aligned}
\Delta \bar{\epsilon}_m^s(r, \Delta r) &= R_m^s(r - \Delta r)\left(1 - e^{-b_m \rho_e' \Delta r}\right) \\
&\quad + \left(T(r)\rho(r)\mathrm{d}A\Delta r\Omega_m \frac{B_m}{b_m} - \Delta R_m^s(r)\right)
\end{aligned} \tag{4.65}$$

where:

$$\Delta R_m^s(r) = T(r)\rho(r)\mathrm{d}A\Omega_m \frac{B_m}{\rho_e'(r)b_m^2}\left(1 - e^{-b_m \rho_e'(r)\Delta r}\right) \tag{4.66}$$

The average total radiant energy deposited into this element is the sum of the deposited primary and scattered radiant energy (Equations 4.64 and 4.65)

$$\begin{aligned}
\Delta \bar{\epsilon}_m(r, \Delta r) &= R_m^p(r - \Delta r)\left(1 - e^{-a_m \rho_e' \Delta r}\right) \\
&\quad + R_m^s(r - \Delta r)\left(1 - e^{-b_m \rho_e' \Delta r}\right) \\
&\quad + \Delta \epsilon_m^{\text{self}}(r, \Delta r)
\end{aligned} \tag{4.67}$$

where $\Delta\epsilon_m^{\text{self}}(r, \Delta r)$ is the energy deposited due to self-absorption

$$\Delta\epsilon_m^{\text{self}}(r, \Delta r) = T(r)\rho(r)\mathrm{d}A\Delta r\Omega_m \left(\frac{A_m}{a_m} + \frac{B_m}{b_m}\right) - [\Delta R_m^p(r) + \Delta R_m^s(r)] \quad (4.68)$$

The first term in the above equation represents the freshly released total radiant energy into the element on the cone axis, while the second term represents the part of this radiant energy exiting this element.

The above calculations for an element at position r on a cone axis can be generalized to a 3D dose voxel at position $\mathbf{r}$ that intersects all cone axes. The average dose deposited into the voxel is the sum of the dose contributions from all cone axes

$$\bar{D}(\mathbf{r}) = \frac{\sum_{m'=1}^M \Delta\bar{\epsilon}_{m'}(\mathbf{r}, \Delta r_{m'})}{\Delta m} \quad (4.69)$$

where M is the number of cones, $\Delta r_{m'}$ is the total path length across the dose voxel along the cone axis m', and Δm is the mass of the voxel. It should be noted that $\Delta r_{m'}$ depends on the direction of the cone axis relative to the dose voxel. For more details on the dose calculation including self-absorption, the reader is referred to the publication on application of the CCC algorithm in brachytherapy (Ahnesjö et al. 2017).

When the step size Δr approaches zero, we can show that the self-absorption given by Equation 4.68 becomes zero using a first-order Taylor expansion in Δr for $\Delta R_m^p(r)$ (Equation 4.62) and $\Delta R_m^s(r)$ (Equation 4.66). The energy deposited into an *infinitesimal* volume element $\mathrm{d}A\mathrm{d}r$ on the cone axis at position r is then given by

$$\begin{aligned}
\Delta\epsilon_m(r) &\approx R_m^p(r)\left(1 - e^{-a_m\rho_e'\mathrm{d}r}\right) + R_m^s(r)\left(1 - e^{-b_m\rho_e'\mathrm{d}r}\right) \\
&\approx \rho_e'(r)\mathrm{d}r[a_m R_m^p(r) + b_m R_m^s(r)]
\end{aligned} \quad (4.70)$$

When we sum up the dose contributions from all cone axes, we obtain the point dose formula as described in the original CCC publication (Ahnesjö 1989)

$$\begin{aligned}
D(\mathbf{r}) &= \frac{\sum_{m'} \Delta\epsilon_{m'}(\mathbf{r})}{\Delta m} \\
&= \frac{\rho_e'(\mathbf{r})}{\rho(\mathbf{r})} \frac{1}{\mathrm{d}A} \sum_{m'}[a_{m'} R_{m'}^p(\mathbf{r}) + b_{m'} R_{m'}^s(\mathbf{r})]
\end{aligned} \quad (4.71)$$

where the infinitesimal cross section $\mathrm{d}A$ will be cancelled out during the calculation, since the radiant energies $R_{m'}^p$ and $R_{m'}^s$ are also proportional to $\mathrm{d}A$, as they are released from elements on cone axes with the same cross-sectional area $\mathrm{d}A$ (Equations 4.56 and 4.57).

4.3.5.9 Justification of the CCC Approximation

At first sight, the collapsed cone approximation seems unphysical. However, due to the fact that the point dose kernel falls off quickly by an inverse square law and exponential attenuation, and due to smearing effects from multiple interaction points and cones, the approximation works well for most clinical cases. As shown in Figure 4.20, near the interaction voxel where the kernel has the highest contribution, collapsing radiant energy to the cone axis introduces no loss of accuracy because the cone elements are typically much smaller than the dose voxels, while the geometric approximation gets worse at larger distances where the kernel contribution is spread out to more dose voxels but with much smaller magnitude. There is also a smearing or lateral "equilibrium" effect.

For example, as shown in Figure 4.21, the energy released from voxel A should deposit energy to both voxels B and B', but in the collapsed cone approximation, the energy that should have been deposited to voxel B', is deposited onto voxel B instead. For a homogeneous medium and similar primary energy fluence at nearby voxels A and A', we can see that the missing energy to B' from A will be compensated by extra energy received from A'. It is important to note that energy conservation is strictly preserved in this approximation. For strongly inhomogeneous medium, this approximation may introduce some errors. For general clinical practice with typical tissue distributions, this approximation turns out to be generally acceptable, and the method is more accurate than 2D convolution methods, especially in the presence of inhomogeneities(Fogliata et al. 2007).

4.3.5.10 Advantages and Limitations of the CCC

The collapsed cone convolution method has been tested and validated by many studies (Ahnesjö 1989; Knöös et al. 2006). It produces very accurate results for dose distribution in inhomogeneous media when charged particle equilibrium (CPE) is present. With loss of CPE, the accuracy is degraded somewhat, but the correct trend is still predicted by the CCC using a first-order approximation for inhomogeneity correction with density-scaling along cone axes.

Combining total dose calculated by Equation 4.69 with recursions given by Equation 4.58 and Equation 4.60, we can calculate 3D dose distribution very efficiently. Using the recursion formula, along each cone axis, we can calculate the dose contribution to N^3 dose voxels with only N^3 calculations. If we have M cone axes, we just need MN^3 calculations to calculate the whole 3D dose volume (compared to N^7 calculations using the original 3D convolution method). This is a huge time savings for typical 3D dose calculations. For example, for $N = 100$ dose voxels in one dimension and $M = 100$ cone axes, $N^7 = 10^{14}$, while $N^3M = 10^6 \times 100 = 10^8$. If each computation takes 1 ns, then 10^{14} takes about 28 hours, while 10^8 takes only 0.1 seconds. The CCC computation speed is similar to that of the fast Fourier transform method described in Section 4.3.1, but it calculates 3D dose distribution

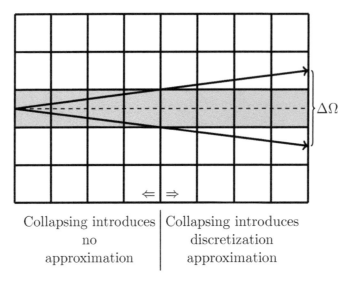

Collapsing introduces | Collapsing introduces
no | discretization
approximation | approximation

Figure 4.20: In the collapsed cone approximation, no approximation is made near the interaction voxel where kernel contribution is high and condensed, and cone elements are smaller than dose elements (voxels). The geometric approximation is getting worse further away but the kernel contribution is getting much smaller. Adapted from reference (Carlsson and Ahnesjö 2000).

more accurately in a heterogeneous medium. Note that we introduced the CCC algorithm using the analytical kernels from the original publication (Ahnesjö 1989). The estimate of the CCC calculation speed is based on this formulation. This estimate may not apply to some other implementations such as those used in helical tomotherapy (Lu et al. 2005; Chen et al. 2011).

The major approximations of the CCC are i) collapsed cone approximation and ii) rectilinear relative electron density scaling for inhomogeneity in the medium. The collapsed cone approximation may cause some inaccuracy near interfaces of inhomogeneities, since the energy equilibrium between adjacent cones, shown in Figure 4.21, will potentially be disrupted. The density scaling approximation may also cause inaccurate dose calculations near the interface of inhomogeneities, since the approximation is less tenable for non-slab geometries between the interaction-source voxel and dose-receiving voxel, or in tissues with very different effective atomic numbers.

4.4 COMMISSIONING REQUIREMENTS

As we can see from the general convolution-superposition formula given by Equation 4.7, in order to calculate the dose distribution in a patient, we need to know the mass density distribution, TERMA, and polyenergetic kernels. To calculate TERMA, we need to calculate mass attenuation coefficients and primary energy fluence. All of

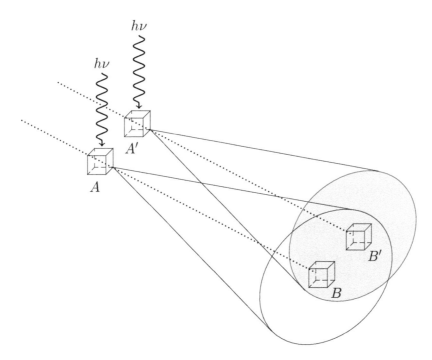

Figure 4.21: The energy released from voxel A should deposit energy to both voxels B and B', but in the collapsed cone approximation, the energy that should have been deposited to voxel B' is deposited onto voxel B. This misplacement of energy is approximately compensated by the same effect for the energy released by voxel A'. Energy is thereby conserved in this approximation. Adapted from reference (Ahnesjö 1989).

these parameters depend on the energy spectrum of the beam. This leads us to the commissioning requirements for beam modelling.

To confirm or infer the energy spectrum for the beam, we need measurements of depth dose curves for different field sizes. The energy spectrum can be pre-calculated with the Monte Carlo method or fitted by the data (Mohan and Chui 1985a; Papanikolaou et al. 1993; Van Dyk et al. 1993). Combining the beam spectrum and dose spread kernels based on the Monte Carlo simulation, we can then calculate the polyenergetic kernel by summing up the kernel contributions from each energy bin in a weighted fashion as given by Equation 4.8. The polyenergetic kernel can be computed in advance, stored, and then be reused for dose calculations for every x-ray beam energy.

The mass attenuation coefficients inside a patient can be obtained by a lookup table index by CT numbers and the energy spectrum of the beam. To obtain primary energy fluence Ψ_0 at the surface of the patient, we need to model the beam including any beam modifiers such as physical wedges, dynamic wedges, and multi-leaf collimator. This requires measured dose profiles with different field sizes, different beam modifiers, and measured at different depths. The parameters used to describe the primary energy fluence will then be fitted to maximize agreement between predicted and measured beam profiles.

Finally, we need to confirm or correct the beam model for MU prediction (monitor unit to Gray) for fulfilling the dose prescription in absolute dosimetry (Gray). This requires measurements of relative output factors with different field sizes at a reference depth. For more details on commissioning of a treatment planning system, the reader is referred to the AAPM Task Group Report 106 (Das et al. 2008).

4.5   DOSE ACCURACY PERFORMANCE

In this section, we will compare the dose calculation accuracy by various convolution methods with the benchmark results from direct measurements or calculations performed by the Monte Carlo method for specific geometric phantoms with inhomogeneous media and more realistic clinical examples.

4.5.1   Phantom Studies

One of the advantages of phantom studies is to have a well designed simple reproducible geometry to test the limitations of various dose calculation algorithms with heterogeneous media. One common simple phantom has a slab geometry (García-Vicente et al. 2003), as shown in Figure 4.22. It consists of solid water and low density cork slabs simulating unit-density tissue and a lung heterogeneity in a patient.

Garcia-Vicente et al. (García-Vicente et al. 2003) performed measurements with film and an ion chamber in the phantom to test fast Fourier transform convolution and multigrid superposition algorithms (an implementation of superposition algorithm not covered in this chapter) for dose calculation in low-density media. A sample result is shown in Figure 4.23. A mean overestimation of about 10% by fast Fourier transform convolution (FFTC) algorithm was found in the calculated dose in the low-density material.

Direct measurements with phantoms is labor-intensive with very limited options for geometry and potential inaccuracy of measured dose in heterogeneous absorbers. A more convenient method is to use Monte Carlo methods, provided that the code is pre-verified extensively against reliable dosimetry. Fogliata et al. (Fogliata et al. 2007) did a comprehensive comparison of various dose calculation algorithms in commercial treatment planning systems (TPS) with Monte Carlo calculations in the presence of simple geometric heterogeneities as shown in Figure 4.24.

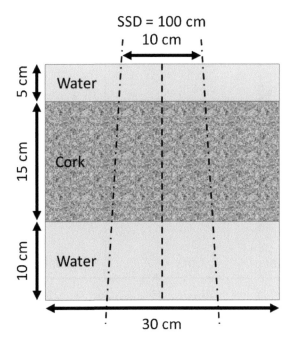

Figure 4.22: Phantom with solid water (blue) and low-density cork (tan) slabs used by Garcia-Vicente et al. for testing FFTC techniques. Adapted with permission (García-Vicente et al. 2003).

Two groups of algorithms were identified by the study (Fogliata et al. 2007) and were classified in Chapter 3. One group is Type-A algorithms where invariant kernels are used in convolution dose calculations such as fast Fourier transform convolution in XiO TPS (FFTC-XiO), pencil-beam convolution in Eclipse (PBC-ECL) and in Helax-TMS (PB-TMS) TPS. The other group is Type-B algorithms where 3D inhomogeneity is taken into account by density scaling anisotropically such as the anisotropic analytical algorithm in Eclipse (AAA-ECL), multigrid superposition-convolution in XiO TPS (MGS-XiO), and collapsed cone in Helax-TMS (CC-TMS) and in Pinnacle (CC-PIN). Effectively, Type-A algorithms do not account for the effect of heterogeneity on the electron transport, while Type-B algorithms approximate the effect of heterogeneity on the electron transport using equivalent path length methods. Therefore, in general, Type-B algorithms agree with the Monte Carlo method better than Type-A algorithms, especially in situations without charged particle equilibrium. For example, comparison of percent depth dose curves between various dose calculation algorithms and MC simulations for a 6 MV x-ray beam with field size 2.8×13 cm^2 in the phantom (Figure 4.24) is shown in Figure 4.25. Results show significant dose difference between Type-A algorithms and the Monte Carlo method, especially inside low density *ultra-light lung* with $\rho = 0.035$ g cm^{-3}, while Type-B algorithms perform much better.

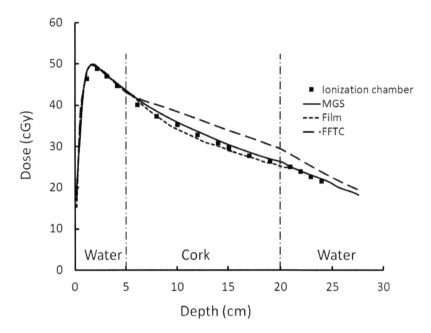

Figure 4.23: Comparison of 6 MV x-ray absolute depth dose curves by film and ion-chamber measurements with fast Fourier transform convolution (FFTC) and multi-grid superposition (MGS) on the water-cork phantom by Garcia-Vicente et al. Adapted with permission (García-Vicente et al. 2003).

A more realistic phantom (CIRS Thorax phantom, CIRS Inc., Norfolk, VA, USA) as shown in Figure 4.26 was used in audits in Europe by the International Atomic Energy Agency (IAEA) on treatment planning system calculations for 3D conformal radiotherapy (Gershkevitsh et al. 2014). The audit was carried out in 60 radiotherapy centres in Europe, comparing TPS dose calculations with measurements by ion chamber in the phantom.

Eight 3D conformal plans were used, including single field, tangential field, four-field box, plans with blocks, oblique incidence, asymmetrically wedged fields, and non-coplanar field. There are 10 specific measurement points in the phantom as shown in Figure 4.26. The agreement criteria vary from 2% to 5%, depending on the plan and the reference measurement point.

The audit found that significant errors could originate from dose calculation algorithms in addition to errors from other sources, such as errors in TPS input data and inaccuracies in beam modeling and calibration. The audit also found significant differences between Type-A and Type-B algorithms. As shown in Figure 4.27, percentage of measurements failing agreement criteria are significantly higher for Type-A algorithms compared to Type-B algorithms for all energy groups. The

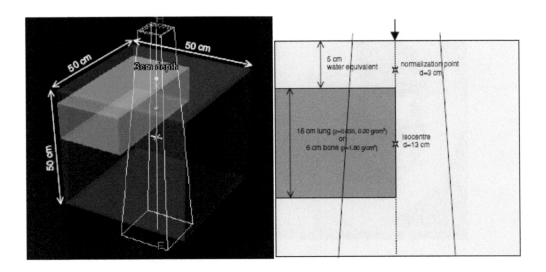

Figure 4.24: Simple geometric phantom with water, lung or bone regions. Reproduced with permission (Fogliata et al. 2007).

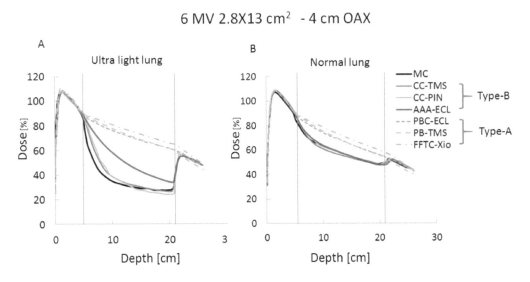

Figure 4.25: Comparison of percent depth dose curves between various dose calculation algorithms and Monte Carlo simulations (black solid lines) for a 6 MV x-ray beam with field size 2.8×13 cm^2 in the simple geometric phantom shown above. Ultra light lung (A) has density $\rho = 0.035$ g cm^{-3}, while normal lung (B) has density $\rho = 0.2$ g cm^{-3}. Adapted with permission (Fogliata et al. 2007).

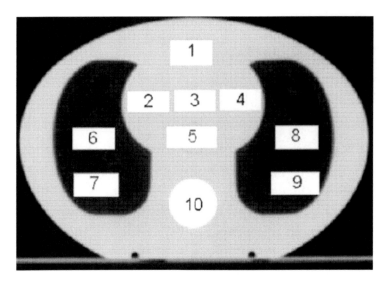

Figure 4.26: CIRS Thorax phantom with 10 measurement positions. Reproduced with permission (Gershkevitsh et al. 2014).

higher discrepancies for Type-A algorithms using Co-60 might be because centres using older Co-60 units were equipped with antiquated TPS using simpler Type-A algorithms, together with problems associated with inaccurate beam modeling and calibration. There is also a general trend that the discrepancies for plans using linear accelerators increase with energy. This is related to the fact that the effect of lateral electron transport increases with energy, while Type-A algorithms do not account for the effect, Type-B algorithms account for it only approximately in the presence of heterogeneity.

The phantom studies have demonstrated that significant errors can be introduced by dose calculation algorithms, especially dose distributions calculated in low density media with high energy beams by Type-A algorithms. The above studies are limited by phantom geometry with simple 3D conformal plans.

4.5.2   Clinical Examples

Results from phantom studies may not reflect real clinical applications. Knöös et al. (Knöös et al. 2006) performed a comprehensive evaluation of dose calculation algorithms using the Monte Carlo simulation as the gold standard for 3D conformal plans in four common radiotherapy sites: prostate, head and neck, breast, and lung cancer. Multiple centres participated in the study with TPS using different dose calculation algorithms, ranging from correction-based (Type-A) to model-based algorithms (Type-B).

Dose calculation accuracy depends not only on the dose calculation algorithm, but also on the beam modeling and calibration; therefore, it is difficult to compare

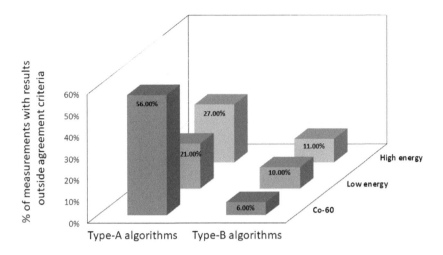

Figure 4.27: Percentage of point-dose measurements with results outside agreement criteria for different algorithm types and energy groups. Adapted from reference (Gershkevitsh et al. 2014).

clinical results among centres with different linear accelerators which were commissioned and calibrated differently. To isolate errors from dose calculation algorithms alone, the study (Knöös et al. 2006) performed the following two tasks:

1. The dose per MU was normalized by the reference dose rate measured with SSD = 100 cm at a depth of 10 cm by each institution.

2. A commercial TPS (Oncentra MasterPlan, Nucletron) with a pencil-beam (Type-A) and a collapsed-cone convolution (Type-B) algorithm was commissioned for an idealized linear accelerator, also called virtual linear accelerator. Benchmark data were created by the Monte Carlo simulations for this virtual linear accelerator. It should be noted, in this study, the Monte Carlo method and the collapsed-cone convolution method calculate the absorbed dose in the *medium*, while the pencil-beam algorithm calculates the dose to a *water* voxel inside a heterogeneous medium.

The mean doses to the planning target volume by Type-A and Type-B algorithms for different sites are summarized in Figure 4.28 and Figure 4.29. From these box plots we see that differences between algorithms for the PTV mean doses in the study are small for most situations. The differences between Type-A and Type-B algorithms are larger for the lung case, where the difference in median value for the PTV mean dose is 2.6% for 6 MV plans and 3.5% for 15 or 18 MV x-ray plans. The influence of electron transport is more significant for higher energy photons incident on lung. High energy beams are not recommended for treatment of small lesions in

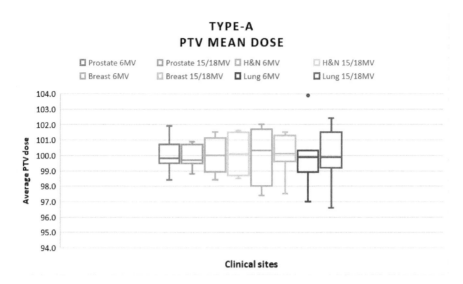

Figure 4.28: Box plot for PTV mean doses by Type-A algorithms for different clinical sites. The dose is normalized to be 100% for the average PTV mean dose by Type-A algorithms. The chart was produced from data by Knöös et al. (Knöös et al. 2006).

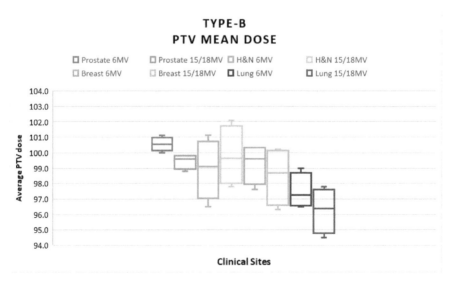

Figure 4.29: Box plot for PTV mean doses by Type-B algorithms for different sites. The dose is normalized to be 100% for the average PTV mean dose by Type-A algorithms. The chart was produced from published data (Knöös et al. 2006).

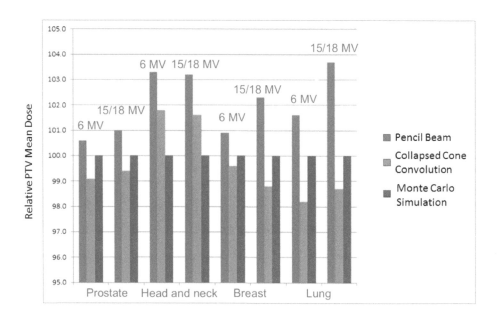

Figure 4.30: Comparison of the Monte Carlo simulations with the dose calculations by the pencil beam and the collapsed-cone convolution algorithms for PTV mean doses of different clinical sites with 6 and 15 or 18 MV beams. The PTV mean doses are normalized to 100% for the Monte Carlo simulation. The chart was produced from published data (Knöös et al. 2006).

lung, unless the effects of electron disequilibrium can be properly accounted for and considered in the clinical dose-volume prescription.

The results of the Monte Carlo benchmark (Knöös et al. 2006) with the virtual linear accelerator are summarized in Figure 4.30. It compares PTV mean doses calculated by the pencil-beam and the collapsed-cone convolution algorithms with the Monte Carlo simulations for different clinical sites with 6 and 15 or 18 MV x-ray beams. The doses are normalized to 100% for PTV mean doses calculated by the Monte Carlo simulations. It shows that the Monte Carlo simulations agree with both the pencil-beam and the collapsed-cone convolution algorithms within 1% for prostate case, where tissue heterogeneity is not significant. For all other sites, the collapsed-cone convolution algorithm generally agrees better with the Monte Carlo simulations (within 2%) than the pencil-beam algorithm, especially for lung sites with 15 or 18 MV x-ray beams.

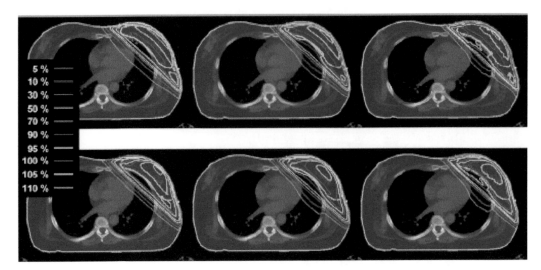

Figure 4.31: Comparison of dose distributions for a breast cancer treatment among pencil-beam (left), collapsed-cone (middle), and Monte Carlo (right) methods for 6 MV (upper row) and 15 or 18 MV (lower row) beams. The PTV contour is displayed in light blue. Reproduced with permission (Knöös et al. 2006).

The limitation of Type-A algorithms is again well illustrated by the dose distributions in lung for a breast case as shown in Figure 4.31. The Monte Carlo results (right panels) show that the beam penumbra is increased inside lung due to lateral electron disequilibrium, especially for high energy beams (right lower panel). This effect is approximated by Type-B algorithm (middle panels) with density scaling, but is missed by Type-A algorithm (left panels). The effects of missing tissue scattering and lung heterogeneity are also reflected in the reduction of PTV mean dose by the Monte Carlo method compared to the Type-A algorithm for the breast case shown in Figure 4.30. It also shows the over corrections for the dose inside lung heterogeneity by the collapsed-cone convolution (Type-B) algorithm for the breast case.

The study (Knöös et al. 2006) shows that both Type-A and Type-B dose calculation algorithms provide sufficient accuracy for clinical sites without significant tissue heterogeneity, such as pelvic radiotherapy for prostate cancer. It also shows that Type-B algorithms provide improved dose calculation accuracy for clinical sites with more tissue heterogeneity, such as head-and-neck, breast, and lung sites. The study was limited to 3D conformal plans; more significant effects are expected for IMRT or VMAT with many small field apertures, especially in low density medium such as lung. Additional studies have been reported more recently over concerns of inaccuracy of dose-calculation algorithms for stereotactic body radiation therapy applications (Disher et al. 2012; Lebredonchel et al. 2017).

4.6 ASSUMPTIONS AND LIMITATIONS

The 3D convolution-superposition dose calculation is generally divided into a two-step process. The first step calculates the energy released by primary photons (i.e. TERMA) at each interaction site. The second step yields the 3D dose distribution by superimposing dose spread kernels resulting from all the energy released by primary photons. In this section we summarize assumptions and limitations of the various implementations of superposition-convolution methods. The major assumptions are:

1. The dose spread kernels can be pre-calculated for an all-water absorber and reshaped in heterogeneous tissue by density-scaling to achieve considerable speed gain over Monte Carlo methods.

2. Effects of atomic number and tissue heterogeneity can be accounted for in the TERMA with attenuation coefficients but only approximated in kernels by density scaling.

3. Effects of polyenergetic beams can be included rigorously in the TERMA calculation but they are approximated *via* polyenergetic kernels.

These assumptions result in some limitations of the algorithms. For the first assumption, in principle, we can always divide dose calculation into the two steps. However, in order to take advantage of convolution-superposition method, we need to make repeated use of pre-calculated kernels. For heterogeneous medium, the TERMA of primary photons can be calculated accurately by ray-tracing through the CT images using effective attenuation coefficients. The major limitation comes from the reuse of pre-calculated kernels with approximations for heterogeneous medium.

The density scaling for TERMA calculation (i.e. ray-tracing) is a good approximation if photon interactions are dominated by Compton scattering with minimal atomic-number effects. The density scaling for kernels is a good approximation only for kernels corresponding to primary charged particles and first Compton scattered photons in a slab geometry between the source voxel and the target voxel. The density scaling along raylines from each interaction site may cause significant dose calculation errors near the interface of heterogeneity as illustrated in Figure 3.11 and in phantom or clinical cases discussed in the previous section. Therefore, the density-scaling assumption is generally valid for megavoltage beams in soft tissue with minimal variations of atomic numbers, such as radiotherapy of prostate cancer.

Polyenergetic beams from a medical linear accelerator create another challenge in defining the kernels with consideration of the x-ray spectrum. This spectrum changes with each location in the patient due to localized beam hardening or softening. In principle, the superposition should be performed with different spectra and different kernels at each location. This would forfeit some of the speed advantage of the method. Fortunately, the beam hardening corrections can be pre-calculated for a water medium and applied to a patient as a reasonable approximation (Ahnesjö 1989).

Besides the above general assumptions and limitations, there are additional assumptions made for various fast implementations. Performing faster convolution integral in Equation 4.9 instead of more general superposition integral in Equation 4.7 requires spatial invariance of kernels. This adds limitations to the application of the algorithms to heterogeneous media. The fast Fourier transform convolution (FFTC) algorithm is one of the fast implementations of the convolution method. As discussed in the previous section, the FFTC may have significant dose calculation errors in low density medium as shown in Figure 4.23.

The pencil-beam convolution algorithm assumes that pre-calculated pencil-beam kernels for a uniform water medium can be used for a patient anatomy. It can only account for inhomogeneities with density scaling along the depth and orthogonal lateral directions *separately*. The algorithm can account for beam shape and modifiers easily, but has difficulty accounting for missing tissue scattering or tissue heterogeneity, as was shown in Figure 4.10.

The singular-value decomposition (SVD) algorithm (Bortfeld et al. 1993) assumes that the pencil-beam kernel can be decomposed into three separate terms, each as a product of depth dependent function and a radial dependent function. This assumption is supported mathematically by the singular-value decomposition theorem. The SVD method has advantage of fast dose calculation and it is widely used for interactive IMRT/VMAT optimization on commercial treatment planning systems such as Pinnacle (Philips Medical Systems) and RayStation (RaySearch Laboratories). However, it has the same limitations in accounting for heterogeneity in a medium as the general pencil-beam algorithms. After an IMRT/VMAT plan is optimized, the final dose is usually recomputed with more sophisticated dose calculation algorithms, such as the collapsed cone convolution (CCC) algorithm. The use of different algorithms for optimization and final dose computation may, however, lead to a suboptimal treatment plan.

The AAA algorithm assumes that the energy is transported along the depth first and then subsequently transported laterally. To account for medium heterogeneity, it assumes that the density scaling can be performed along the depth and lateral directions *independently*. Obviously this assumption does not correspond to the real physical process and pathways, but it does provide a mathematical model to improve conventional 2D pencil-beam algorithms for heterogeneous media.

The CCC algorithm divides 3D space into a discrete set of scattering cones for every interaction point. It assumes that the radiant energy released into a cone can be effectively collapsed onto the cone axis. This collapsed energy is then *linearly* transported, attenuated, and deposited along the cone axis. This approximation speeds up the dose calculation significantly, because instead of tracking radiant energy transport everywhere around the interaction site, the energy transport is only tracked along specified cone axes. In addition to the limitations associated with the general convolution-superposition methods, the collapsed cone approximation may cause inaccurate dose prediction far away from a heterogeneity, as shown in Figure 4.20 where the energy equilibrium between adjacent cones may be imperfect, as shown in Figure 4.21.

4.7 SUMMARY

In this chapter we have discussed some major dose calculation algorithms based on the general convolution-superposition principle. This is currently the most widely used method in commercial software. We have also reviewed their clinical applications and limitations in Sections 4.5 - 4.6. A brief summary of the algorithms with their key assumptions and computation efficiency is provided in Table 4.1.

The major advantage of convolution-superposition algorithms is their computation efficiency compared to the more accurate Monte Carlo or Boltzmann transport methods. An estimate for the number of calculations for each algorithm is listed in Table 4.1. If dose spread kernels are assumed to be spatially invariant, the superposition integral becomes the convolution integral and the fast Fourier transform can be used to speed up the calculation significantly. For 3D convolution (FFTC) with N dose voxels in one dimension, it takes the order of $3N^3\log_2 N$ calculations, while for 2D convolution with pencil-beam kernels, it takes the order of $2N^3\log_2 N$ calculations. When the SVD method is used to factor out the depth dependence of the kernels, the number of calculations can be further reduced to $6N^2\log_2 N$. As discussed in previous sections, for more accurate dose calculations with heterogeneous medium, kernels need to be scaled locally using radiological path length. This results in the increase of the number of calculations to about N^5 for the AAA algorithm and MN^3 for the CCC algorithm when M cones are used for each dose voxel.

The actual computation time depends on the software and hardware configuration of treatment planning systems. For example, for a VMAT head-and-neck plan using two 360-degree arcs on a Pinnacle planning system (Philips) with an 8-core Intel 2.9 GH Xeon processor, it takes about 30 seconds to calculate 3D dose distribution with 10^6 voxels if the SVD algorithm is used, but it takes about 4.5 minutes with the CCC algorithm. On a RayStation planning system (RaySearch Lab) using parallel calculation with a GPU processor, it takes only 18.2 seconds to calculate the same plan with the same dose matrix using the CCC algorithm. On an Eclipse planning system (Varian) equipped with distributed calculations using 13 frame agent servers, it takes 35 seconds to calculate the same plan with the same dose matrix using the AAA algorithm when 30 CPUs were used in parallel.

In general, the convolution-superposition methods provide acceptable dose calculation accuracy for megavoltage x-rays in a medium without significant variation of heterogeneities such as radiotherapy of prostate cancers in pelvic region. There exist trade-offs between the speed and the dose calculation accuracy among various fast implementations. The SVD method combined with the fast Fourier transform is one of the fastest algorithms, therefore it is widely used in real-time IMRT/VMAT optimization, while the final dose distribution is typically calculated by either the CCC or the AAA algorithm.

The major limitation of convolution-superposition algorithms is the inaccuracy of dose calculation with heterogeneous media using density-scaling approximation and ignoring atomic-number effects. This may cause significant errors for dose calculated with high-density material (e.g. hip prosthesis) and near the interface of heterogeneous medium. One should be cautious about the dose calculated under the condition of severe lateral charged-particle disequilibrium, such as stereotactic ablative radiotherapy for lung cancer using high energy photon beams with very small field size in low density medium (Disher et al. 2012; Bibault et al. 2015).

Despite some limitations, the convolution-superposition dose calculation algorithms plays a very important role in treatment planning systems due to a good compromise between calculation efficiency and accuracy for most clinical applications, especially for IMRT/VMAT optimization, and real-time plan adaptation with image-guided radiation therapy (IGRT).

To overcome any residual limitations of superposition-convolution methods, more accurate dose calculation algorithms are being implemented in treatment planning systems with fast computer technology. For example, the Boltzmann transport method has recently been implemented as Acuros XB in Eclipse TPS (Varian). Its speed for 3D computations is comparable to that of the AAA 2.5D algorithm for clinical VMAT plans. VMAT using dynamic arcs has been gradually replacing fixed-gantry IMRT for many clinical sites due to its conformal dose distribution, achieved with more degrees of freedom in beam directions. Similar gains have been made using GPU or CPU acceleration for execution of Monte Carlo codes. The computation time for Monte Carlo methods is independent of number of beams, depending mainly on the *aggregate* number of particles passing through a dose matrix voxel; this is ideal for taking advantage of parallel processing techniques. The accuracy of Monte Carlo and Boltzmann methods depends on discretization and ultimately is limited by the validity of input physical data sets. For Monte Carlo codes, the number of particle histories used in the simulation determines the precision of dose calculation results. With incessant improvements in computer speed and computations shared across multiple processors, the Monte Carlo (Chapter 5) and Boltzmann transport methods (Chapter 6) will play a progressively more important role in future clinical applications.

Table 4.1: Summary of dose calculation algorithms. N is the number of dose voxels in one dimension. M is the number of cones used in the CCC algorithm.

Algorithm Name	Treatment Planning System (TPS)	Key Assumptions	Efficiency	Key References
FFTC	Old TPS	Point kernels are spatially invariant	$3N^3\log_2 N$	(Boyer and Mok 1986) (Mohan and Chui 1987) (Zhu and Boyer 1990) (Wong et al. 1996)
Pencil Beam	Old TPS	Convolution can be reduced from 3D to 2D	$2N^3\log_2 N$	(Mohan and Chui 1986) (Ahnesjö et al. 1992)
SVD	Pinnacle 16 RayStation 7	Pencil-beam kernels can be decomposed into three terms as products of depth-dependent and radial-dependent functions	$6N^2\log_2 N$	(Bortfeld et al. 1993)
AAA	Eclipse 15.5	The radiant energy is transported along the depth first and then laterally	N^5	(Ulmer and Harder 1995) (Ulmer et al. 2005) (Tillikainen et al. 2008)
CCC	Pinnacle 16 RayStation 7 Monaco 5 Oncentra	The radiant energy released into a cone can be effectively collapsed onto the cone axis	MN^3	(Ahnesjö 1989) (Ahnesjö and Aspradakis 1999)

BIBLIOGRAPHY

Ahnesjö, A. (1984). Application of transform algorithms for calculation of absorbed dose in photon beams. In *Proceedings of the VIIIth International Conference on the Use of Computers in Radiation Therapy*, pp. 17–20.

Ahnesjö, A. (1987). Invariance of convolution kernels applied to dose calculations for photon beams. In *Proceedings of the IXth International Conference on the Use of Computers in Radiation Therapy*, pp. 99–102.

Ahnesjö, A. (1989). Collapsed cone convolution of radiant energy for photon dose calculation in heterogeneous media. *Medical Physics 16*(4), 577–592.

Ahnesjö, A., P. Andreo, and A. Brahme (1987). Calculation and application of point spread functions for treatment planning with high energy photon beams. *Acta Oncologica 26*, 49–56.

Ahnesjö, A. and M. Aspradakis (1999). Dose calculations for external photon beams in radiotherapy. *Physics in Medicine and Biology 44*, R99–R155.

Ahnesjö, A., M. Saxner, and A. Trepp (1992). A pencil beam model for photon dose calculation. *Medical Physics 19*(2), 263–273.

Ahnesjö, A., B. van Veelen, and A. Tedgren (2017). Collapsed cone dose calculations for heterogeneous tissues in brachytherapy using primary and scatter separation source data. *Computer Methods and Programs in Biomedicine 139*, 17–29.

Ahnesjö, A., L. Weber, A. Murman, M. Saxner, I. Thorslund, and E. Traneus (2005). Beam modeling and verification of a photon beam multisource model. *Medical Physics 32*(6), 1722–1737.

Bibault, J., X. Mirabel, T. Lacornerie, E. Tresch, N. Reynaert, and E. Lartigau (2015). Adapted prescription dose for Monte Carlo algorithm in lung SBRT: Clinical outcome on 205 patients. *PLoS One 10*(7), 1–10.

Bjarngard, B. (1987). On Fano's and O'Connor's Theorems. *Radiation Research 109*(2), 184–189.

Bortfeld, T., W. Schlegel, and B. Rhein (1993). Decomposition of pencil beam kernels for fast dose calculations in three-dimensional treatment planning. *Medical Physics 20*(2), 311–318.

Boyer, A. and E. Mok (1985). A photon dose distribution model employing convolution calculations. *Medical Physics 12*(2), 169–177.

Boyer, A. and E. Mok (1986). Calculation of photon dose distributions in an inhomogeneous medium using convolutions. *Medical Physics 13*(4), 503–9.

Boyer, A., Y. Zhu, L. Wang, and P. Francois (1989). Fast Fourier transform convolution calculations of x-ray isodose distributions in homogeneous media. *Medical Physics 16*(2), 248–253.

Breitman, K., S. Rathee, C. Newcomb, B. Murray, D. Robinson, C. Field, H. Warkentin, S. Connors, M. MacKenzie, P. Dunscombe, and G. Fallone (2007). Experimental validation of the Eclipse AAA algorithm. *Journal of Applied Clinical Medical Physics 8*(2), 76–92.

Carlsson, A. and A. Ahnesjö (2000). The collapsed cone superposition algorithm applied to scatter dose calculations in brachytherapy. *Medical Physics 27*(10), 2320–2332.

Chen, Q., M. Chen, and W. Lu (2011). Ultrafast convolution/superposition using tabulated and exponential kernels on GPU. *Medical Physics 38*(3), 1150–1161.

Cooley, J. and J. Tukey (1965). An algorithm for the machine calculation of complex Fourier series. *Mathematics of Computation 19*(90), 297–301.

Das, I., C. Cheng, R. Watts, A. Ahnesjö, J. Gibbons, X. Li, J. Lowenstein, R. Mitra, W. Simon, and T. Zhu (2008). Accelerator beam data commissioning equipment and procedures: Report of the TG-106 of the Therapy Physics Committee of the AAPM. *Medical Physics 35*(9), 4186–4215.

Disher, B., G. Hajdok, S. Gaede, and J. Battista (2012). An in-depth Monte Carlo study of lateral electron disequilibrium for small fields in ultra-low density lung: Implications for modern radiation therapy. *Physics in Medicine and Biology 57*(6), 1543–59.

Field, C. and J. Battista (1987). Photon dose calculations using convolution in real and Fourier space: assumptions and time estimates. In *Proceedings of the Ninth International Conference on the Use of Computers in Radiation Therapy*, pp. 103–106.

Fogliata, A., G. Nicolini, E. Vanetti, A. Clivio, and L. Cozzi (2006). Dosimetric validation of the anisotropic analytical algorithm for photon dose calculation: fundamental characterization in water. *Physics in Medicine and Biology 51*(6), 1421–1438.

Fogliata, A., E. Vanetti, D. Albers, C. Brink, A. Clivio, T. Knöös, G. Nicolini, and L. Cozzi (2007). On the dosimetric behaviour of photon dose calculation algorithms in the presence of simple geometric heterogeneities: comparison with Monte Carlo calculations. *Physics in Medicine and Biology 52*(5), 1363–1385.

Gagné, I. and S. Zavgorodni (2007). Evaluation of the analytical anisotropic algorithm in an extreme water-lung interface phantom using Monte Carlo dose calculations. *Journal of Applied Clinical Medical Physics 8*(1), 33–46.

García-Vicente, F., Á. Miñambres, I. Jerez, I. Modolell, L. Pérez, and J. Torres (2003). Experimental validation tests of fast Fourier transform convolution and multigrid superposition algorithms for dose calculation in low-density media. *Radiotherapy and Oncology 67*(2), 239–249.

Gershkevitsh, E., C. Pesznyak, B. Petrovic, J. Grezdo, K. Chelminski, M. Do Carmo Lopes, J. Izewska, and J. Van Dyk (2014). Dosimetric inter-institutional comparison in European radiotherapy centres: Results of IAEA supported treatment planning system audit. *Acta Oncologica 53*(5), 628–636.

Hasenbalg, F., H. Neuenschwander, R. Mini, and E. Born (2007). Collapsed cone convolution and analytical anisotropic algorithm dose calculations compared to VMC++ Monte Carlo simulations in clinical cases. *Physics in Medicine and Biology 52*(13), 3679–3691.

Johns, H. and J. Cunningham (1983). *The Physics of Radiology* (4th ed.). Charles C. Thomas, Springfield Illinois.

Kawrakow, I. and D. Rogers (2003). The EGSnrc code system: Monte Carlo simulation of electron and photon transport. Manual NRCC PIRS-701, National Research Council of Canada, Ottawa, Canada.

Knöös, T., E. Wieslander, L. Cozzi, C. Brink, A. Fogliata, D. Albers, H. Nyström, and S. Lassen (2006). Comparison of dose calculation algorithms for treatment planning in external photon beam therapy for clinical situations. *Physics in Medicine and Biology 51*(22), 5785–5807.

Lebredonchel, S., T. Lacornerie, E. Rault, A. Wagner, N. Reynaert, and F. Crop (2017). About the non-consistency of PTV-based prescription in lung. *Physica Medica 44*, 177–187.

Lu, W., G. Olivera, M. Chen, P. Reckwerdt, and T. Mackie (2005). Accurate convolution/superposition for multi-resolution dose calculation using cumulative tabulated kernels. *Physics in Medicine and Biology 50*(4), 655–680.

Mackie, T. (1983). Photon dose calculation using a convolution method that includes non-local electron energy deposition. *Medical Physics 10*(4), 536.

Mackie, T., A. Bielajew, D. Rogers, and J. Battista (1988). Generation of photon energy deposition kernels using the EGS Monte Carlo code. *Physics in Medicine and Biology 33*(1), 1–20.

Mackie, T., P. Reckwerdt, T. McNutt, M. Gehring, and C. Sanders (1996). Photon beam dose computations. In T. Mackie and J. Palta (Eds.), *Teletherapy: Present and Future*, Chapter 6, pp. 103–135. Madison, Wisconsin: Advanced Medical Publishing.

Mackie, T., J. Scrimger, and J. Battista (1985). A convolution method of calculating dose for 15-MV x rays. *Medical Physics 12*(2), 188–96.

McDermott, P. (2016). Convolution/superposition dose computation algorithms. In *Tutorials in radiotherapy physics: Advanced topics with problems and solutions*, Chapter 3, pp. 121–170. Boca Raton, Florida: CRC Press, Taylor & Francis Group.

Mohan, R. and C. Chui (1985a). Energy and angular distributions of photons from medical linear accelerators. *Medical Physics 12*(5), 592–597.

Mohan, R. and C. Chui (1985b). Validity of the concept of separating primary and scatter dose. *Medical Physics 12*(6), 726–730.

Mohan, R. and C. Chui (1986). Differential pencil beam dose computation model for photons. *Medical Physics 13*(1), 64–73.

Mohan, R. and C. Chui (1987). Use of fast Fourier transforms in calculating dose distributions for irregularly shaped fields for three-dimensional treatment planning. *Medical Physics 14*(1), 70–77.

O'Connor, J. (1957). The variation of scattered x-rays with density in an irradiated body. *Physics in Medicine and Biology 1*(4), 305.

O'Connor, J. (1984). The density scaling theorem applied to lateral electronic equilibrium. *Medical Physics 11*(5), 678–680.

Oliver, P. and R. Thomson (2017). A Monte Carlo study of macroscopic and microscopic dose descriptors for kilovoltage cellular dosimetry. *Physics in Medicine and Biology 62*(4), 1417–1437.

Papanikolaou, N., T. Mackie, C. Meger-Wells, M. Gehring, and P. Reckwerdt (1993). Investigation of the convolution method for polyenergetic spectra. *Medical Physics 20*(5), 1327–1336.

Sharpe, M. and J. Battista (1993). Dose calculations using convolution and superposition principles: The orientation of dose spread kernels in divergent x-ray beams. *Medical Physics 20*(6), 1685–1694.

Sievinen, J., W. Ulmer, and W. Kaissl (2005). AAA photon dose calculation model in Eclipse. Manual, Varian Medical Systems, Palo Alto, California.

Tillikainen, L., H. Helminen, T. Torsti, S. Siljamäki, J. Alakuijala, J. Pyyry, and W. Ulmer (2008). A 3D pencil-beam-based superposition algorithm for photon dose calculation in heterogeneous media. *Physics in Medicine and Biology 53*(14), 3821–3839.

Tillikainen, L., S. Siljamäki, H. Helminen, J. Alakuijala, and J. Pyyry (2007). Determination of parameters for a multiple-source model of megavoltage photon beams using optimization methods. *Physics in Medicine and Biology 52*(5), 1441–1467.

Ulmer, W. and D. Harder (1995). A triple Gaussian pencil beam model for photon beam treatment planning. *Journal of Medical Physics 5*(1), 25–30.

Ulmer, W., J. Pyyry, and W. Kaissl (2005). A 3D photon superposition/convolution algorithm and its foundation on results of Monte Carlo calculations. *Physics in Medicine and Biology 50*(8), 1767–1790.

Van Dyk, J., R. Barnett, J. Cygler, and P. Shragge (1993). Commissioning and quality assurance of treatment planning computers. *International Journal of Radiation Oncology Biology Physics 26*(2), 261–273.

Van Esch, A., L. Tillikainen, J. Pyykkonen, M. Tenhunen, H. Helminen, S. Siljamäki, J. Alakuijala, M. Paiusco, M. Iori, and D. Huyskens (2006). Testing of the analytical anisotropic algorithm for photon dose calculation. *Medical Physics 33*(11), 4130–4148.

Varian (2016). Eclipse photon and electron reference guide 15.1. Manual, Varian Medical Systems, Palo Alto, California.

Wong, E., Y. Zhu, and J. Van Dyk (1996). Theoretical developments on fast Fourier transform convolution dose calculations in inhomogeneous media. *Medical Physics 23*(9), 1511–1521.

Woo, M. and J. Cunningham (1990). The validity of the density scaling method in primary electron transport for photon and electron beams. *Medical Physics 17*(2), 187–94.

Zhu, Y. and A. Boyer (1990). X-ray dose computations in heterogenous media using 3D FFT convolution. *Physics in Medicine and Biology 35*(3), 351–368.

Upper Photo - First EGS course at National Research Council of Canada (1986).
Lower Photo - Monte Carlo School held at Ettorre Majorana Centre (Italy, 1987).
Attendees included Martin Berger, Stephen Seltzer, Ralph Nelson, Ted Jenkins,
Dave Rogers, Alex Bielajew, Art Boyer, Rhade Mohan, Jeff Williamson, Alan
Nahum, David Raeside and Pedro Andreo.
Photos provided by Dave Rogers.

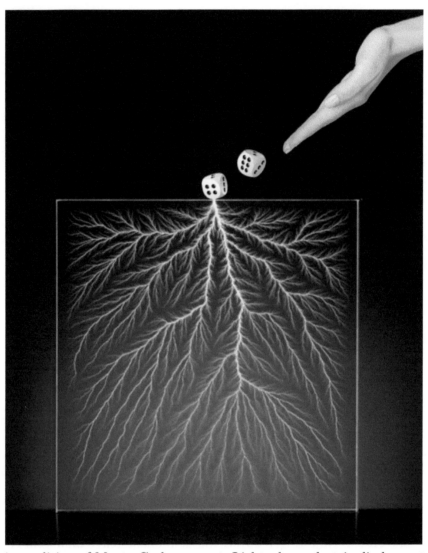

Artistic rendition of Monte Carlo concept. Lichtenberg electric discharge tree of megavoltage electrons deposited in an acrylic plastic slab.
Design and photo courtesy of John Patrick, Steve Sawchuk and Mike Walker
(mike@mikewalkerphoto.com).

CHAPTER 5

Stochastic Radiation Transport Methods

Stephen Sawchuk

University of Western Ontario and London Health Sciences Centre

CONTENTS

5.1 MONTE CARLO METHOD AND MEDICAL PHYSICS

S TOCHASTIC or Monte Carlo methods (MCMs) are commonly used to model a succession of multiple random events and they have been successfully applied to diverse fields such as science and socio-economics. MCMs can be programmed to simulate the conduct of scientific experiments or emulate scenarios which are either impossible, extremely difficult or too expensive to perform in the real world. Moreover, artificial conditions can be introduced or suppressed to model "what if" situations that help identify the relative importance of a host of competing variables. In general, computer simulations are very versatile, and they can prove to be more efficient, cost-effective, and safer than execution in actual situations. For example, computer-aided design of complex experimental apparatus avoids costs in iterative engineering of the instrument. After the equipment is built, additional simulations can help identify points of failure or corrections of experimental data. This is especially true in the field of medical physics, specifically for utilization in radiation dosimetry, including dose measurements and computations. MCMs are used for simulation of particle interactions in detectors and tissue, culminating in accurate predictions of particle fluences and radiation dose distributions. A prime example has been the steady progress achieved in absolute international radiation dosimetry through improved design of various instruments, smarter interpretation of detector signals, and application of correction factors to measured data (Rogers 2006). Currently, it is commonly accepted by most medical physicists that the MCMs are the gold standard amongst dose calculation algorithms, although recent developments in Boltzmann techniques (Chapter 6) are yielding competitive performance.

The Monte Carlo stochastic approach is very different from that used in other algorithmic approaches, namely *deterministic* methods including the convolution-superposition algorithm (Chapter 4) and solution of a system of Boltzmann transport equations (Chapter 6). The distinguishing feature is the *stochastic* aspect of MCMs arising from the *statistical sampling* of cross sections describing interaction events for photons and secondary particles (Chapter 2). In the terminology of MCMs, these cross sections are regarded as probability density distributions that are *randomly sampled*, rather than used as direct energy propagation operators in deterministic transport equations.

Some early applications of MCMs involved obtaining numerical solutions of definite integrals in one dimension, such as estimation of the area under a curve using randomly-placed points (recall Figure 3.14). In MCMs, the integral is approximated by counting the fraction of points landing under the curve, as opposed to above the curve. A frequently-cited example is the determination of the value of π (Nelson et al. 1985) by counting the fraction of random points that fall within a circle enclosed in a square (https://academo.org/demos/estimating-pi-monte-carlo/). Building upon this approach, random sampling methods lead to a mechanism for propagating radiation and integrating (literally) the energy deposited in tissue voxels. The mathematical solution for the average dose deposited at a voxel in the patient by a radiation beam is indeed a complex multi-dimensional integral. It reflects all possible permutations and combinations of particle pathways and energy transfers that lead to local energy deposits. This integral can be evaluated by random sampling without expressing it analytically. The secret to obtaining realistic results is to identify all relevant interaction mechanisms and apply accurate individual probability density functions using assumptions that are acceptable for the intended scope of clinical applications. Typical outputs from MCMs include the entire frequency distribution of scored variables, such as particle fluences. More often only the average value and variance of frequency distributions are needed and are calculated for a finite number of simulated photon histories. As long as accurate probability density functions are applied, multiple sequential events will yield end results that converge to a desired accuracy limited by the number of radiation-depositing particles.

The underlying processes of radiation physics are fortunately well understood for applications to megavoltage energy transport and distribution. For example, by knowing the total interaction cross section and corresponding linear attenuation coefficient for photons traversing an absorbing material, the path lengths to interaction sites become exponentially distributed (recall Figure 2.9). Similarly, the probability that a type of scattering process will occur at the selected interaction site depends on a current random state variable, such as incoming energy. The redirection and energy loss of the inbound photon is chosen by sampling the angular differential cross section for inelastic scattering. Simulating a sequence of free paths and interaction events, one particle shower at a time, over a large number of instances builds up the dose to a matrix of tissue voxels in the patient.

To recap, the Monte Carlo technique has been used to solve radiation transport problems in many branches of medical physics including x-ray imaging, nuclear medicine, radiation protection, and radiation therapy:

- interactions within an imaging detector, patient-scattered radiation, pixel signal interpretation, collimator design, patient dose from diagnostic imaging

- shielding design of radiation therapy rooms, treatment planning, dose verification

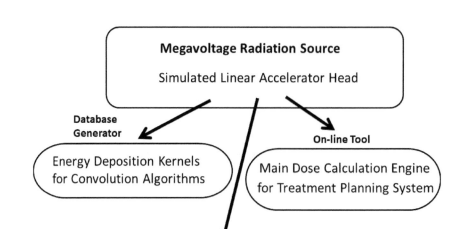

Figure 5.1: Applications of Monte Carlo methods to radiation therapy.

We now direct attention on propagation of radiation fluences incident on cancer patients treated with megavoltage x-rays and the subsequent delivery of energy to human tissue voxels.

5.1.1 Monte Carlo Applications to Radiation Therapy

We will firstly describe the general methodology of MCMs and the radiation transport simulation process. Then we will focus on specific applications to radiotherapy and state the advantages and disadvantages compared with deterministic strategies. Finally, we will describe aspects of implementation of MCMs in practical treatment planning systems (TPS), from modelling of the radiation source during commissioning to subsidiary applications as a dose verification tool. Figure 5.1 presents a summary of current applications.

5.1.2 Goal and Scope of Chapter

This chapter highlights the basic concepts and aims to provide an intuitive explanation of MCMs, structured in such a way as to facilitate comparisons with other dose computation algorithms. It is important from a clinical implementation standpoint that MCMs should be treated as any other dose algorithm within a treatment planning system. Valid implementation requires the clinical physicist to understand the fundamentals and assumptions of the algorithm including any hidden limitations associated with commissioning (Chetty et al. 2007). Ideally, the reader of this chapter should be equipped with an undergraduate knowledge of mathematics and statistics, and a graduate course in radiological physics. This chapter can be used for self-study in graduate medical physics programs or residency programs. It may also be of value to staff involved with support of commercial radiotherapy planning software.

It is a great challenge to write and clearly explain the methodology of this algorithm. The current stage of evolution of MCMs is not only complicated but there is an enormous span of existing resources from past and recent publications. Indeed, there are many excellent reference articles which the reader is encouraged to explore beyond the scope of this introductory chapter. Some of the most important and popular techniques are conveniently summarized and referenced within this chapter.

5.2 MONTE CARLO METHOD IN THEORY

We present the method with its theoretical foundation first, followed by practical implementation considerations.

5.2.1 Genesis and History

Randomness is inherent and ubiquitous in nature; understanding gained from natural randomness can lead to human discovery. The first MCMs are attributed to the Compte de Buffon in solving of the needle problem in the 1700s (Bielajew 2013; Vassiliev 2016). The simulation models the random dropping of needles with a given length onto a grid of parallel lines, and predicts the expected fraction of needles that will intersect any one of these lines. The beginning of computerized MCMs and applications to radiation transport began in the 1930s and 1940s. In 1934, Fermi used them to study the diffusion of neutrons in nuclear reactions but he did not publish his method (Vassiliev 2016). Von Neumann describes this statistical technique originally suggested by Ulam and and the name **Monte Carlo** method is attributed to *Metropolis* (Metropolis 1987). It was also used to solve neutron diffusion problems arising at the Los Alamos Laboratory in the late 1940s. (Bielajew 2013; Vassiliev 2016; Andreo et al. 2017).

5.2.2 Definition of Monte Carlo Method

The MCM has often been described simply as the *process of conducting a mathematical experiment to estimate the expected outcome of a stochastic process by sampling from the governing probability densities*. Generally, problems and applications are cast in the mathematical form of differential or integro-differential equations that are difficult to solve in analytical closed form. The MCMs with appropriate probability densities for the specific application are used for statistical *inference of the solution* to these complex equations (Metropolis 1987; Vassiliev 2016).

5.2.3 Anecdotal Application of Monte Carlo Simulation

The MCM clearly relies on *a-priori* knowledge of **accurate** probabilities of individual events. To highlight the importance of this foundation, we describe a simple example from the past popular television game show, "Let's Make A Deal". The game consists of the host named Monty Hall (Monte might have been a more suitable spelling in our context), a contestant, and 3 separate doors hiding unknown prizes on the stage. The valuable prize, such as a vintage custom Les Paul electric guitar, is hidden behind only one of the doors. The other two doors are hiding undesirable junk prizes. We know from basic statistics that the probability of *initially* choosing the door with the valuable prize is 1 in 3, and that of getting either of the joke prizes is 2 in 3.

Steps in Playing the Monty Hall Game:

- Step 1: Monty Hall asks the contestant to pick one of the three doors.

- Step 2: Contestant chooses a door but the door is *not* opened. The prize behind this door remains unknown to the contestant.

- Step 3: Monty then reveals one of the junk prize doors (avoiding the valuable prize).

- Step 4: Then the contestant is asked whether to either keep the original door choice or switch to one of the unopened doors.

What are the odds of winning the valuable prize if the contestant either sticks with the original choice or switches doors? It might not be what you intuitively expect! (Vos Savant 1990).

- Step 5: The contestant makes the choice and the chosen door is opened to reveal the real prize or junk prize. The theoretical odds of winning are 2 in 3 in favour of switching doors, but only 1 in 3 to sticking with the initial door. Probabilistically the odds are twice in one's favour by switching doors to win the prize. This may appear unintuitive and if the reader is not convinced, then consider performing a Monte Carlo simulation of this game (Gill 2011) with the following steps shown in Table 5.1 and please also see the demo (https://youtube/Xp6V_lO1ZKA).

Table 5.1: Steps of Monte Carlo algorithm for the Monty Hall game

1	Draw a random number [1,2,3] to set the door hiding the valuable prize. This is a well kept secret.
2	Draw another random number [1,2,3] to simulate the contestant's initial choice
3	The computer has internal information as to what is behind each door. The computer determines which door not selected by the contestant is a joke prize door. There are 2 possible scenarios. Scenario One: Both doors have joke prizes. Then the computer randomly selects between the two doors and reveals the choice. Scenario Two: One door is a joke door and the other has the major prize. Then the computer just reveals the door with the joke prize.
4	Draw a random number to simulate the choice the contestant has between sticking with original choice or switching doors.
5	If sticking is chosen and it matches with the prize door then the tally *sticking* is incremented by one
6	If sticking is chosen and does not match the prize door then the tally for *switching* is incremented.
7	Similarly, if switching is chosen and it matches the prize door then the tally for *switching* is incremented.
8	If switching is chosen and the prize door does not match then the tally for *sticking* is incremented.
9	The tallies for the *sticking* and *switching* strategies should match the total number of runs of the simulation.
10	The fraction of *sticking* tally over *total* tally gives the probability of winning by sticking to the contestant's initial choice.
11	The fraction of *switching* tally over the *total* tally gives the probability of winning by switching from the contestant's initial choice.

Table 5.2: Probabilities in a single Monty Hall game

Item	Door 1	Door 2	Door 3
Hidden prizes	junk	junk	prize
odds	1/3	1/3	1/3
step 1 and 2	choose		
step 3 reveal		junk	
steps 4 and 5 - stick or switch?	stick	junk	Switch to win!
Win probability after choosing door #1	1/3	2/3	

Table 5.2 explains the distributions for an example of a prize placed behind Door 3 and an initial selection of Door 1. If programmed as in the above example, the theoretical odds (2/3 for switching *versus* 1/3 for sticking) will be approached after a large number of trials. The correct initial door pick is right with a 1/3 probability; switching would cause a loss in 1/3 of trials. If the initial pick is wrong (2/3 probability), switching will favour a win with 2/3 probability!

5.2.4 Main Ingredients of Monte Carlo Method

5.2.4.1 Pseudo-Random Numbers

Principle #8: " Monte Carlo simulations mimic radiation experiments by rolling the dice and utilizing realistic probabilities describing particle interactions"

Random number generators (RNGs) are the essential core element of MCMs. The art and science of RNGs is a field of study unto its own (Butler 1956; McGrath and Irving 1975; Andreo 1991). Truly random numbers can be generated from hardware devices that, for example, monitor random natural events such as Poisson-distributed counts of particles emitted and detected from a radioactive source. However, computer systems implement pseudo-RNGs which can produce uncorrelated sequences of numbers with very long periods before the sequence repeats itself. An advantage of using computed RNGs is that debugging of Monte Carlo codes is facilitated by the ability to run exactly the same sequence of events in a simulation when initiated by the same random number seed. Most Monte Carlo software packages either have access to external pseudo-RNGs with excellent properties or incorporate the RNG directly within the simulation. Because of the wide range of RNGs and implementations, only the basic concepts will be described in this section.

A pseudo RNG delivers random numbers through an iterative equation such as:

$$I_{i+1} = \alpha I_i + c \text{ modulo } m \quad \text{for} \quad i = 0, 1, 2 ... N_{\text{period}}$$

where I_0 is the random number seed, α is the multiplier, c is the increment, m is the modulus and N_{period} is the period of the generator - the length of the string of numbers before the sequence repeats itself exactly. The *modulo*

Table 5.3: Popular well-tested random number generators (RNGs)

Name	References
RANMAR	EGS4 (Nelson et al. 1985), XVMC (Fippel 1999), EGSnrc (Kawrakow 2000)
RANLUX	EGSnrc (Kawrakow 2000; Luscher 1994)

operator yields the remainder after dividing the number by m. For example, if $I_0 = 987654321, \alpha = 987, c = 999$, and $m = 19$ we have:

$$I_1 = (987654321 \times 987 + 999) \text{ modulo } 19 = 4$$

A reliable random number generator has the following properties: a long period, uniformly distributed numbers, a sequence that is uncorrelated, and portability. It delivers the same sequence on different computer platforms (Fippel 2013a). There exists a battery of statistical tests that can be run to ensure that these characteristics are achieved. Ideally the period is a very large number especially when consuming many numbers in simulating numerous interactions with multiple branching decision points. Note that if the RNG period is exceeded during a simulation it may not necessarily have major impact on results unless the repetition occurs at the start of a fresh particle history. Table 5.3 provides an abbreviated list of well-tested RNGs used in contemporary MC codes and documentation is available from the CERN accelerator facility (https://root.cern.ch/doc/v612/).

From this point on, we assume that pseudo-RNGs produce a sequence of uniformly distributed real numbers on the normalized interval between $[0, 1]$. The notation of square brackets beside the limit indicates inclusion of the limit: whereas, round brackets indicate non-inclusion. For example, the notation $[0, 1]$ means that both 0 and 1 are included and $(0, 1]$ means that 0 is not included and 1 is included in the range. We will assume that the RNGs have been thoroughly tested and do not introduce any bias in the simulation results.

5.2.4.2 Probability Distributions

Probability distribution functions (pdf) are the main engines of MCMs, fuelled by random number generators. Many types of probabilty distributions are utilized and within this section we will describe two that are pertinent to simulations in medical radiation physics: uniform and Gaussian distributions.

The uniform distribution is the most utilized distribution in random sampling and it is the starting point for most sampling techniques (Hammersley and Handscomb 1964). It describes a constant frequency of values over the range of random variables where all are equally likely to occur. In Figure 5.2, x values ranging from

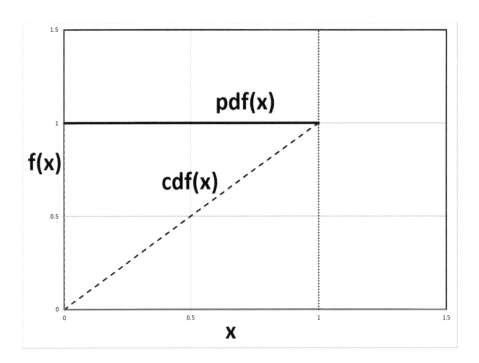

Figure 5.2: Uniform probability density function (pdf). The variable limits are $x = 0$ to $x = 1$ for each function. The probability density function (pdf) is the horizontal line at $y = 1$ and the cumulative density function (cdf) is shown as a dotted sloped line from $y = 0$ to $y = 1$.

finite limits 0 to 1.0 form a constant horizontal line at a particular probability level of 1.0. In this special case, the total area under the probability density function (pdf) is equal to unity which means that the pdf is normalized to unity by area.

Taking the uniform distribution to be a probability density function (pdf) between more general limits [a,b] has some very useful properties. First, the pdf of the random variable x is defined as follows:

$$\text{pdf}(x) = \frac{1}{b-a}, \qquad a \le x \le b \tag{5.1}$$

and it has an associated cumulative distribution function (cdf) given by

$$\text{cdf}(x) = \int_a^x \text{pdf}(x')\,dx' = \frac{1}{b-a} \int_a^x dx' = \frac{x-a}{b-a}$$

$$\tag{5.2}$$

From basic statistics (Chatfield 1975), the expectation value of x is given by

$$\langle x \rangle = \int_a^b x \, \mathrm{pdf}(x) \, \mathrm{d}x = \int_a^b \frac{x}{b-a} \, \mathrm{d}x = \frac{x^2}{2(b-a)} \Big|_a^b = \frac{b+a}{2}$$

and the variance is

$$\sigma_x^2 = \int_a^b (x - \langle x \rangle)^2 \, \mathrm{pdf}(x) \mathrm{d}x$$

$$= \int_a^b x^2 \, \mathrm{pdf}(x) \, \mathrm{d}x - \langle x \rangle^2 \int_a^b \mathrm{pdf}(x) \, \mathrm{d}x$$

$$= \frac{1}{b-a} \int_a^b x^2 \, \mathrm{d}x - \frac{(a+b)^2}{4}$$

$$= \frac{(b-a)^2}{12}$$

Substituting in for $a = 0$ and $b = 1$ for the special case shown in Figure 5.2, we define the random variable $x = \eta$ for $\eta \in [0, 1]$

$$\mathrm{pdf}(\eta) = 1, \qquad 0 \le \eta \le 1 \tag{5.3}$$

with normalized cdf given by

$$\mathrm{cdf}(\eta) = \eta \qquad \text{with } \mathrm{cdf}(0) = 0 \text{ and } \mathrm{cdf}(1) = 1 \tag{5.4}$$

The expectation value and variance are then given by

$$\langle \eta \rangle = \frac{1}{2} \qquad \text{and } \sigma_\eta^2 = \frac{1}{12} \tag{5.5}$$

From now on in this chapter, random numbers drawn from a uniform distribution between the special limits of 0 and 1 are referred to as a uniformly distributed random numbers with the acronym *udrn*.

5.2.4.3 Gaussian Distribution

Another important pdf is the Gaussian function which is bell-shaped with tails extending laterally to $\pm\infty$. This function is often used to add artificial noise to simulated radiation sensors or to study the effect of uncertainties of input parameters, for example. The cdf and pdf for a Gaussian distribution are given by the following expressions:

$$\mathrm{cdf}_{\mathrm{Gauss}}(x) = \int_{-\infty}^x \mathrm{pdf}_{\mathrm{Gauss}}(x') \, \mathrm{d}x'$$

$$\mathrm{pdf}_{\mathrm{Gauss}}(x) = \frac{1}{\sigma\sqrt{2\pi}} \exp \left\{ -\frac{(x-\mu)^2}{2\sigma^2} \right\} \tag{5.6}$$

One derives the mean, μ and variance, σ^2 as follows:

$$\langle x \rangle = \int_{-\infty}^{\infty} x \, \mathrm{pdf}_{\mathrm{Gauss}}(x) \, \mathrm{d}x$$

$$= \int_{-\infty}^{\infty} (x' + \mu) \, \mathrm{pdf}_{\mathrm{Gauss}}(x') \, \mathrm{d}x'$$

$$= \mu \int_{-\infty}^{\infty} \mathrm{pdf}_{\mathrm{Gauss}}(x') \, \mathrm{d}x' + \int_{-\infty}^{\infty} x' \, \mathrm{pdf}_{\mathrm{Gauss}}(x') \, \mathrm{d}x'$$

$$= \mu + \frac{1}{\sigma\sqrt{2\pi}} \int_{-\infty}^{\infty} x' \exp\left\{-\frac{x'^2}{2\sigma^2}\right\} \mathrm{d}x'$$

$$= \mu - \frac{\sigma}{\sqrt{2\pi}} \exp\left\{-\frac{x'^2}{2\sigma^2}\right\}\Bigg|_{-\infty}^{+\infty}$$

$$= \mu$$

$$\int_{-\infty}^{+\infty} (x - \mu)^2 \, \mathrm{pdf}_{\mathrm{Gauss}}(x) \, \mathrm{d}x = \int_{-\infty}^{+\infty} x'^2 \, \mathrm{pdf}_{\mathrm{Gauss}}(x') \, \mathrm{d}x'$$

$$= \frac{1}{\sigma\sqrt{2\pi}} \int_{-\infty}^{\infty} x'^2 \exp\left\{-\frac{x'^2}{2\sigma^2}\right\} \mathrm{d}x'$$

$$= -\frac{\sigma^2}{\sigma\sqrt{2\pi}} x' \exp\left\{-\frac{x'^2}{2\sigma^2}\right\}\Bigg|_{-\infty}^{+\infty} + \frac{\sigma^2}{\sigma\sqrt{2\pi}} \int_{-\infty}^{\infty} \exp\left\{-\frac{x'^2}{2\sigma^2}\right\} \mathrm{d}x'$$

$$= \sigma^2$$

The substitution $x' = x - \mu$ is used for derivations of both parameters. Integration-by-parts was used for the variance integral (Thomas and Finney 1979).

5.2.4.4 Evaluating Integrals

The area under the curve of any one-dimensional function, $f(x)$, between finite limits a and b is expressed as

$$I = \int_a^b f(x) \, \mathrm{d}x \tag{5.7}$$

The simplest Monte Carlo procedure to evaluate this integral begins by drawing random variables from the uniform distribution using Equation 5.2 and setting

$$\mathrm{cdf}(x) = \frac{x - a}{b - a} = \eta \tag{5.8}$$

Inverting in terms of x yields the following:

$$x = a + (b - a)\eta$$

or

$$x_i = a + (b - a)\eta_i \text{ for } i = 1, 2..., N \tag{5.9}$$

Equation 5.7 is intentionally re-written to highlight that this is actually a discrete process of selecting N points. The result is that x is scaled to fit the limits $[a, b]$ using the random variable $\eta \in [0, 1]$. In this simple case of using random sampling to evaluate the integral I, Equation 5.7 is approximated by

$$I \approx \frac{(b-a)}{N} \sum_{i=1,[\text{pdf=udrn}]}^{N} f(x_i) \tag{5.10}$$

where the subscripted term in the summation [pdf=udrn] indicates random numbers from a uniform distribution. We have seen how random sampling works using a uniformly distributed random number to solve a definite integral by obtaining the area under the curve in one dimension. This can be extended to many dimensions as for complicated systems such as the ones used in radiation transport problems. This and other sampling approaches can be used to produce any probability density frequency distribution shape. Using the uniform pdf is very simple but not necessarily the most efficient distribution shape to sample from. The section regarding Importance Sampling (Section 5.2.5.3) discusses this aspect.

5.2.4.5 Simulation Results and Uncertainty

This section serves as a refresher of some important statistical properties for calculating uncertainties in the results of Monte Carlo simulations (Chatfield 1975; Salvat et al. 2008). Given a random variable x which is distributed according to the probability density function pdf(x), the expected value of $f(x)$ over the range $[a, b]$ is given by the standard definition as follows:

$$\langle f(x) \rangle = \int_a^b f(x) \, \text{pdf}(x) \, \mathrm{d}x \tag{5.11}$$

where in general the pdf is a normalized function according to

$$\text{pdf}(x) = \frac{\text{pdf}'(x)}{\int_a^b \text{pdf}'(x) \, \mathrm{d}x}$$

The variance is given by

$$\text{var}[f(x)] = \int_a^b (f(x) - \langle f(x) \rangle)^2 \, \text{pdf}(x) \, \mathrm{d}x$$

$$= \int_a^b f^2(x) \text{pdf}(x) \, \mathrm{d}x - \langle f(x) \rangle^2 \tag{5.12}$$

One can then write any one-dimensional integral into the form

$$I = \int_a^b A(x) \, \mathrm{d}x = \int_a^b f(x) \text{pdf}(x) \, \mathrm{d}x$$

where:

$$f(x) = \frac{A(x)}{\text{pdf}(x)}$$

The area under the curve and expectation value are then interconnected as follows:

$$\langle f(x) \rangle = \int_a^b f(x)\text{pdf}(x)\,\mathrm{d}x = I$$

and the variance using Equation 5.12 can be written as

$$\text{var}[f] = \int_a^b f^2(x)\,\text{pdf}(x)\,\mathrm{d}x - I^2 \tag{5.13}$$

The Monte Carlo estimate of the integral I is given by the sample mean as follows:

$$\bar{f} = \frac{1}{N} \sum_{i=1,\text{pdf}(x_i)}^{N} f(x_i) \tag{5.14}$$

where the random samples of x_i are taken from $\text{pdf}(x_i)$. Provided that the second moment of $f(x)$ exists (or is finite) and then taking the limit as $N \to \infty$ while recalling the central limit theorem in Equation 5.14 (Feller 1967) gives

$$\lim_{N\to\infty} \bar{f} = \langle f \rangle = I$$

with the variance given by

$$\text{var}[f(x)] = \lim_{N\to\infty} \left\{ \frac{1}{N} \sum_{i=1}^{N} [f(x_i)]^2 - \left[\sum_{i=1}^{N} \frac{f(x_i)}{N} \right]^2 \right\} \tag{5.15}$$

Independent Monte Carlo simulations generally produce variable estimates of $\bar{f}$. Thus the estimate $\bar{f}$ is considered itself to be a random variable, with an unknown pdf and its expected value and variance are given by the following (Chatfield 1975; Salvat et al. 2008):

$$\langle \bar{f} \rangle = \langle f \rangle$$

$$\text{var}[\bar{f}] = \frac{1}{N} \text{var}[f(x)]$$

Thus, the standard error of $\bar{f}$, namely σ, is given by

$$\sigma \equiv \sqrt{\text{var}[\bar{f}]} = \sqrt{\frac{\text{var}[f(x)]}{N}} \tag{5.16}$$

providing the statistical uncertainty of the Monte Carlo estimate $\bar{f}$.

Equation 5.16 very importantly sets forth the mechanism for decreasing the uncertainty in Monte Carlo results, keeping in mind the escalation in computational effort that this function would entail. Thus to reduce the statistical uncertainty in a scored quantity by 10-fold would require a 100-fold increase in the sample size N (Salvat et al. 2008). This power law clearly sets a bottleneck to the accuracy that can be attained with limited computer resources. Often radiation transport problems require on the order of 10^6 incident photon histories to achieve dose accuracy of a few % in typically-sized voxels in the patient; a reduction of uncertainty by a factor of 10 requires an increase to 10^8 incident histories.

The pdf of $\bar{f}$ is a *Gaussian distribution* with mean $\langle f(x) \rangle$ and standard deviation σ in the limit $N \to \infty$ and noting that the central limit theorem (James 1980) ensures that the summation will converge to a stable and correct expectation value.

$$\mathrm{pdf}(\bar{f}) = \frac{1}{\sigma \sqrt{2\pi}} \exp \left\{ -\frac{(\bar{f} - \langle f \rangle)^2}{2\sigma^2} \right\}$$

Then the result of a scored quantity in a Monte Carlo simulation may be represented by $\bar{f} \pm$ an uncertainty based on, σ, to envelope the true expectation value, $\langle f(x) \rangle$. Uncertainty reporting varies with different MC software packages use a range from 1σ, 2σ, or even 3σ, as in PENELOPE, which yields confidence intervals of 68.3%, 95.4%, 99.7%, respectively (Salvat et al. 2008).

It is important to compare error bounds between MCMs and deterministic numerical quadrature techniques. Scored quantities or tallies in Monte Carlo simulations are average values with an uncertainty window. This is different from other dose calculation methods that yield the expectation values directly without statistical uncertainty but with potential inaccuracy due to discretization. Quadrature techniques such as the trapezoidal rule (TR) and Simpson's rule (SR) (Thomas and Finney 1979) for numerically evaluating definite integrals in multi-dimensional space, namely s, are a good example. The discretization results in small step sizes that are called nodes. Equation 5.17 is a discretized numerical scheme for solving the definite one-dimensional integral of $f(x)$. Picture the curve $f(x)$ with the area of one panel (trapezoid) being the width (or step) $\Delta x = (x_{m+1} - x_0)/m$ multiplied by the average of the lengths, $f(x_i)$ and $f(x_i+1)$, at 2 ends of a node pair. The total area is then given by the sum of all these individual panels. The closer the spacing (i.e. smaller panel strips), the more nodes there are over the range of integration and the more accurate is the result. The node sizes are determined by m, the number of subdivisions within the integration region along one of the dimensions. Provided that the second derivative of the integrand is continuous, the error terms are given by

$$\int_{x_0}^{x_{m+1}} f(x)\,\mathrm{d}x = \sum_{i=1}^{m+1} \frac{(f(x_{i-1} + f(x_i))}{2} + O\left(\Delta x^2\right) \tag{5.17}$$

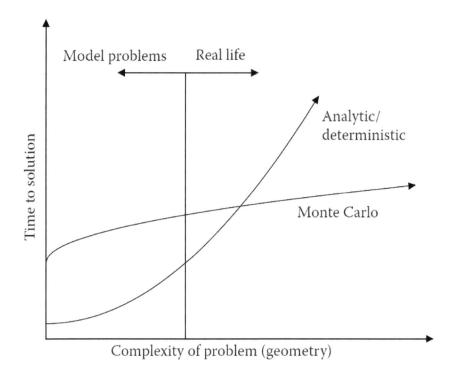

Figure 5.3: Comparison of computational efficiency of stochastic and deterministic methods. Reproduced with permission (Bielajew 2013).

Note that the error estimate is on the order of Δx^2. The error terms for these basic integration techniques are of second order and fourth order, respectively, as follows:

$$O(\Delta x^2) \approx O(m^{-2}) \text{ for TR and } O(\Delta x^4) \approx O(m^{-4}) \text{ for SR}$$

The number of nodes in one dimension is given by $N = m + 1$ and in multiple dimensions, s, that error becomes $N = m + 1)^s$. In terms of the number of nodes, N, the error terms in s-dimensions are given by

$$O(m^{-2}) = O(N^{-2/s}) \text{ for TR and } O(m^{-4}) = O(N^{-4/s}) \text{ for SR}$$

For an absolute error of the order 10^{-2} for TR or 10^{-4} for SR, the number of nodes required is on the order of 10^s. This computational penalty is referred to as the "curse of dimensionality" (Niederreiter 1992). Comparing this to MCMs where the uncertainty is determined by $O(N^{-1/2})$, it is clear that MCMs have an advantage over second order quadratures when $s > 4$ and $s > 8$ for fourth order schemes. In brief, results of MCMs are less dependent on dimensionality of the problem. A more rigorous treatment of computational efficiency is detailed by Bielajew (Bielajew 2013; Borgers 1998). In Figure 5.3, the computational burden is shown as a function of dimensional complexity for each method.

5.2.5 Sampling of Interaction Events

5.2.5.1 Direct Inversion Method

The direct method of sampling is based on integrating the pdf analytically or with a numerical scheme, equating the result to a uniformly distributed random number (udrn), and then inverting the expression to select the random variable. The formal concept is depicted as follows (Salvat et al. 2008). A cumulative distribution function(cdf), $\mathrm{cdf}(x)$, of the probability density function (pdf), $\mathrm{pdf}(x)$, is a monotonically increasing function of x between finite limits [a,b]. We write

$$\mathrm{cdf}(x) = \int_a^x \mathrm{pdf}(\alpha)\, \mathrm{d}\alpha, \qquad a \le x \le b$$

The cumulative function is normalized (i.e. area under the curve is unity) as follows:

$$\mathrm{cdf}(b) = \int_a^b \mathrm{pdf}(x)\, \mathrm{d}x = 1$$

$$0 \le \mathrm{cdf}(x) \le 1 \qquad \text{for } a \le x \le b$$

Defining a new random variable, η, by the following transformation and limits:

$$\eta = \mathrm{cdf}(x) \qquad a \le x \le b \qquad \text{for } 0 \le \eta \le 1$$

and applying the fundamental theorem of calculus (Thomas and Finney 1979)

$$\frac{\mathrm{d}\eta}{\mathrm{d}x} = \frac{\mathrm{d}}{\mathrm{d}x}[\mathrm{cdf}(x)] = \frac{\mathrm{d}}{\mathrm{d}x}\int_a^x \mathrm{pdf}(\alpha)\, \mathrm{d}\alpha = \mathrm{pdf}(x)$$

$$\therefore \ \mathrm{d}\eta = \mathrm{pdf}(x)\, \mathrm{d}x$$

The random variable x corresponds to η together with the pdf of each random variable, namely, $\mathrm{pdf}(x)$ and $\mathrm{pdf}_\eta(\eta)$ respectively and it defines the following correspondence:

$$\mathrm{pdf}_\eta(\eta)\, \mathrm{d}\eta = \mathrm{pdf}(x)\, \mathrm{d}x = \ \mathrm{d}\eta$$

again using the fundamental theorem result. Thus,

$$\mathrm{pdf}_\eta(\eta) = 1$$

The above simply reflects the fact that all values of random variable η between the limits of $0 \le \eta \le 1$ are equally likely. Now we stipulate that the inverse of $\mathrm{cdf}(x)$, $\mathrm{cdf}^{-1}(\eta)$ exists uniquely so that we can write

$$\eta = \mathrm{cdf}(x) \text{ and then } x = \mathrm{cdf}^{-1}(\eta)$$

Thus for every value, $\eta \in [0, 1]$ we obtain a corresponding selected random variable $x \in [a, b]$ that is distributed according to the differential $\mathrm{pdf}(x)$ (Salvat et al. 2008). Then by sampling, η_i, a udrn between [0,1] and utilizing cdf^{-1}, we can write

$$x_i = \mathrm{cdf}^{-1}(\eta_i): \ for: \ i = 1, 2, 3, ..., N$$

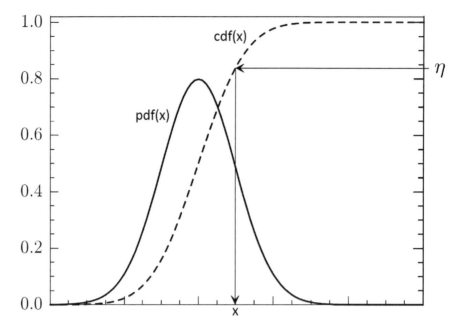

Figure 5.4: Example of a cdf that is inverted to select a value of x for each randomly drawn value of η. The slope of the cdf(x) bunches up more values in the region where the slope of cdf(x) is steeper. The resultant distribution of x values is pdf(x).

The x_i are members of the distribution pdf(x_i) and the summation is entirely over values of x sampled from the distribution pdf(x). This is a direct method of inversion and hence coined the direct, inverse, or transformation sampling method in one dimension. This method relies on the ability to integrate the pdf(x) and invert the resulting cdf(x) function (see Figure 5.4) using a *single* random number call.

A simple numerical example of using the direct method is demonstrated with the linear distribution function $f(x) = \frac{2x}{99}$ for $x \in [1, 10]$. Given this normalized definite integral in one dimension,

$$I = \int_1^{10} f(x)\,dx = \int_1^{10} \frac{2x}{99}\,dx = 1 \tag{5.18}$$

one easily defines the cdf

$$\text{cdf}(x) = \int_1^x \frac{2x'}{99}\,dx' = \frac{x^2 - 1}{99} \qquad \text{with cdf}(10) = 1$$

Now draw a udrn $\eta \in [0, 1]$ and equate it to the cdf as follows:

$$\eta_i = \text{cdf}(x_i) = \frac{x_i^2 - 1}{99} \quad \text{for } i = 1, 2, 3, ...N$$

Inverting for x_i gives

$$x_i = \sqrt{99\eta_i + 1} \qquad (5.19)$$

For large N, one builds up the linear distribution of $f(x) = \text{pdf}(x) = 2x/99$. We already know the solution to this definite integral without using Monte Carlo simulation and can verify the convergence to the true value. Pretending the solution is not known analytically, the *integral* (5.18) can be solved simply using a uniform distribution. By drawing a udrn, $\eta \in [0, 1]$ and noting that in this case the pdf$(x_i) = 1/9$, we obtain the following:

$$\eta = \int_1^x \frac{1}{9} \, dx' = \frac{x - 1}{9}$$

and after inversion, we obtain

$$x_i = 9\eta_i + 1$$

The MC estimate of the integral 5.18 is given by

$$\frac{1}{N} \sum_{i=1, [\text{pdf}(x_i)=1/9]}^{N} \frac{\frac{2x_i}{99}}{\left(\frac{1}{9}\right)} = \frac{2}{11N} \sum_{i=1, [\text{pdf}(\eta_i)=1]}^{N} (9\eta_i + 1)$$

In the limit as $N \to \infty$,

$$\lim_{N \to \infty} \frac{2}{11} \sum_{i=1, \text{pdf}=1}^{N} \left[\frac{9\eta_i}{N} + \frac{1}{N} \right] = \frac{2}{11} \left[\frac{9}{2} + 1 \right] = 1 = I$$

noting that

$$\lim_{N \to \infty} \frac{1}{N} \sum_{i=1, \text{pdf}=1}^{N} \eta_i = \frac{1}{2}$$

This result follows from the characteristics of a uniform distribution given by Equation 5.5. Also note the change in the pdf index in the lower limit of the summation.

A more relevant practical example of direct sampling used in dose computations describes the uniform selection of the azimuthal angle ϕ for Compton scattering of unpolarized photons which is axisymmetric. Given that the range is $\phi \in [0, 2\pi]$, drawing the udrn $\xi \in [0, 1]$, the normalized uniform distribution of angles is given intuitively by rescaling the random numbers to span the angular limits, as shown more formally below:

$$\xi = \text{cdf}(\phi) = \frac{1}{2\pi} \int_0^\phi d\phi' = \frac{\phi}{2\pi}$$
$$\text{giving } \phi = 2\pi\xi \qquad (5.20)$$

The direct method is also applicable to discretized functions in which case a pdf is essentially a histogram or a tabulated function, such as an energy spectrum. The example presented below uses tabulated values of a Gaussian pdf and direct inversion to solve for the random variable using a numerical technique. The Gaussian is already a normalized pdf so that a udrn $\eta \in [0, 1]$ is drawn and equated to

$$\eta = \mathrm{cdf}_{\mathrm{Gauss}}(x) = \frac{1}{\sigma\sqrt{2\pi}} \int_{-\infty}^{x} \exp\left\{-\frac{(x' - \mu)^2}{2\sigma^2}\right\} \, \mathrm{d}x'$$

Then a root-finding approach such as the Newton-Raphson method is applied to the following equation:

$$f(x) = 0 \tag{5.21}$$

where:

$$f(x) = \mathrm{cdf}_{\mathrm{Gauss}}(x) - \eta_i = \frac{1}{\sigma\sqrt{2\pi}} \int_{-\infty}^{x} \exp\left\{-\frac{(x' - \mu)^2}{2\sigma^2}\right\} \, \mathrm{d}x' - \eta = 0 \tag{5.22}$$

$$f'(x) = \frac{\mathrm{d}f(x)}{\mathrm{d}x} = \frac{1}{\sigma\sqrt{2\pi}} \exp\left\{-\frac{(x - \mu)^2}{2\sigma^2}\right\}$$

The root, x, of Equation 5.21 or more explicitly Equation 5.22 is found iteratively applying the following equation for each sampled η:

$$x_{n+1} = x_n - \frac{f(x_n)}{f'(x_n)}$$

with initial estimate x_0. The solution x is obtained for each η when convergence is reached between the difference of iterates, i.e. the following convergence criterion is satisfied:

$$|x_{n+1} - x_n| \leq \text{ user defined tolerance value}$$

This technique can be costly in computational resources for a large number of iterations, but it expands the applicability of the inversion method to discretized functions, when analytical expressions of the $\mathrm{cdf}(x)$ are not available.

5.2.5.2 Rejection Method

The rejection method, attributed to Von Neumann, can be used as an alternative when the inverse of the cdf is difficult to obtain in analytical closed form (Hammersley and Handscomb 1964). Another density function $q(x)$ on interval $[a, b]$ will be used to compare with the pdf. It is chosen such that it is equipped with the following properties: (1) easy to sample and possibly using the direct method (2) similar in shape to $\mathrm{pdf}(x)$ and (3) $k \geq \mathrm{pdf}(x)/q(x)$ where $k > 1$ is a positive bounding constant. In fact the closer that $q(x)$ mimics $\mathrm{pdf}(x)$, the more efficient is the sampling process because it leads to fewer rejections of drawn random numbers. The graphical representation is shown in Figure 5.5. The procedure is as follows:

Step 1. Begin by drawing a sample of the udrn $\eta_i \in [0, 1]$:

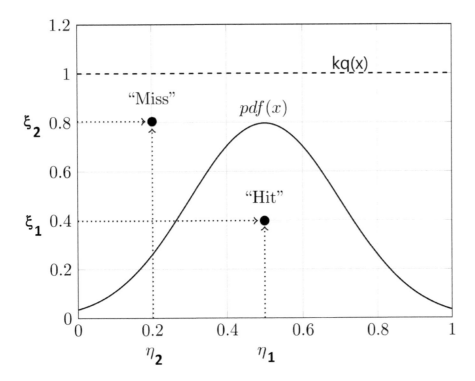

Figure 5.5: Use of rejection sampling. Perform random sampling directly using $q(x)$ to obtain x-coordinate. Test with udrn ξ_i which is the y-coordinate. Use hit or miss coordinates to build up probability density pdf(x). Note that $kq(x) \geq \text{pdf}(x)$.

Step 2.

$$\eta_i = \int_a^{x_i} q(x)\, dx \tag{5.23}$$

Step 3. Solve and invert for x_i in terms of η_i from Equation 5.23.
Step 4. Draw another udrn, ξ_i uniformly distributed on $[0, 1]$.
Step 5. Perform the following rejection test:

$$\text{If } \xi_i < \frac{\text{pdf}(x_i)}{kq(x_i)}$$

Accept η_i and thus x_i. Otherwise reject and go back to Step 1.

The successful x_i values that are accepted will populate the density function pdf(x). Ideally, pdf(x) on $[a, b]$, is normalized to pdf$'(x)$ such that pdf$'(x) = \text{pdf}(x)/\text{pdf}_{\max}(x)$ making the maximum value of pdf$'(x) = 1$, in line with the standard range of random numbers.

A simple numerical example is given of the uniform distribution $q(x) = 1/(b-a)$, and the functions $\text{pdf}(x)$ and $k = \text{pdf}_{max}(x)\,(b-a)$:

$$pdf(x) = \frac{1}{3} + x^2 \qquad x \in [0,3]$$

and

$$q(x) = \frac{1}{3} \qquad x \in [0,3] \text{ and } k = 28$$

Step 1. Draw udrn, $\eta_i \in [0,1]$.
Step 2 and Step 3. This gives $\eta_i = x_i/3$ and invert to $x_i = 3\eta_i$.
Step 4. Draw udrn $\xi \in [0,1]$.
Step 5. Test rejection with the following condition:

$$\text{If } \xi < \frac{\frac{1}{3} + x_i^2}{(28)(\frac{1}{3})} \left\{ = \frac{\text{pdf}(x_i)}{kq(x_i)} \right\}$$

Accept η_i and thus x_i. Otherwise reject and go back to Step 1.

In general, if the values of the $\text{pdf}(x)$ are highly modulated over an enormous range, then this process becomes very inefficient because many random draws will be rejected in the low probability regions. Using importance sampling (Section 5.2.5.3) is one way to remedy this issue. The overall efficiency for this technique is given by

$$\epsilon = \int_a^b \text{pdf}(x)\,\mathrm{d}x = \int_a^b \frac{\text{pdf}(x)}{kq(x)}q(x)\,\mathrm{d}x = \frac{1}{k}$$

This technique is frequently used in MCMs more for convenience rather than efficiency since it is *less efficient* than the direct method that only requires a single random number draw. It is important to note that in practice, the direct method, however, cannot be used because many pdfs are not analytically integrable. If analytic solutions do exist, they may not be easily invertible or require too many time-consuming function evaluations. However, in some cases discrete forms such as histograms or lookup tables can be used to avoid these slow evaluations. It is worth noting that in the early implementation of electronic MCMs, the rejection technique was almost abandoned because of poor efficiency (Hammersley and Handscomb 1964). Now, with faster computers, this issue has dissipated somewhat and it offers an advantage of simple trouble-shooting of suspicious events. Often these sampling techniques are combined with other methods that help boost the efficiency as is the case for variance reduction techniques (see Section 5.3.5).

We now demonstrate a clever two-part rejection technique applied to a linear function

$$f(x) = x \text{ with } x \in [0,1]. \tag{5.24}$$

This technique is used in part for the random sampling applied to the Klein-Nishina cross section for Compton scattering (Nelson et al. 1985). The reason for using it is related to its improved efficiency over the direct method that one would expect to use. Recall the previous example of Equation 5.18. Applying the substitution

$$x' = \frac{x-1}{10-1} \, ,$$

$$I = \int_1^{10} \frac{2x}{99} \, dx = \frac{2}{11} \left\{ 9 \int_0^1 x' \, dx' + 1 \right\}$$

leaves a simple linear function x in the integrand. The direct method of sampling a linear function was already presented using the example of $\text{pdf}(x) = 2x/99$ and Equation 5.19 resulted in the evaluation of a square root of the udrn.

To elaborate on this special technique without losing generality, we use a simple linear function $f(x) = x$ over the range $x \in [0,1]$. Using the $\text{pdf}(x) = 1$, on a simpler integral I we have

$$I = \int_0^1 f(x) \, dx = \int_0^1 \frac{f(x)}{\text{pdf}(x)} \text{pdf}(x) \, dx = \int_0^1 \frac{(x)}{1} (1) \, dx$$

The direct sampling method is applied to $\text{pdf}(x) = 1$ and the rejection part is used on $f(x) = x$ as the rejection border. Therefore, draw the udrn $\eta \in [0,1]$ and equate and invert as follows:

$$\eta = \int_0^x dx' = x \qquad \eta, x \in [0,1]$$

Draw another udrn $\xi \in [0,1]$ and test. If $\xi > f(x) = x = \eta$ then reject x and start again at the beginning by sampling η. Otherwise accept x.

In the above rejection step, normally, one would continue to re-sample the udrn η and continue until the above rejection test is false so that finally the value $x = \eta$ is accepted. The computer could churn through many udrn's, both η and ξ, until settling on a value. This is not considered to be efficient. One realizes a trick based on symmetry in the linear function x to make this process more effective. In this specific case, each udrn (η or ξ in this case) is drawn from the same standard random sampling $\text{pdf}(x) = 1$ yielding the $\text{cdf}(x) = x$, which is also equal to the rejection function $f(x) = x$. This fact is the basis for this trick. The two udrn's that are ultimately drawn are ultimately equated (cdf) or compared ($f(x)$) to the same linear function x indicating the symmetry. Furthermore, another important feature to realize is that the udrn's are each independent so that the order in which they are drawn does not matter. To further elaborate on this special case, consider the two scenarios below:

Rejection Scenario 1

1. Sample $\eta \in [0, 1]$
2. Then $\eta = x$
3. Sample $\xi \in [0, 1]$
4. Test if $\xi > f(x) = x = \eta$ from step 2
5. if NOT true accept $\eta = x$ i.e. realize that $\xi < \eta$

Since both udrn's η and ξ are independent and their order of drawing does not matter, then the above routine can be switched to:

Rejection Scenario 2

1. Sample $\xi \in [0, 1]$
2. Then $\xi = x$
3. Sample $\eta \in [0, 1]$
4. Test if $\eta > f(x) = x = \xi$ from step 2
5. if NOT true accept $\xi = x$ i.e. realize that $\eta < \xi$

Thus in Rejection Scenario 1, η is drawn and then ξ and if $\xi \leq \eta$ one accepts $x = \eta$. In Rejection Scenario 2, ξ is drawn and then η and if $\eta \leq \xi$ one accepts $x = \xi$. What this means is that one can draw a pair of udrn's, i.e. $\eta \in [0, 1]$ and $\xi \in [0, 1]$ and take the *maximum* of the two values as the accepted $x-$value. In other words,

$$x = max(\eta_i, \xi_i) \tag{5.25}$$

Only two udrn's need to be drawn! Note that drawing 2 udrns in this way and taking the maximum of them is more efficient computationally than drawing one udrn and evaluating a square root expression.

5.2.5.3 Importance Sampling

Importance sampling is a technique that uses a surrogate pdf and it can be used for reducing the variance in scored quantities. It has some similarity with rejection methods, with regards to the shape of a surrogate pdf function. Consider in general such a function, $h(x)$. Then,

$$F(x) = \int_a^x f(x') \, dx' = \int_a^x \frac{f(x')}{h(x')} h(x') \, dx' \qquad h(x') \neq 0 \tag{5.26}$$

Under optimal circumstances, random sampling performed using pdf= $h(x)$ instead of pdf= $f(x)$ can lead to a more efficient sampling sequence. The optimal qualities for such an $h(x)$ are that it should closely resemble $f(x)$ in areas where $f(x)$ is most relevant to the problem at hand. It is less complex, more convenient, and more efficient to use. Frequently, it is difficult to find an $h(x)$ that will satisfy all of these optimal criteria but when it is possible, there are clear gains in efficiency.

Firstly draw a random sample from the pdf $h(x)$ and then construct the distribution for $f(x)$ by using $f(x)/h(x)$ instead of $f(x)$. Thus, there is compensation for the fact that we are using the surrogate pdf, $h(x)$, for random sampling. To clarify, consider the definite integral

$$I = \int_0^1 f(x)\,\mathrm{d}x = \int_0^1 \frac{f(x)}{h(x)} h(x)\,\mathrm{d}x \tag{5.27}$$

Recognize the expectation value as follows:

$$\left\langle \frac{f}{h} \right\rangle = \int_0^1 \frac{f(x)}{h(x)} h(x)\,\mathrm{d}x \tag{5.28}$$

Applying the central limit theorem to the Monte Carlo estimate of (5.27) or (5.28) for pdf $= h(x) = 1$, the solution is given by

$$\int_0^1 f(x)\,\mathrm{d}x = \lim_{N\to\infty} \left\{ \frac{1}{N} \sum_{i=1,[\text{pdf}=h(x)=1]}^{N} f(x_i) \right\} = I \tag{5.29}$$

and for a general $h(x)$

$$\int_0^1 f(x)\,\mathrm{d}x = \lim_{N\to\infty} \left\{ \frac{1}{N} \sum_{i=1,[\text{pdf}=h(x)]}^{N} \frac{f(x_i)}{h(x_i)} \right\} = I \tag{5.30}$$

Thus in the limit for large N, the expectation value approaches I, regardless of which pdf is used.

For a generalized $h(x)$, consider the variance given by

$$\mathrm{var}\left[\frac{f(x)}{h(x)} \right] = \int_0^1 \left(\frac{f(x)}{h(x)} - I \right)^2 h(x)\,\mathrm{d}x = \int_0^1 \left(\frac{f^2(x)}{h^2(x)} - I^2 \right) h(x)\,\mathrm{d}x$$

$$= \int_0^1 \left(\frac{f^2(x)}{h^2(x)} \right) h(x)\,\mathrm{d}x - I^2 \tag{5.31}$$

In the limit as $N \to \infty$ the variance becomes

$$\mathrm{var}\left[\frac{f(x)}{h(x)} \right] = \lim_{N\to\infty} \left\{ \frac{1}{N} \sum_{i=1,[h(x)]}^{N} \frac{f^2(x_i)}{h^2(x)} - I^2 \right\}$$

It is clear that the choice of $h(x)$ can effect the variance greatly (Salvat et al. 2008), whereas, the expectation value (5.29 and 5.30) is unaffected in the limit.

Consider a simple numerical example of reducing the variance, as follows:

$$I = 3 \int_0^1 x^2 \, dx \text{ with } f(x) = 3x^2$$

It is easy to verify that the solution is $I = 1$. Reformulate this integral to demonstrate importance sampling for two different normalized pdf functions $h(x) = 1$ and $h(x) = 2x$, as follows:

$$I = \int_0^1 f(x) \, dx = \int_0^1 \frac{f(x)}{h(x)} h(x) \, dx$$

Draw a random sample from the uniform distribution, $h(x) = 1$. The variance is given by

$$\sigma^2_{h=1} = \int_0^1 (f^2(x) - 1)^2 \, dx = \int_0^1 (3x^2 - 1)^2 \, dx = \frac{4}{5}$$

On the other hand drawing a random sample from $pdf = h(x) = 2x$ with

$$H(x) = \int_0^x h(u)du = 2 \int_0^x x \, dx = x^2 \text{ noting that } H(1) = 1$$

the variance is given by

$$\sigma^2_{h=2x} = \int_0^1 \left(\frac{f(x)}{h(x)} - I \right)^2 dH(x) = \int_0^1 \left(\frac{3x^2}{2x} - 1 \right)^2 (2x) \, dx = \frac{1}{8}$$

We have demonstrated that the mean value is the same in the limit as $N \to \infty$. However, from above it is clear that the variances are quite different. In this case, $h(x) = 2x$ appears to remove more of the fluctuation away from $f(x) = 3x^2$ over the range $0 \leq x \leq 1$ than does $h(x) = 1$ and the variance appears to be dramatically reduced. The variance ratio in the limit indicates a reduction by a factor of 6.4, i.e.:

$$\frac{\sigma^2_{h=1}}{\sigma^2_{h=2x}} = \frac{\frac{4}{5}}{\frac{1}{8}}$$

For a balanced perspective on the efficiency aspect, however, it is also important to realize that computationally the inversion of $h(x) = 2x$ requires evaluating a square root of the udrn in calculating I, which is less efficient than the case for $h(x) = 1$ which involves calculating the square of the udrn.

5.2.5.4 Composite Sampling Method

The mixed method is used heavily in Monte Carlo simulations of radiation and uses partitioning of the pdf into different regions. The technique involves partitioning of the pdf for discrete sampling and then continuing with the direct method and rejection methods (Butcher 1961; Hammersley and Handscomb 1964).

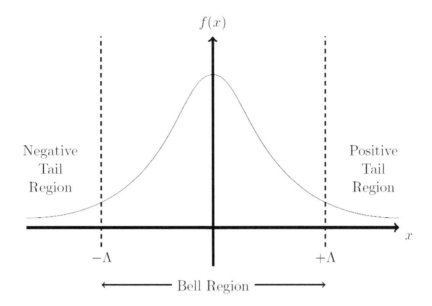

Figure 5.6: Gaussian pdf $= f(x)$ represented by two sections for the random sampling technique (Butcher 1961). Λ separates the two tails from the bell region.

It is worth focusing on a familiar example and different approach to sampling a Gaussian pdf. Most references cite the elegant Box-Muller method (Box and Muller 1958; Muller 1959) for sampling utilizing polar coordinates. In this case, the Gaussian pdf (see Equation 5.6) with mean $\mu = 0$ and standard deviation $\sigma = 1$ is partitioned into 2 parts: the middle region of the Gaussian or the bell portion, and the symmetric tail sections one negative and one positive, as in Figure 5.6. For k partitions one uses the following general formulae:

$$\text{pdf}_{\text{Gauss}}(x) = \sum_{i=1}^{k} \alpha_i g_i(x) h_i(x)$$

$$g_i(x) \leq 1$$

$$\int_{limit_1}^{limit_2} h_i(x) \, \mathrm{d}x = 1$$

The constants, α_i, are used in a random selection process to determine which partition, i, is chosen in a manner similar to using branching ratios for selecting x-ray interactions. The parameter p_i forms the following branching ratios:

$$p_i = \frac{\sum_{j=1}^{i} \alpha_j}{\sum_{j=1}^{k} \alpha_j}$$

For this case, where $k = 2$, there are two branches to select from. Then $h_i(x)$ is sampled directly to obtain x followed by the rejection technique applying the $g_i(x)$ to accept or reject x. In this example (Hammersley and Handscomb 1964; Butcher 1961), partition 1 is given by

$$\alpha_1 = \Lambda\sqrt{\frac{2}{\pi}} \qquad g_1(x) = \exp\left[-\frac{x^2}{2}\right] \qquad h_1(x) = \frac{1}{2\Lambda} \qquad |x| < \Lambda$$

and partition 2 is given by

$$\alpha_2 = \frac{1}{\lambda}\sqrt{\frac{2}{\pi}}\exp\left[\frac{\lambda^2}{2} - \lambda\Lambda\right] \qquad g_2(x) = \exp\left[-\frac{1}{2}[|x| - \lambda]^2\right]$$

$$h_2(x) = \frac{\lambda}{2}\exp\left[-\lambda[|x| - \Lambda]\right] \qquad |x| > \Lambda$$

Randomly draw udrn $\eta_1 \in [0, 1]$ to select a partition.

If $\eta_1 < \dfrac{\alpha_1}{\alpha_1 + \alpha_2}$ is true then go to partition 1, *or* $\qquad\qquad$ (5.32)

If $\dfrac{\alpha_1}{\alpha_1 + \alpha_2} < \eta_1 < 1$ is true then go to partition 2 $\qquad\qquad$ (5.33)

For partition 1, use the direct method and draw another udrn $\eta_2 \in [0, 1]$. Then equate and invert:

$$\eta_2 = \frac{1}{2\Lambda}\int_{-\Lambda}^{x} dx' = \frac{x + \Lambda}{2\Lambda} \text{ so that } x = 2\Lambda\eta_2 - \Lambda \qquad\qquad (5.34)$$

Then draw another udrn $\eta_3 \in [0, 1]$ and test using rejection function, $g_1(x)$.

$$\text{If } \eta_3 < \exp\left[-\frac{x^2}{2}\right]$$

then accept x of Equation 5.34. Otherwise go back to the very beginning and randomly re-select η_1 and reuse Equations 5.32 and 5.33 to determine the main partition.

Partition 2 involves the tails of the function which are clearly separated. Care is taken for selection and this will require yet another partition, i.e., the negative tail is partition 2a and the positive tail is partition 2b. The range $(-\infty, -\Lambda]$ is for partition 2a and note that in this region we have

$$\frac{\lambda}{2}\int_{-\infty}^{x}\exp\left[-\lambda(-x' - \Lambda)\right]dx' = \frac{\lambda}{2}\int_{-x}^{\infty}\exp\left[-\lambda(x' - \Lambda)\right]dx'$$

$$= \frac{1}{2}\exp\left[-\lambda(-x - \Lambda)\right]$$

At $x = -\Lambda$,

$$\frac{\lambda}{2} \int_{-\infty}^{-\Lambda} \exp\left[-\lambda(-x' - \Lambda)\right] \, \mathrm{d}x' = \frac{1}{2}$$

On the other hand, the range for partition 2b is $[\Lambda, \infty)$ and note that in this region we have

$$\frac{\lambda}{2} \int_{-\infty}^{-\Lambda} \exp\left[-\lambda(-x' - \Lambda)\right] \, \mathrm{d}x' + \frac{\lambda}{2} \int_{\Lambda}^{x} \exp\left[-\lambda(x' - \Lambda)\right] \, \mathrm{d}x'$$

$$= 1 - \frac{1}{2} \exp\left[-\lambda(x - \Lambda)\right] \tag{5.35}$$

At $x = \Lambda$, Equation 5.35 equals $1/2$ and as $x \to +\infty$ it approaches 1.

With all of this in mind, for partition 2, draw a udrn $\eta_2 \in [0, 1]$ and if $\eta_2 \leq 1/2$ then the negative tail region partition 2a is selected and apply the direct method and inversion to give

$$\eta_2 = \frac{1}{2} \exp\left[-\lambda(-x - \Lambda)\right] \qquad \text{so that } x = \frac{1}{\lambda} \ln[2\eta_2] - \Lambda \tag{5.36}$$

Note that x is taking on negative values for $\eta_2 \in [0, 1/2]$. Otherwise if $1/2 < \eta_2 \leq 1$, then the positive tail region partition 2b is selected. Here, use the direct method and inversion to obtain

$$\eta_2 = 1 - \frac{1}{2} \exp\left[-\lambda(x - \Lambda)\right] \text{ so that } x = \Lambda - \frac{1}{\lambda} \ln\left[2(1 - \eta_2)\right] \tag{5.37}$$

and note that positive values of $x \in [\Lambda, \infty)$ for $\eta_2 \in [1/2, 1]$ are acquired.

Finally for this partition draw another udrn $\eta_3 \in [0, 1]$ and test using rejection function, $g_2(x)$.

$$\text{If } \eta_3 < \exp\left[-\frac{1}{2}\left[|x| - \lambda\right]^2\right]$$

then accept x for partition 2a in Equation 5.36 or partition 2b in Equation 5.37 noting whether x is from the negative region or the positive region. Otherwise go back to the very beginning and randomly re-select η_1 and reuse Equations 5.32 and 5.33 to determine the main partition.

5.3 MONTE CARLO METHOD IN PRACTICE

All particle transport calculations are attempts at solving the linearized Boltzmann Equation either explicitly or implicitly. This equation is an energy-balancing equation, based in interaction cross sections, in phase space as discussed in Chapter 3 and used extensively in Chapter 6. Solutions to this equation are difficult even under the best of circumstances such as in simplified geometries or with untenable assumptions. There is overwhelming difficulty in obtaining analytical solutions especially when realistic geometries and inhomogeneous tissue distributions are present. The net increment of particle counts in phase space, defined by 3D location and particle status, is given by the sum of all particle-production mechanisms, less all destructive mechanisms. The production terms correspond to particles that arise from explicit radiation sources, radioactive decay, or from production of secondary particles scattered *inwardly* to match the phase space bin attributes (location, status). Destructive terms correspond to particles that are absorbed, decay away, or scatter *outwardly* into another phase space bin with different attributes. Monte Carlo methods keep track of all these elements as individual (or grouped) interactions occur. This can best understood by adopting the analogy of riding the particle across sequential steps of a simulation procedure.

5.3.1 Particle Tracking in Phase Space

The concept of a multi-dimensional phase space was introduced in Chapter 3. For Monte Carlo methods, the phase space coordinates in most Monte Carlo codes consist of particle position $r(x, y, z)$, energy E, and direction cosines $(sin(\theta)cos(\phi), sin(\theta)sin(\phi), cos(\theta))$ where θ and ϕ are polar and azimuthal angles, respectively. The Monte Carlo method tracks particles in phase space, including real space, as particle interactions develop. By comparison, deterministic methods treat particles as cohorts of a radiation field that spread energy. This is the case for convolution-superposition algorithms and Boltzmann equation solvers. Random sampling techniques model the free path distance between interaction site and select the types of interactions that can cause absorption, energy losses, directional changes, with spawning of secondary particles. As interactions occur, the phase parameters for the incoming and outgoing particles are updated. The simulation continues until all the particles are either stopped or exit the volume of interest. This occurs, for example, when the particle energy falls below a predetermined energy cutoff value, where it is assumed to be fully absorbed in the medium.

In order to account for all of the particles during the simulation Monte Carlo programs utilize a storage stack with a stack pointer. The particle stack is a storage memory that holds the phase space parameters of particles generated during the simulation so that they can be processed later in sequence. At any point during the simulation the future of any particle is independent of its previous history making this stack procedure valid. The stack is essential because for each interaction it is possible to create one or more additional particles and one must store the newborn phase space parameters as well as revise data for pre-existing particles involved in the transaction. There is a stack pointer for each particle being simulated starting in ascending order of kinetic energy and the particle with the lowest kinetic energy will be processed first. This emptying scheme helps prevent a stack overflow with too many particles stored at once (Nelson et al. 1985). The stack pointer parameter continues to move to the next lowest energy particle for its turn to be simulated until the stack is emptied and the simulation is ready for a fresh history or the simulation has been fully terminated. This entire simulation process is repeated N times in other words for N histories. It is important to note that each particle is not only tracked for phase space parameters, but also other important features such as the particle charge, an identifier for the current region of travel, statistical weight for variance reduction purposes and other user defined parameters to aid in scoring and tracking. Monte Carlo simulations can thus identify regions where particles originate and this unique feature greatly assists in flagging important regions of a simulation or trouble-shooting program bugs.

Simulation packages must prepare the materials database for all of the media that will be used in the simulation. Sometimes these come in the form of formulae describing interaction cross sections or they are generated beforehand in the form of lookup tables that will be accessed during the simulation, such as in the preparatory program Preprocessor Electron Gamma Shower (PEGS) (Nelson et al. 1985). The material data consist of all possible interaction cross sections, stopping power data for charged particles and associated ranges (Chapter 2).

The start of the simulation initializes the transport parameters such as energy thresholds and cutoffs, energies, particle phase space bin counts, details of the geometry, number of histories N, and scoring values. A phase space file can be initiated empty or pre-filled with output data from a previous simulation (e.g. accelerator head model). This information includes spectra of energy, direction, position and statistical weighting. For this type of initial phase space file, the initialization includes, for example, random sampling of the x-ray spectrum as a probability density function. A general flow diagram for the simulation of an photon-electron cascade is shown in Figure 5.7. From left to right, the process starts by reading in the necessary input interaction data for the materials involved and setting fixed transport parameters which define the energy at which the simulation will be stopped (e.g. cut-off energy E_{cut} and P_{cut} for electrons and photons, respectively), threshold energies for the creation of newborn particles ($E_{thresh} = $ AE or AP for delta ray and bremsstrahlung production), along with details of the geometry. Then the simulation proceeds, randomly selecting the possible trajectories and interactions of the incident particle(s) and updating the phase space data and stack pointer for all primary and secondary particles as the stack is emptied. During the simulation, scored quantities are tallied in a plane or volume of interest. A photon history proceeds until all stacked particles have been terminated by energy loss or exiting conditions. History cycles are then initiated until the desired accuracy is achieved in the scored quantity.

Monte Carlo techniques simulate *individual photon* interactions in a discrete manner regardless of their origin (Section 5.3.2). The entire history is described as analogous because it mimics natural reality, simulating one interaction event at a

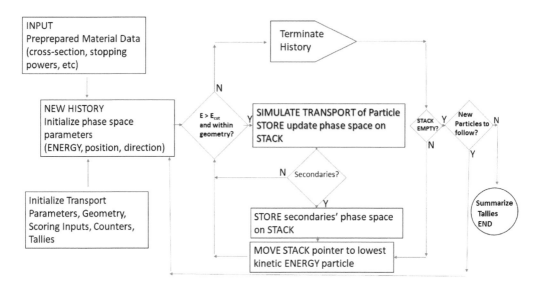

Figure 5.7: General flow for a generic Monte Carlo program. (Andreo et al. 2017)

time. This is feasible because the mean free path for megavoltage photons is such that the string of Compton scattering events in a patient typically involves less than 10 events (Figure 3.17). In principle, other types of particles could be simulated in analog mode. Practically, however, this analog mode of simulation is reserved for uncharged particles where the number of interactions per history is manageable. If the number of interactions escalates by orders of magnitude, the simulation of individual events becomes impractical, causing excessively long computing times; grouping of multiple interactions become necessary, as is the case for megavoltage electrons (Section 5.3.3).

5.3.2 Photon Transport in Analog Mode

The typical Monte Carlo photon simulation begins at the radiation source. The starting parameters for the photon may range from simple to complex. One could simply simulate a mono-directional mono-energetic pencil or a complex multi-particle beam from an accelerator, described by phase space file containing energy spectra, positional coordinates, directional components for all emerging particles. The particle enters the geometry and the distance to its first or next interaction is sampled. The distance between subsequent interaction points is also randomly sampled and is directly linked to total cross section at the current energy for the material being traversed. When the interaction point is reached, the interaction type is selected by random branching depending upon individual interaction cross sections at current energy and medium. Depending upon the type of interaction new particles may be created or absorbed in accordance with differential cross sections. Depending upon their resultant kinetic energy the emerging particles are placed on the stack described earlier, to be emptied in the order from lowest energy to highest energy. Lower energy particles are emptied first since they are most likely to complete their history and avoid overflowing the stack. The lowest energy particle continues in the transport process until it loses enough energy to be below the energy cut-off or leaves the region of interest.

During each step of a given history, the values of interest such as dose are calculated or scored in a matrix of voxels after each history is completed and the result eventually converges to an expectation value. Depending upon the software package, uncertainties are reported in various ways (see equations 5.16, and 5.48, 5.49, 5.50, 5.53). The number of simulated particle histories reduces the statistical uncertainty at the expense of calculation time. The overall process is shown in Figure 5.8.

5.3.2.1 Free Path Selection

An important initial step in simulating radiation transport is determining the distance between interaction sites. This topic was introduced in Chapter 2. While the free path can range from $[0, \infty)$, the probability that a photon will interact between x and $x + dx$ is given by

$$\mathrm{pdf}(x) = \mu \exp\{-\mu\,x\}\,dx$$

The normalized $\mathrm{cdf}(x)$ needed for inversion random sampling is given by

$$\mathrm{cdf}(x) = \frac{\int_0^x \exp[-\mu x']\,dx'}{\int_0^\infty \exp[-\mu x]\,dx} = \mu \int_0^x \exp[-\mu x']\,dx' = 1 - \exp\{-\mu x\}$$

The latter is the probability that an interaction occurs within a geometric distance x. Note that $\lambda = \mu x$ is the dimensionless radiological distance - the geometric distance expressed as a number of mean free paths (mfp $= 1/\mu$). Utilizing the direct inversion technique with the normalized cdf,

$$\mathrm{cdf}(\lambda) = 1 - \exp\{-\lambda\}$$

$$\eta = \mathrm{cdf}(\lambda) = 1 - \exp\{-\lambda\} \quad \text{where } \eta \in [0, 1)$$

Inverting the function leads to

$$\lambda = -\ln(1 - \eta) \tag{5.38}$$

Since $1 - \eta$ is as much a random number as η, we can write

$$\lambda = -\ln(\eta) \quad \text{where } \eta \in (0, 1]$$

This procedure provides a random number of mean free paths to the next interaction site. It changes as the photon energy deteriorates after multiple interactions and the value of μ is updated following each interaction event.

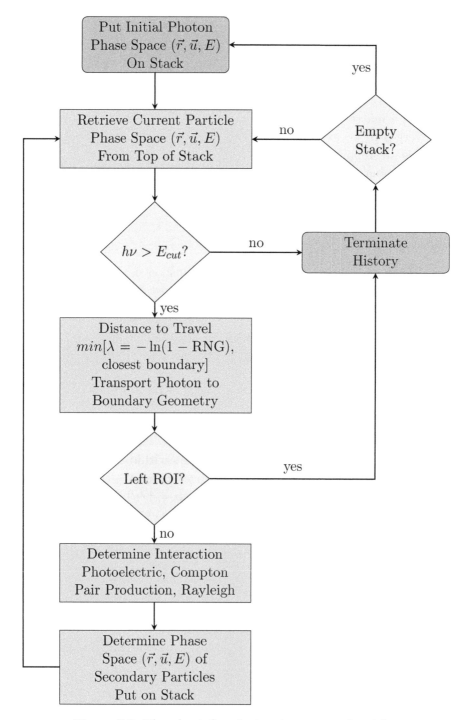

Figure 5.8: Flowchart for photon transport algorithm

5.3.2.2 Selecting the Type of Interaction

When the interaction point is reached by the photon, it is time to decide the type of interaction that will occur. The random sampling in this case is discrete and direct. The pdf is made up of each of the cross sections of the 4 competing processes, normalized to the total cross section. In other words the random sampling procedure will determine which interaction will occur from the pool of possibilities, namely, Rayleigh, photoelectric absorption, Compton scattering or pair production. The cdf is constructed by summing the cumulative components of each process, then partitioned into discrete numerical ranges along a line of unit length.

$$\left[0 \rightarrow \frac{\mu_1}{\mu_{\text{Tot}}}, \frac{\mu_1}{\mu_{\text{Tot}}} \rightarrow \frac{\mu_1 + \mu_2}{\mu_{\text{Tot}}}, \frac{\mu_1 + \mu_2}{\mu_{\text{Tot}}} \rightarrow \frac{\mu_1 + \mu_2 + \mu_3}{\mu_{\text{Tot}}}, \frac{\mu_1 + \mu_2 + \mu_3}{\mu_{\text{Tot}}} \rightarrow 1 \right]$$

The branching procedure starts with drawing a udrn $\eta \in [0, 1]$. The interaction by process i, labelled as μ_i, of m processes can be found by searching for the appropriate index j that encompasses η in the following way:

$$\sum_{j=1}^{i-1} \frac{\mu_j}{\mu_{\text{Tot}}} \leq \eta \leq \sum_{j=1}^{i} \frac{\mu_j}{\mu_{\text{Tot}}} \tag{5.39}$$

where the total probability is given by

$$\sum_{i=1}^{m} \mu_i = \mu_{\text{Tot}} \qquad \text{and} \qquad \sum_{j=1}^{0} \frac{\mu_j}{\mu_{\text{Tot}}} = 0$$

Depending upon the value of η that is randomly selected and where it lands within the interaction intervals, the type of interaction is determined. In the following section we give an example, assuming that a Compton interaction was set to occur.

5.3.2.3 Simulation of Compton Scattering

Generally, random sampling of photon interaction outcomes is performed utilizing the individual differential cross sections in pdf format (Chapter 2). A valuable example is the Compton process - the most prevalent interaction in the megavoltage energy range of radiation therapy. Here we know that an incident photon interacts with atomic electrons in tissue and scatters away with reduced energy and a change of direction. The lost energy is passed onto a Compton recoil electron which is placed on the stack for subsequent processing. The Klein-Nishina cross section governs this process for free atomic electrons with binding energies that are negligible with respect to the energy of the incident photon. In this section we describe the sampling and simulation technique that is utilized by the EGS4 program (Nelson et al. 1985). There is also an alternative valuable description (Raeside 1976) which originated with Kahn (Kahn 1956). The main equations required are the Compton differential

cross section and the Compton momentum-energy conservation equations. We will label Compton scattering reference with "inc" for incoherent scattering for easier readability.

The Klein-Nishina cross section, differential with respect to energy, $h\nu'$, is given by

$$\frac{\mathrm{d}\sigma_{\text{inc}}}{\mathrm{d}(h\nu')} = \frac{X_0 n\pi r_0^2 m_0 c^2}{(h\nu)^2}\left\{\left[\frac{\frac{1}{\alpha^2}}{\epsilon} + 1 - 2\frac{(1+\alpha)}{\alpha^2}\right]/\epsilon + \frac{(1+2\alpha)}{\alpha^2} + \epsilon\right\} \tag{5.40}$$

where:

$$\begin{aligned}
h\nu' &\equiv \text{scatter photon energy (MeV)}\\
h\nu &\equiv \text{incident photon energy (MeV)}\\
X_0 &\equiv \text{radiation length (cm)}\\
n &\equiv \text{electron density (electron}/cm^3)\\
r_0 &\equiv \text{classical electron radius (cm)}\\
c &\equiv \text{speed of light}\\
m_0 &\equiv \text{electron rest mass}\\
m_0c^2 &\equiv \text{electron rest energy (MeV)}\\
\epsilon &\equiv h\nu'/h\nu\\
\alpha &\equiv h\nu/m_0c^2
\end{aligned}$$

The energy of the scattered photon is given by the well-worn expression

$$h\nu' = \frac{h\nu}{1+\alpha\{1-cos(\theta)\}} \tag{5.41}$$

The energy limits of integration over the cross section are given by $h\nu'_{min}$ and $h\nu'_{max}$. The first limit at the scattered photon angle of $\theta = 0$ indicates a grazing hit of the electron. The second limit at angle $\theta = \pi$ corresponds to a direct hit with backscattering of the incident photon.

$$\begin{aligned}
h\nu'_{min} &= h\nu/(1+2\alpha)\\
h\nu'_{max} &= h\nu
\end{aligned}$$

The scattered photon energy is sampled using the relative energy variable $\epsilon = h\nu'/h\nu$ and the interval corresponds to $[\epsilon_0, 1]$ defined by the following:

$$\epsilon_0 = \frac{1}{1+2\alpha} \tag{5.42}$$

The differential cross section (5.40) is repackaged by scaling, factoring, and partitioning in the following way:

$$\frac{\mathrm{d}\sigma_{\text{inc}}}{\mathrm{d}(h\nu')} = \frac{X_0 n\pi r_0^2 m_0 c^2}{(h\nu)^2}\left\{\frac{1}{\alpha^2\epsilon^2} + \frac{1}{\epsilon} - 2\frac{(1+\alpha)}{\alpha^2\epsilon} + \frac{(1+2\alpha)}{\alpha^2} + \epsilon\right\}$$

Note that substitution for the expression for $cos(\theta)$ from Equation 5.41 is used to obtain an expression for $sin^2(\theta)$, and we then obtain the following progression:

$$cos(\theta) = 1 + \frac{1}{\alpha} - \frac{1}{\alpha \epsilon}$$

$$-sin^2(\theta) = cos^2(\theta) - 1 = \frac{1 + 2\alpha}{\alpha^2} - \frac{2(1 + \alpha)}{\epsilon \alpha^2} + \frac{1}{\epsilon^2 \alpha^2}$$

$$\frac{d\sigma_{inc}}{d(h\nu')} = \frac{X_0 n\pi r_0^2 m_0 c^2}{(h\nu)^2} \left\{ \frac{1}{\epsilon} + \epsilon - sin^2(\theta) \right\}$$

$$= \frac{X_0 n\pi r_0^2 m_0 c^2}{(h\nu)^2} \left\{ \frac{1}{\epsilon} + \epsilon - \frac{(1 + \epsilon^2)}{1 + \epsilon^2} sin^2(\theta) \right\} \qquad (5.43)$$

which is partitioned into the following expression (Butcher 1961):

$$\frac{d\sigma_{inc}}{d(h\nu')} = \frac{X_0 n\pi r_0^2 m_0 c^2}{(h\nu)^2} \left\{ \frac{1}{\epsilon} + \epsilon \right\} \left\{ 1 - \frac{\epsilon}{1 + \epsilon^2} sin^2(\theta) \right\}$$

$$\frac{d\sigma_{inc}}{d\epsilon} = \frac{X_0 n\pi r_0^2 m_0 c^2}{h\nu} f(\epsilon)g(\epsilon) \propto f(\epsilon)g(\epsilon)$$

where:

$$f(\epsilon) = \frac{1}{\epsilon} + \epsilon$$

and

$$g(\epsilon) = 1 - \frac{\epsilon}{1 + \epsilon^2} sin^2(\theta)$$

Thus, the above expressions for f and g when multiplied together yield either of the right-hand side expressions of equations (5.43).

Note that in the sampling procedures that follow, one can re-normalize the constants of proportionality and not effect the actual sampling technique. Mixed sampling in multiple stages is performed firstly with function $f(\epsilon)$ to obtain the random variable ϵ and then rejection sampling is carried out with function $g(\epsilon)$ where the latter is used for the rejection decision. The partition for $f(\epsilon)$ is divided into 2 pieces as follows:

$$f(\epsilon) = a_1 f_1(\epsilon) + a_2 f_2(\epsilon)$$

where:

$$a_1 = \ln\left(\frac{1}{\epsilon_0}\right) \qquad f_1(\epsilon) = \frac{1}{\ln\left(\frac{1}{\epsilon_0}\right)} \times \frac{1}{\epsilon} \qquad \epsilon \in [\epsilon_0, 1]$$

and

$$a_2 = \frac{(1 - \epsilon_0^2)}{2} \qquad f_2(\epsilon) = \frac{2\epsilon}{(1 - \epsilon_0^2)} \qquad \epsilon \in [\epsilon_0, 1]$$

Sampling for ϵ begins with the partitioned branching ratios. Choose a udrn $\eta_1 \in [0, 1]$.

$$\text{If } \eta_1 < \frac{a_1}{a_1 + a_2}$$

then use partition 1 and $f_1(\epsilon)$. Then choose another udrn $\eta_2 \in [0, 1]$.

$$\int_{\epsilon_0}^{\epsilon} f_1(\epsilon') \, d\epsilon' = \frac{1}{\ln \frac{1}{\epsilon_0}} \int_{\epsilon_0}^{\epsilon} \frac{d\epsilon'}{\epsilon'} = \frac{1}{\ln \frac{1}{\epsilon_0}} (\ln \epsilon - \ln \epsilon_0) = \eta_2$$

and inverting for the solution, ϵ is then given by

$$\epsilon = \epsilon_0 \exp \left[\eta_2 \ln \frac{1}{\epsilon_0} \right]$$

Otherwise

$$\frac{a_1}{a_1 + a_2} \le \eta_1 \le 1$$

and partition 2 is utilized with $f_2(\epsilon)$. In this case, further partitioning takes place with the following transformation:

$$\epsilon'' = \frac{\epsilon - \epsilon_0}{1 - \epsilon_0} \text{ yielding } \quad d\epsilon'' = \frac{d\epsilon}{1 - \epsilon_0} \tag{5.44}$$

Then transform as follows:

$$h(\epsilon'') = f_2(\epsilon) \frac{d\epsilon}{d\epsilon''} = \frac{2\epsilon}{1 - \epsilon_0^2} (1 - \epsilon_0) = \frac{2}{1 + \epsilon_0} \left[(1 - \epsilon_0)\epsilon'' + \epsilon_0 \right]$$

The function, $h(\epsilon'')$ simplifies in terms of α, utilizing Equation 5.42, to the following:

$$h(\epsilon'') = \frac{2\alpha}{1 + \alpha} \epsilon'' + \frac{1}{1 + \alpha}$$

Re-partitioning into 2 parts again leads to the following:

$$a_1'' = \frac{\alpha}{1 + \alpha} \quad h_1(\epsilon'') = 2\epsilon'' \quad \epsilon'' \in [0, 1]$$

and

$$a_2'' = \frac{1}{1 + \alpha} \quad h_2(\epsilon'') = 1 \quad \epsilon'' \in [0, 1]$$

Now select a udrn $\eta_2 \in [0, 1]$ and

$$\text{If } \eta_2 \le \frac{\alpha}{1 + \alpha} \tag{5.45}$$

is true, then the first partition, $h_1(\epsilon'')$ is chosen. This is a linear pdf in ϵ'' so that a direct sampling technique could be used or a special rejection technique. Here, we select udrn's $\eta_3 \in [0, 1]$ and $\eta_4 \in [0, 1]$ and set $\epsilon'' = max[\eta_3, \eta_4]$. Refer to Equation step 5.24 and the explanation that follows in the rejection technique to Equation 5.25 in Section 5.2.5.2.

If inequality (5.45) does not hold then

$$\frac{\alpha}{1+\alpha} < \eta_2 \leq 1$$

and the second partition, $h_2(\epsilon'') = 1$ is used as a uniform distribution. We then draw a udrn $\eta_3 \in [0, 1]$ and set:

$$\epsilon'' = \eta_3$$

When ϵ'' is accepted, we revert back to ϵ from the original substitution and construct the rejection function.

The next phase of sampling requires the rejection technique utilizing the ϵ that was accepted above through either partition, and calculation of the rejection function, i.e., $g(\epsilon)$. Draw a udrn η_4 and test with the following:

$$\eta_4 < g(\epsilon)$$

If true, accept the ϵ; otherwise reject and start again at $f(\epsilon)$ and η_1.

One arrives at the calculation of $sin^2(\theta)$ by using the substitution for $cos(\theta)$, namely Equation 5.41, as follows:

$$\cos(\theta) = 1 + \frac{(\epsilon - 1)}{\alpha\epsilon} = 1 - t$$

where:

$$t = \frac{(\epsilon - 1)}{\alpha\epsilon}$$

and thus

$$\sin^2(\theta) = 1 - \cos^2(\theta) = \{1 - \cos(\theta)\}\{1 + \cos(\theta)\} = t(2 - t)$$

Finally, secondary photon energies are calculated and an azimuthal angle [see equations (5.20)] is selected uniformly at random in the interval $[0, 2\pi]$.

5.3.3 Electron Transport in Condensed Mode

The role of electron transport is crucial in dose calculation algorithms and it is highlighted in this section as implemented in Monte Carlo modelling. Over the energy range of interest to radiation therapy, i.e. 10 keV to 50 MeV, charged particles suffer an enormous number of single interactions in tissue over very short distances compared to photons (Chapter 2). Thus, charged particle simulations would take much more time to model in analog mode than uncharged particles.

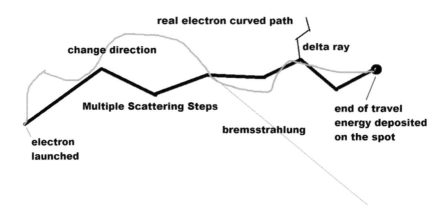

Figure 5.9: Multiple scattering steps along an electron path, with continuous energy deposition close to the track and occasional delta-ray or bremsstrahlung interactions that carry energy beyond the track. Note that in the middle of each multiple scattering step the hinge method is utilized in some of the MC transport codes such as PENELOPE (Salvat et al. 2008).

Until computers increase dramatically in speed a single-scatter model for electrons is impractical for clinical time-sensitive applications. Instead multiple interactions can be statistically grouped together along steps that are considerably longer than a mean free path. This grouping process is known as "Condensing the History (CH)". Condensation of histories for charged particles can be considered a variance reduction technique, described later in Section 5.3.5. It is complicated and it requires many facets working in concert to effectively depict collective multiple-scattering while maintaining accuracy.

The condensation model was invented by necessity (Berger 1963) and specifies Class I and Class II Monte Carlo methods. Class I uses a series of step lengths determined in advance, grouping interactions together with random sampling deferred to the *end of each step* (Andreo et al. 2017). Class II divides the energy transport *along each step* into two distinct mechanisms, separating catastrophic rare events with large energy transfers from milder continuous interactions with frequent small energy exchanges. This mixed procedure uses analog simulation for the major discrete energy loss events, and statistical grouping for continuous softer collisions. Megavoltage electrons are most often simulated with Class II Monte Carlo techniques (e.g. in EGS code). The simulation splits catastrophic hard interactions producing bremsstrahlung or energetic δ-rays from soft interactions with continuous Coulomb slowing down along each track. The trajectories of the primary and catastrophic particles are illustrated in Figure 5.9.

User-defined threshold energies for secondary photon and charged particle events force the division of labour between occasional large and routine small energy exchanges. Hard collisions by Möller or Bhabha scattering by electrons and positrons, respectively, produce particles with energies *above* threshold energy $E_{\text{thresh}} = AE$ (Nelson et al. 1985). These knock-on particles merit follow-up as separate distinct newborn entities. Similarly, bremsstrahlung x-rays produced by inelastic scattering with transfers above the threshold energy $P_{\text{thresh}} = AP$ (Nelson et al. 1985) are treated as newborn photons in analog mode. Softer charged particle collisions and bremsstrahlung x-rays dump their energy locally along the track *via* restricted stopping powers. This energy is deposited within a sausage-shaped region wrapped around the main particle track, while harder events give rise to energy deposited remotely from this core region. The size of this region is determined by threshold energy values. These values should be set to maintain the spatial resolution required for the problem being studied (Fippel 2013a).

Harder interactions for charged particles are simulated similarly to the analog method for photons. The step distance, s, to the next major interaction is randomly sampled as for photons in Equation 5.38. To be more explicit, the process uses

$$s = -\frac{1}{\sigma_{\text{tot}}^{\text{e-}}} \ln(1 - \eta)$$

where $\sigma_{\text{tot}}^{\text{e-}}$ is the total hard interaction cross section including the Moller and bremsstrahlung cross sections. Note that in the above expressions, positrons are not included (i.e. Bhabha scattering) but could be considered in a similar fashion if the prevalence of events is expected to deposit significant dose contributions. When a hard interaction location is reached, random sampling is performed using the restricted differential cross section (or stopping power) at which point the type of hard collision, Möller (electron-electron collision) or Bhabha (positron-electron collision), or *bremsstrahlung* is selected similarly to the random branching procedure used to triage photon events (Equation 5.39).

Within the travel trajectory between catastrophic events, a large number of soft interactions will occur. Consideration is given into combining all minor events using multiple scattering theory that considers a statistical ensemble of particles undergoing scattering events. Within this realm the effects of scattering, energy loss, path length correction, straggling of energy loss and range are treated. Energy loss is accounted for in a continuous manner (recall the csda approximation described in Chapter 2) with the particle depositing energy to the surrounding medium through restricted collisional and radiative stopping powers. These quantities restrict attention to the vicinity of an electron track, consistent with AE and AP parameters. The small angular deflections that occur in one step are combined in composite multiple scattering angle. Care must be taken in selecting a step size. The distance travelled between two steps should be optimized such that the total number of steps or energy loss is small enough to be accurate but large enough that the modelled multiple collision theories remain physically valid in predicting energy loss and angular scattering. Thus there is a balance between short path lengths to maximize accuracy and resolution *versus* larger steps to promote computational efficiency and risk validity of theory (Andreo et al. 2017). Most MC algorithms limit the maximum step size that can be taken in a multiple scattering step. This is usually based on a percentage maximum energy loss parameter such as specified by the parameter ESTEPE used in EGS4, for example (Nelson et al. 1985). Then the restricted stopping power and mass density of the surrounding medium are used to calculate the physical step taken in real space (Fippel 2013a).

Where a particle ends up after a step is predicted by multiple scattering theory of elastic and semi-elastic interactions, as in a random walk problem. An appropriate multiple scattering model must be used for random sampling of multiple angular deflections to provide a realistic stepping move. Some of the models that are used include Fermi-Eyges, Molière or Goudsmit-Saunderson approaches. For example, Fermi-Eyges uses a 2D Gaussian distribution giving the probability that the electron is scattered within the solid multiple scattering angular section, $(\theta + d\theta), (\phi + d\phi)$, where:

$\theta \equiv$ is the azimuthal multiple scattering angle
$\phi \equiv$ is the polar multiple scattering angle

The Gaussian probability density model works well for small cumulative scattering angles, but large angle scattering is underestimated. This gives way to the improved Molière theory of scattering adopted by EGS4 (Nelson et al. 1985), but still suffering to some extent from the large angle scatter issue. Goudsmit-Saunderson scattering theory was put into effect in EGSnrc (Kawrakow 2000), PENELOPE (Salvat et al. 2008), and MCNP (Briesmeister 2000), producing multiple scattering angles closer to reality. The overall flowchart for charged particle transport is summarized in Figure 5.10.

In reality, the real path of each charged particle is a continuous tortuous curved path ultimately resulting in a displacement that is lateral to the straight step taken. The straight step length that is simulated in the multiple scattering step is corrected to reflect this meandering path. The nature of these corrections will vary depending upon the simulation package that is used. Corrections are applied to the energy loss, energy straggling, path length and path length straggling. They are also applied for situations where the crossing of boundaries of different media comes into effect.

Energy Loss An energy loss is associated with each CH step. Energy loss is accounted for in a continuous (csda) manner with the particle depositing energy to the surrounding medium governed by both the restricted collisional and radiative

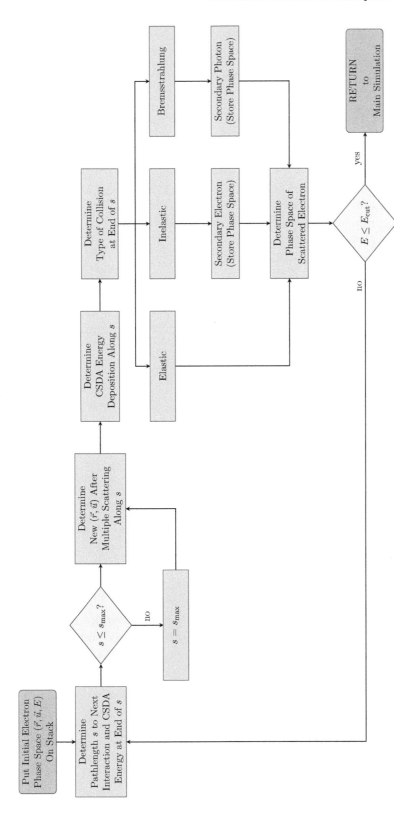

Figure 5.10: Flowchart highlighting major steps of electron transport in condensed history Class II algorithm. Adapted with permission (Andreo et al. 2017).

stopping powers, omitting the hard energy exchanges, as follows:

$$\left(\frac{dE}{ds}\right)_{col} = S_{col}(\bar{r}, E) - N(\bar{r}) \int_{E_{thresh}}^{E} E' \sigma_{col}(\bar{r}, E, E') \, dE'$$

$$\left(\frac{dE}{ds}\right)_{rad} = S_{rad}(\bar{r}, E) - N(\bar{r}) \int_{P_{thresh}}^{E} (h\nu)' \sigma_{rad}(\bar{r}, E, h\nu') \, d(h\nu') \qquad (5.46)$$

where:

$$N(\bar{r}) \equiv \text{number of scattering targets per volume at } \bar{r}$$
$$S_{col}(\bar{r}, E) \equiv \text{unrestricted collisional stopping power}$$
$$S_{rad}(\bar{r}, E) \equiv \text{unrestricted radiative stopping power}$$
$$\sigma_{rad}(\bar{r}, E, h\nu') \equiv \text{collisional cross section}$$
$$\sigma_{col}(\bar{r}, E, E') \equiv \text{radiative cross section}$$
$$s \equiv \text{path length of travel}$$

The actual step length, s, is determined using Equation 5.47:

$$s = -\int_{E_{initial}}^{E_{step}} \frac{dE}{\left(\frac{dE}{ds}\right)_{col} + \left(\frac{dE}{ds}\right)_{rad}} \qquad (5.47)$$

where:

$$E_{step} \equiv E_{initial} - \Delta E$$
$$E_{initial} \equiv \text{is the initial energy of the electron at beginning of the step}$$
$$\Delta E \equiv \text{is the energy lost during the step}$$

Energy Straggling For soft interactions, energy loss straggling is the fluctuation of energy about the mean energy loss predicted solely by stopping powers, $E_{step} = E_{initial} - \Delta E$, in the above Equation 5.47 of the energy loss distribution. The mean energy loss per multiple scattering step in this equation is constant for the same starting energy, $E_{initial}$. It is clear that there is indeed a fluctuation of energy loss about the mean in the explicit treatment of hard interactions. Thus, energy loss straggling is not emulated in this way unless the threshold energies are set small enough to reduce the portion of simulation that concerns average soft interactions. In this way most of the energy straggling is taken care of by the portion that simulates hard interactions in analogue manner and *soft straggling* is less important. The caviat, again, is that computation time will increase. Some MCM packages perform random sampling about an energy distribution such as a Gaussian as in the work of Bohr (Bohr 1948), Landau (Landau 1944) or Vavilov (Vavilov 1957). Energy straggling is not considered as important as range straggling (Fippel 2013a) described next.

Range Straggling   The electron traverses a multiple scattering step in a straight line and at the end of this step the multiple scattering angle is randomly chosen. To emulate the curved path of a real electron, a path length correction is also applied with a lateral displacement which also takes into account the fluctuations about the mean value of the actual range and lateral displacement. This is called range straggling for soft interactions and is independent of the electron energy loss (Fippel 2013a). There are several techniques to deal more accurately with these effects. The *random hinge* method utilizing random sampling *between* multiple scattering steps rather than only at the end of each step yields more accurate corrections (Salvat et al. 2008; Kawrakow 2000; Fippel 2013a) (see Figure 5.9). The algorithm in EGSnrc (Kawrakow 2000) uses first and second order spatial moments (Fippel 2013a).

Boundary Transitions   Radiation transport simulation near boundary interfaces of differing media must be carefully treated to ensure that correct path lengths, lateral deflections, dose deposition, and regional tracking are maintained. However, "for arbitrarily-shaped material interfaces, an exact theory of multiple scattering does not exist" (Fippel 2013a). Figure 5.11 shows electron tracks, one simulated by using CH steps and a second one a sample of a real curved electron path near and within two different adjacent media. In this example, the simulated track does not reflect the fact that material II exists. Being far enough away from interfaces where the multiple scattering steps are small relative to the distance to the nearest interface will cause minimal issues. However, when step sizes are of the order of distance to the boundary, ambiguity of materials crossed by the electron will arise. There are programs that have been developed that monitor the neighbourhood for the closest approach to boundaries of different media. When a distance to a boundary is on the order of the multiple scattering step size, the step size is progressively shortened on approach. Great progress was achieved with PRESTA (Bielajew and Rogers 1986) in EGS4 (Nelson et al. 1985). EGSnrc (Kawrakow 2000) utilizes a user-defined parameter d_{near} to flag nearness to a boundary and switches the electron scattering algorithm from multiple scattering to single scattering mode. Fortunately, in radiation therapy planning applications, the boundary-crossing issues have generally been resolved (Fippel 2013a).

5.3.4   Tallying of Results

The calculated quantities of interest in a Monte Carlo simulation for radiation therapy will depend upon the specific application. For example, accelerator head modelling will score fluences or energy spectra crossing a predefined planar surface. On the other hand, a radiation treatment planning system will score dose deposition in a 3D matrix of voxels. The planes or voxels of interest within the simulation geometry are flagged for scoring or tallying of these quantities. For a set number of histories, the averages of these quantities will be calculated together with their uncertainties. The user has the option to continue the simulation if the uncertainty

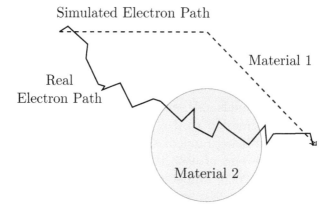

Figure 5.11: Two electron paths are indicated. One is a simulated piece-wise straight path along a multiple scattering step and the other is a more realistic curved path. Boundary crossing algorithms progressively shorten step length on approach to a boundary to help resolve ambiguity of the materials being traversed (Fippel 2013a).

level is unacceptable. For N histories the mean of scoring quantity x is given by

$$\bar{x} = \frac{1}{N} \sum_{i=1}^{N} x_i$$

and in practice it is convenient to calculate the standard error, using the *sample standard* deviation of $\bar{x}$ in the form

$$s_{\bar{x}} = \sqrt{\frac{1}{N(N-1)} \left\{ \left(\sum_{i=1}^{N} x_i^2 \right) - \frac{1}{N} \left(\sum_{i=1}^{N} x_i \right)^2 \right\}} \qquad (5.48)$$

Note that the quantities $\sum x_i$ and $\sum x_i^2$ are summed over each history and that at the end of the simulation Equation 5.48 is used for the final tally. This is usually referred to as the history-by-history method. Also recall that Equation 5.48 yields the uncertainty in $\bar{x}$ and not the spread in the x_i.

For m independent runs of the simulation, the grand total number of histories is

$$N = \sum_{k=1}^{m} N_k$$

Each independent run of N_k histories has a mean value $\bar{x}_k$ with its sample standard deviation of the mean, $s_{\bar{x}_k}$. From these, the weighted mean value and the standard uncertainty for total N are obtained as follows (Andreo et al. 2017):

$$\bar{x} = \frac{\sum_{k=1}^{m} \frac{\bar{x}_k}{s_{\bar{x}_k}^2}}{\sum_{k=1}^{m} \frac{1}{s_{\bar{x}_k}^2}} \qquad \text{and } s_{\bar{x}} = \frac{1}{\sqrt{\left\{\sum_{k=1}^{m} \frac{1}{s_{\bar{x}_k}^2}\right\}}} \qquad (5.49)$$

If the following is true,

$$\frac{1}{s_{\bar{x}_k}^2} \propto N_k$$

then one can use the following simpler expressions instead (Andreo et al. 2017):

$$N = \sum_{k=1}^{m} N_k$$

$$\bar{x} = \sum_{k=1}^{m} \frac{N_k}{N} \bar{x}_k$$

$$s_{\bar{x}} = \sqrt{\sum_{k=1}^{m} \left(\frac{N_k}{N}\right)^2 s_{\bar{x}_k}^2} \qquad (5.50)$$

A significant amount of computational time can be spent with the history-by-history method. Summations of complicated functions after each history can slow things down considerably, although there are some smart computational remedies that have been developed (Sheikh-Bagheri et al. 2006; Sempau et al. 2000). Currently, EGS (Nelson et al. 1985; Sheikh-Bagheri et al. 2006) also uses this approach but previously used the batch method. In the batch approach, the simulation is divided up into n equal batches of N/n histories per batch. To ensure that the central limit theorem can be utilized, n should be large enough and usually ranges between 10 to 40. Then the sum over the jth batch is given by

$$x_j = \sum_{i=1}^{N/n} x_{i,j} \text{ for } j = 1, n \qquad (5.51)$$

where $x_{i,j}$ is the ith quantity of interest for batch j and x_j is the sum for the jth batch. The sample mean is

$$\bar{x} = \frac{1}{N} \sum_{j=1}^{n} x_j \qquad (5.52)$$

The sample variance for x_j is given by

$$s_{x_j}^2 = \frac{1}{n-1} \sum_{j=1}^{n} \left\{x_j^2 - n\bar{x}^2\right\}$$

and the variance on the sample mean $\bar{x}$ is given by

$$s_{\bar{x}}^2 = \frac{s_{x_j}^2}{n} \tag{5.53}$$

In order to combine the results of m independent runs of the batch method or, in other words, using the batch method m different times for m different runs, Equations 5.50 can be used. In utilizing batch method Equations 5.51 to 5.53 for 1 of m independent runs, one must be careful to correspond each N in Equation 5.52 with each N_k in Equations 5.50, each $\bar{x}$ in Equation 5.52 with each $\bar{x}_k$ in Equations 5.50, and also each $s_{\bar{x}}^2$ in Equation 5.53 with each $s_{\bar{x}_k}^2$ in Equations 5.50. With all of this in mind, one should always report the type of statistics used in simulations to remove any ambiguity because of the different uncertainty specifications that exist (Sechopoulos et al. 2018).

5.3.5 Variance Reduction Techniques

Variance reduction techniques (VRTs) can be defined as "statistical methods that improve the efficiency of a MC calculation without affecting the accuracy of the physics and assuring that statistical estimators remain unbiased" according to Andreo et al. (Andreo et al. 2017). It is one of the major components for improving efficiency of a simulation. Variance reduction techniques have evolved as Monte Carlo codes have progressed and become specialized for radiation therapy applications. They can be quite sophisticated and are applied to many aspects of a simulation, from accelerator head modelling to calculating dose distributions within the patient. Their utilization will accelerate the clinical implementation of Monte Carlo methods.

The common overall goal is to reduce uncertainty in the scored quantity to an acceptable tolerance level. The standard error of the mean for a particular number of total histories is a measure based on the variance as we have seen in the previous section with equations (5.16 and 5.31). In reporting results, the uncertainty is often quoted in terms of the standard error of the mean and it is proportional to the inverse of $\sqrt{N}$. The brute force method to decrease the uncertainty by a factor of k is to increase the number of histories by a factor of k^2. With unlimited resources and computing power, this effort would be reasonable. In practice, the specialist attempts to reduce the uncertainty for a given computational burden (i.e. smaller uncertainty for fixed N investment), by employing smarter statistical strategies instead of a brute-force increase in N. Alternatively, one could say that the goal of variance reduction is to maintain the same uncertainty, but at reduced computational labour (i.e. smaller N with fixed uncertainty). Effective MCMs rely on reducing the coefficient of the $\sqrt{\frac{1}{N}}$. These strategies include distorting the input data in such a way as to reduce the variance of the parent distribution such as importance sampling or other techniques that do not necessarily require exponential increase in the number of histories (Hammersley and Handscomb 1964).

There exist scenarios where regions of interest are infrequently traversed by simulated particles resulting in poorer statistics or higher uncertainty. VRT strategies can be employed to improve the statistics in these zones. Examples are the modelling of the response within an ionization chamber and the dose distribution outside of the primary radiation fields. There are methods to speed up calculations using hardware and smart software optimization but these are not strictly referred to as variance reduction. Condensed history techniques (Section 5.3.3) are not strictly considered to be VRTs unless it can be mathematically proven that they are equivalent to explicitly simulated analog transport, maintaining fidelity to the underlying physics (Andreo et al. 2017).

It is useful to define the efficiency, ϵ, of a calculation based on the desired uncertainty and the computational effort required to achieve it. The formal definition is

$$\epsilon = \frac{1}{s^2(N)\,T(N)}$$

where s^2 is the estimate of the variance σ^2 of the calculated quantity and T is the expended CPU time. Thus, in order to increase the efficiency one must decrease either quantity or both. Note that the efficiency is essentially independent of the number of histories, N, because s^2 varies as $1/N$ and T varies as N (Fippel 2013b).

In Section 5.2.5.3, importance sampling was demonstrated to be an effective tool for variance reduction. Four additional commonly-used techniques will now be discussed in this section.

Interaction Forcing   The purpose of forcing photon interactions is to artificially boost the interaction probability of a specific type of interaction so that it occurs more frequently than the natural frequency, in order to improve the statistics of results affected by upstream interactions. The photon is forced to interact between current location and the location where it may deposit dose. Figure 5.12 depicts the geometry with ray AB of the photon projected through the region where it might not interact at all. The interaction is forced to occur between points A and B at a hypothetical point I. The number of mean free paths (L) between points A and B must pre-calculated, as follows:

$$L = \sum_{A}^{B} \mu_i r_i$$

where μ_i and r_i are the linear attenuation coefficient and ray segment through the voxel i respectively. The number of mean free paths, $\lambda \in [0, L]$ is randomly selected by drawing a udrn $\eta \in [0, 1]$ and equating it to the following cdf to give

$$\eta = \int_0^\lambda \frac{\exp\{-\lambda'\}}{1 - \exp\{-L\}}\,\mathrm{d}(\lambda') = \frac{1 - \exp\{-\lambda\}}{1 - \exp\{-L\}}$$

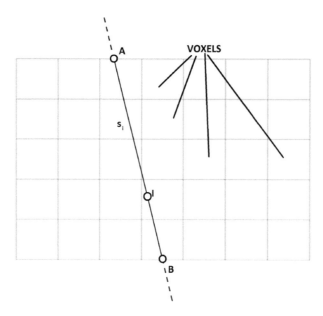

Figure 5.12: The photon enters at point A in the geometry and may exit along ray line AB through point B *without* any intervening interactions. The photon can be forced to interact at point I but statistical weighting corrections must then be applied afterwards to "keep the game fair".

with the number of mfp's given by

$$\lambda = -\ln\left(1 - \eta(1 - \exp\left\{-L\right\})\right)$$

Notice that for a real interaction $L \to \infty$ approaches the proper number of mean free paths λ given by Equation 5.38.

Following the ray trace along AB to interaction point, I, with the number of mean free paths $\lambda = \sum_A^I \mu_i r_i$ gives the required distance of travel. All secondary particles involved in the forced interactions are given a reduced weight to compensate for the artificial physics. Statistical weights are tracked in the stack along with the phase space states for each particle. Weight factors are used as a correction factor for the statistical variance reduction technique applied. In this case, the particle weight, is changed to $weight_{\text{current}} = weight_{\text{previous}} \times weight_{\text{forcingVRT}}$ where $weight_{\text{forcingVRT}} = 1 - \exp\left\{-L\right\}$. This ensures that scored quantities such as dose deposition are reduced by that weight in order to fairly score the dose quantity. This is the way in which weighting factors are applied and it is important to note that $weight = 1$ is the normal default for all particles at the start of simulations.

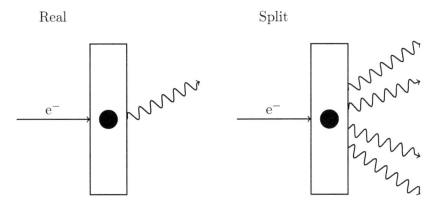

Figure 5.13: Particle splitting variance reduction technique. Interaction site of accelerated electron striking a target and producing either a real single or many virtual bremsstrahlung photons.

Energy Deposition Kernels   An important example of applying the strategy of forced interactions can be found in an EGS user code (Nelson et al. 1985), heavily used in bootstrapping convolution-superposition algorithms (Chapter 4). The user code SCASPH is used to produce energy deposition kernels that serve as a startup database for dose calculation using these algorithms. The kernels are generated for a homogeneous sphere of water exposed to incident monoenergetic photon pencil beams ranging from 0.1 to 50 MeV. The incident photons are *forced to interact at the center* of a large sphere instead of by random path selection. The energy deposited by secondary charged particles resulting from primary, first-scattered, second-scattered, multiply-scattered and bremsstrahlung and annihilation photons are scored in neighbouring voxels. Chapter 2 (Section 2.6.1) provides more details on the nature of this kernel data set.

Particle Splitting   The most common technique is uniform photon splitting in which one photon is split into n_{split} photons with particle weights adjusted so that the result is not biased. An example of this is used for bremsstrahlung splitting in which the accelerated electrons striking the target in a linear accelerator produce bremsstrahlung photons, as shown in Figure 5.13. Because this interaction is rare compared to delta ray production, the bremsstrahlung cross section is independently sampled n_{split} times to produce n_{split} independent bremsstrahlung photons each of reduced statistical weighting $1/n_{\text{split}}$. There is a resultant energy exchange that should occur between the initiating electron and only one bremsstrahlung photon in reality. The photon chosen is handled by randomly selecting one from the set of n_{split} possible photons. This together with the statistical weighting adjustment ensures that the physics is played fairly and energy is conserved, on average. Ultimately the splitting game adds more artificial *bremsstrahlung* photons per $N_{\text{histories}}$ accelerated electrons which reduces the uncertainty in simulation for the same $N_{\text{histories}}$ of electrons without playing the splitting game.

Russian Roulette In Russian roulette, a single bullet is loaded in a 6-chamber cylinder of a hand gun. The probability of it being fired in a random spin of the cylinder is 1 in 6. In a Monte Carlo simulation, the discarding of inconsequential particles with minimal impact on results can be accomplished by implementing an elimination of particles with a Russian roulette strategy. Computing time is conserved by avoiding the tracking of particles that will not contribute to the scoring goals of the simulation and hence their elimination will not bias the results (Fippel 2013b; Andreo et al. 2017; Salvat et al. 2008). The mechanism begins by assigning a survival probability, P_{survive}, to those unwanted particles intended for discarding. A udrn $\eta \in [0, 1]$ is drawn and compared with the latter. If $\eta < P_{\text{survive}}$ the particle survives and its statistical weighting is adjusted to $weight_{\text{NEW}} = weight_{\text{current}} \div P_{\text{survive}}$ to play the game fairly. Otherwise, the particle is discarded in such a way that it is no longer tracked and no longer contributes to any of the scoring.

Particle splitting and Russian roulette techniques both create efficiency but from opposing mechanisms. Splitting is applied to increase the number of particles going into a region of interest while Russian roulette is invoked to remove particles moving away from it (Salvat et al. 2008). Often these are played *successively* as in the case of the embellished number of *bremsstrahlung* photons created in conjunction with photon splitting in the LINAC target and the secondary electrons that they create along their path. The "weighting" game that is played for the successive mechanisms is *bremsstrahlung* photons with $weight = 1/n_{\text{split}}$ interact by Compton interaction in flattening filter launching a recoil electron which initially has the same weight as the parent photon. However, in surviving the game of Russian roulette the electron's weight is adjusted to $weight_{\text{adjusted}} = 1/n_{\text{split}} \times 1/P_{\text{survive}}$. Note that by setting $P_{\text{survive}} = 1/n_{\text{split}}$ the surviving electron's statistical weight becomes $weight = 1$ (Fippel 2013b). It is important to validate the discarding of particles. There may be a possibility that electrons can still produce bremsstrahlung photons before being discarded.

Range rejection A particle that is far enough away from any region of interest and has little opportunity of being transported there may be stopped prematurely and discarded. This technique is most often applied to charged particles with limited finite range, as opposed to photons with potentially long free paths. The mechanism requires use of the current range and extrapolation to the closest distance of approach to the region of interest. If the range is smaller than the nearest boundary, then the particle's current energy is deposited on the spot in the current region and the particle may be discarded. Simulation time is saved by avoiding the tracking of this particle.

The strategy to abandon particles because they no longer can contribute substantially to the simulation result has a similarity with both range rejection, and imposing terminal energy cut-offs such as E_{cut} and P_{cut} (not to be confused with condensed history energy thresholds). These cut-offs force termination of particle histories with local dumping of their terminal energy. Usually this is applied to but is not limited to charged particles and caution must be exercised not to bias the physical results. The energy cutoffs can be set *regionally* and the choices depend upon the local media because this has an impact on the local cross section and range of travel. This could be applied to electrons travelling through voxels with a low probability of creating *bremsstrahlung* while running out of energy. Thus, efficiency is gained using either technique leaving only a small risk of biasing the physics.

There are other techniques that speed up calculations and simulations. They involve exploitation of geometric symmetries, using pre-calculated lookup tables rather than computing quantities on the fly, the reciprocity theorem (see Chapters 3 and 4), and parallel computing (Chapter 7). Special cases may involve manipulating input interaction data that can negate certain interactions that occur in nature but are irrelevant to the problem at hand. For example, analysing the behaviour of electron interactions while suppressing the production of *bremsstrahlung* can be informative (Sawchuk et al. 1992). Here the total collisional cross section and stopping power were modified to include only the Moller collisional cross section. In other words, the bremsstrahlung cross section was artificially nullified. This technique is more efficient than actually simulating with the real cross sections and the subsequent tracking of energy deposition by bremsstrahlung photons.

There are an enormous number of VRTs available in simulation software. Those already discussed are common and many newer including hybrid techniques have been developed. More thorough descriptions are available in other works (Fippel 2013b; Chetty et al. 2007; Rogers et al. 1995; Kawrakow et al. 2004; Kawrakow 2005; Jenkins et al. 1988). Because of the heavy reliance on VRTs to save computer resources, it is important to verify that their application does not bias the physical results for the problem under investigations. VRTs can be tested by comparison with known analytical, calculated or measured results including the results from *full* simulation runs *without using the VRT*. Validating VRTs must use a sufficient number of histories to discern statistically significant differences in expectation values (Chetty et al. 2007). It should be noted that a VRT implemented to reduce the uncertainty on one set of parameters can at the same time increase the uncertainty in another set (Salvat et al. 2008). One must be aware of the impact of this cross-talk effect, bearing in mind the aims of the simulation process.

5.4 RADIATION TREATMENT PLANNING

Monte Carlo simulations for radiotherapy take on a multitude of roles. Applications include accelerator beam modelling, treatment planning both on-line and off-line, treatment device modelling and design, and quality assurance (QA) of treatment plans whether direct patient calculations or indirect in phantoms.

A complete simulation from the accelerator treatment head to a portal imager on the exit side of the patient involves a long succession of continuous interactions. This begins with simulating the initial accelerated electrons incident on the x-ray target continuing with the electron-photon shower through the accelerator head onto field-shaping devices through to the patient and finally traversing onto a portal detector system. For convenience, this can be broken up into sub-stages (Chetty et al. 2007) where at the end of each stage, a phase space file captures all particle information crossing an interim scoring surface. This information is stored in such a way to be reused as input to begin the next stage of simulation. A phase space file incorporates the state of all simulated particles crossing the scoring surface and also may include other information that might be pertinent, such as previous interaction type or site prior to reaching the scoring plane. There can be 3 main divisions of the accelerator-patient-imaging space. Stage 1 simulates transport of a shower through the components that apply to the patient treatment generally. Here, the succession goes through the electron exit window, target, primary beam collimator, flattening filter, monitor chamber and mirror. An actual beam model can be formed from phase space 1. Stage 2 is more specific to customized treatments of patients where field openings with and without patient specific filters (physical wedges or compensators), movable components such as multi-leaf collimator (MLC) leaves, auxiliary collimators, and gantry angle are specific to individual treatment. Stage 3 involves simulation through the patient or phantom space. This is usually combined with stage 2 because each stage is specifically dependent on patient-specific parameters. Stage 4 involves the exiting beam onto a detector device such as a portal imager. Figure 5.14 depicts the succession of components and the three scoring planes for the phase spaces files.

The main advantages of staging the source simulations is computational time-saving by not having to repeat unnecessary fundamental simulation of upstream stages. This would mostly involve stage 1, as stage 2 would vary widely because of custom treatment settings and devices per patient. On the other hand, there are disadvantages pertaining to the burden of storage that is required to house large amounts of pertinent phase space information and also longer retrieval times to access data (Chetty et al. 2007; Sempau et al. 2001). Another disadvantage is residual uncertainty in the phase space data that can propagate to downstream stages (Sempau et al. 2001). Reduction of uncertainty usually requires increasing the number of histories, creating yet larger phase space files and delays in data readout.

5.4.1 Modelling of the Radiation Source

"A beam model in treatment planning is any algorithm that delivers the location, direction, and energy of particles to the patient dose-calculating algorithm" according to TG-105 (Chetty et al. 2007). To build the model requires a set of accurate detailed information regarding the geometry of all of the head components and test

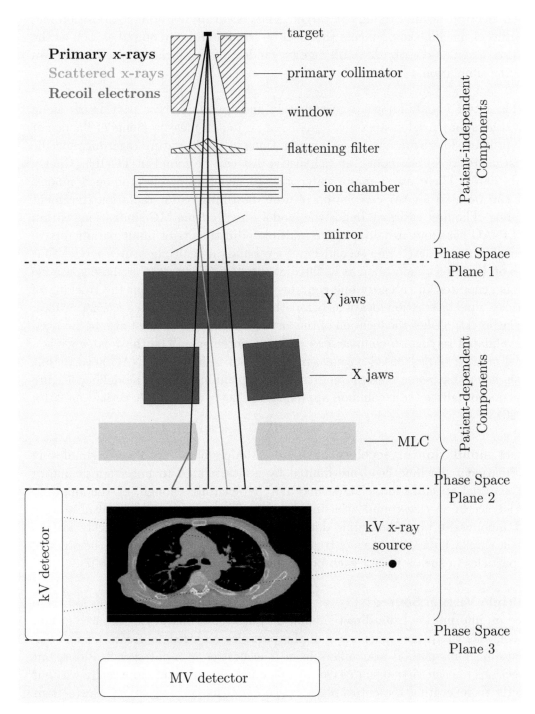

Figure 5.14: Monte Carlo simulation of particles crossing the accelerator head, with locations of phase space planes.

of sensitivity of simulation results to these parameters, including the initial accelerated electron beam striking the target. Thus accurate material composition and densities of machine components that will be irradiated are required as well as the position, angular distribution and energy window of the incident electrons beginning the simulation.

The latter is only known approximately and therefore these parameters along with others may need systematic adjustment to match measurements of fluence or dose profiles. There are numerous investigations in the literature regarding specific adjustments to source model; an exhaustive list can be found in TG-105 (Chetty et al. 2007). There are 3 major categories to generate accelerator head models that can be used for the calculation of dose distributions for radiation treatment planning. The first category builds the model directly from MC simulations within the LINAC head but usually require multiple adjustments of input parameters to match measured data. The second category builds a multiple source model that relies on the direct simulation as in the first category. However, the phase space file that is generated in category one contains more information from the tracking of particles' final interaction locations before they cross the phase space scoring surface. The latter establishes the location of the virtual sources which must also be tweaked and validated relying on comparisons with measurements. The third category is a model created purely based on measurements, not reliant on LINAC head details and is not unlike some conventional treatment planning system modelling. Figure 5.15 conceptualizes these common approaches (Chetty et al. 2007; Verhaegen 2013; Fix 2013).

Direct simulation of accelerator head Models based on directly simulating the treatment machine head are limited by issues related to uncertain geometry and material specifications. These may not be accurate enough, or vendors may restrict access to this confidential information. There is also a burden of storing large amounts of phase space file data and being able to access it quickly for the patient simulation calculations of treatment planning. There can be on the order of 10^9 particles in such a file for each beam configuration (Chetty et al. 2007).

Multiple Virtual Sources These models create virtual sources and group them based on information from direct simulation phase space data. Each source grouping contains the phase space state of the particles required to create fluence distributions. This method also suffers from inaccuracies of input specifications, but corrections to each virtual source can be made individually and empirically without directly re-generating the entire phase space file (Chetty et al. 2007). Corrections are reliant on matching downstream predictions using the virtual source model and comparing to to measured data sets, such as percentage depth dose and beam profiles, for example.

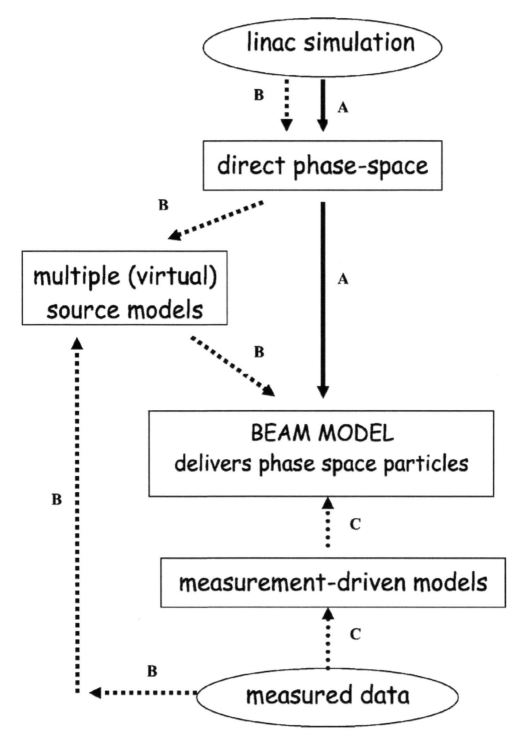

Figure 5.15: Schematic for Modelling the Radiation Source. Reproduced with permission (Chetty et al. 2007).

Empirical Models These models are driven purely by measured dose data and were common in early dose calculation models. Depending upon the basis of the beam model, whether deterministic or stochastic, calculation parameters are fitted to produce a match to a standard set of measured data. These models are not subject to the issues of uncertain LINAC head specifications and a direct simulation phase space is not used (Chetty et al. 2007). However, the selected parameters may not be unique and are optimized for dose distributions in water. They may not produce reliable results in heterogeneous tissue.

5.4.2 Modelling of the Patient

Modelling of the patient-dependent portions needs to be specifically addressed. It not only includes simulation of beam transport through the individual patient but also the customizable components of a LINAC such as secondary collimation including MLCs, beam filters, gantry angle with possible dynamic delivery. Accurate CT data conversion to mass density and atomic composition, such as using the mapping of Hounsfield units to electron densities (see Chapter 3) and utilizing tissue composition data from ICRU Report 46 is also extremely important for accurate dose calculation. Reading in DICOMRT CT patient data and utilizing automatic segmentation with specified ranges of Hounsfield Units (HU) works well for speed and convenience in allocating the material data for simulation. One must ensure appropriate calibration of the CT unit and also be aware and correct for CT image artifacts (Schlegel et al. 2006; Fippel 2006).

There must be careful optimization of voxel sizes in scoring the patient dose distribution in regions of intense dose gradients. In Monte Carlo simulations, the smaller the voxel, the greater the uncertainty. Increasing the number of histories can resolve this issue but this comes at the expense of increased computational time which could make clinical treatment planning impractical. Increasing the voxel size leads to lower spatial resolution and dose blurring effects. High statistical fluctuations in calculated dose distributions may cause difficulties related to using points or a single voxel for normalization, prescriptions or monitor unit calculations. Instead it is highly recommended to use dose-volumes for prescribing doses or isodoses (Chetty et al. 2007). These can be addressed by applying smoothing or de-noising algorithms to Monte Carlo generated dose distributions, many of which are borrowed from image-smoothing experience. These are usually applied to voxels receiving lower doses with high uncertainty.

5.4.3 Clinical Commissioning and Dose Calibration

The acceptance and commissioning requirements for a Monte Carlo radiation treatment planning system (RTPS) is similar to that of any other treatment planning system. Any RTPS must have its beam models constructed, acceptance-tested and commissioned before clinical use. The testing aims to verify and validate each section of the model (see again Figure 5.15) by comparing to measurement data and calculation from other RTPSs including other MC system calculations guided by an expert user. Specific aspects of unique Monte Carlo systems should be addressed because there are many varieties with different approaches to modelling and data input requirements. Vendors should provide a guide to acceptance and commissioning. Also, the recent guideline document from the AAPM, "AAPM Medical Physics Practice Guideline 5.a" addresses commissioning and quality assurance of the *beam modelling* and *calculation* portion of a modern RTPS for photon and electron beams, including the 3 algorithm types described in this book (Smilowitz et al. 2015). Guidance can be found on minimal requirements for data collection and dose accuracy tolerances in homogeneous and heterogeneous phantoms.

Recently, the AAPM Research Committee Task Group 268 (Sechopoulos et al. 2018) addresses a checklist for reporting Monte Carlo results, but this can also be used for checking and thoroughly verifying that any MC algorithm, including one devoted to treatment planning, is understood by the user. A checklist is provided to validate important aspects such as scoring, statistics and uncertainties, CT data utilization and dose grid, and data smoothing algorithms.

It is important to note the method that is often used to calculate absolute dose and MU settings from Monte Carlo treatment plans. Monitor units and the absolute dose distribution per beam for any Monte Carlo simulation can ultimately be obtained with the aid of a calibration factor. This beam-specific factor is calculated from a *separate* Monte Carlo simulation that mimics the beam in question under its standard beam calibration conditions such as at a reference depth with a standard field size. Once this *calibration run* is performed with high precision, the average dose at the reference voxel per incident electron *striking the target* of the accelerator can be related back to the typical machine calibration of 1 cGy / MU. This establishes the calibration factor that can then be used for other MC simulations to convert relative doses to absolute doses for the same beam model as follows:

$$1\,\mathrm{cGy/MU} = \alpha\,\mathrm{cGy/incident\ electron}$$

where the reference voxel scored an average dose per incident electron of α cGy per incident electron in the Monte Carlo calibration simulation.

The absolute dose to a voxel for *another* simulation of the same beam with a voxel scoring of:

$$\beta\,\mathrm{cGy/incident\ electron}$$

can be calculated using:

$$\beta\,\mathrm{cGy/incident\ electron} \times \frac{1\mathrm{cGy/MU}}{\alpha\,\mathrm{cGy/incident\ electron}} = \frac{\beta}{\alpha}\,\mathrm{cGy/MU}$$

Note that in the literature the calibration factor may be stated as the number of incident particles to deliver 1 MU with a corresponding uncertainty value for the reference voxel. In the above example, this is 1 MU delivered under reference conditions with $1/\alpha$ incident particles where $\alpha << 1$. To ensure that this is done properly and avoid discrepancies the beam model should ideally relate back to the number of incident electrons striking the target. For example, some phase space beam models such as the BEAMnrc code (Rogers et al. 2011) record the number of source (incident electrons striking target) electrons but others do not, making it difficult to obtain this value. Backscatter into the monitor chamber is an important confounding aspect to consider in some accelerators. Many of these calibration and normalization issues are very well documented and explained in the work of Popescu et. al (Popescu et al. 2005).

5.4.4 Monte Carlo Codes used Clinically

The most recent review article (Brualla et al. 2017) provides an update on available Monte Carlo codes, classified according to their application to either commissioning, treatment planning, or dose quality assurance (QA). Table 5.4 lists the salient features of each algorithm. Since that review, PRIMO has been updated for IMRT and VMAT treatment planning using a fast DPM infrastructure. The surge in computer performance is expected to bring new products to market. Examples include Radify developed initially at McGill University and now commercialized as RADCALC-3DMC (Lifeline Software Incorporated), ProSoma Core for dose verification (Med-Com GmbH, Darmstadt, Germany), and software for MRIdian treatment planning (ViewRay Technologies Incorporated).

Many of the Monte Carlo packages that are related to treatment planning are rooted in only a few key parent programs such as the family of EGSnrc, BEAMnrc and DOSXYZnrc; the family of PENELOPE and PENFAST; the family of ETRAN and MCNP; the family of VMC and XVMC and a few others. For example, the Monaco Monte Carlo RTPS particle simulation system from the LINAC head is constructed as a virtual energy fluence model created with 3 virtual sources. There is a primary photon source near the bremsstrahlung target (Fippel et al. 2003), a secondary photon source near the flattening filter (Fippel et al. 2003) and a third source of contaminant electrons virtually located down stream of the filter (Sikora and Alber 2009). Each source has a Gaussian shaped distribution. The latter was initiated from a phase space simulation using BEAMnrc (Rogers et al. 2011). The resulting beam model utilizes 11 parameters and is commissioned with a series of depth dose measurements and beam profiles of varying field sizes depending upon the parameter being adapted (Fix 2013). The approach to modelling radiation transport is approximately 100 times faster through the MLC's. The patient calculation

Table 5.4: Monte Carlo software packages (Brualla et al. 2017)

System	Application		Monte Carlo Software			
	Treatment Planning	QA*	LINAC Head	Patient Dose	Self-Contained*	Software Platform
CARMEN	✓		BEAMnrc	DOSXYZnrc		MATLAB®
CERR		✓		VMC++/DPM		MATLAB®
Corvus	✓		PSF	PEREGRINE	C	
Eclipse	✓		PSF	MMC	C	
eIMRT		✓	BEAMnrc	DOSXYZnrc	✓	
iPLAN RT	✓		Model	XVMC	C	
IsoGray	✓		PENELOPE	PENFAST	✓	
MCDE	✓		BEAMnrc	DOSXYZnrc		GRATIS
MCDOSE	✓		Model	EGS4		FOCUS
MCV		✓	EGS4	EGS4		Pinnacle
MCVS		✓	BEAMnrc	DOSXYZnrc		CERR
MMCTP		✓	BEAMnrc	EGSnrc		
Monaco	✓		Model	XVMC	C	
MSKCC		✓	Model	EGS4	✓	
Oncentra	✓		Model	VMC++		
Pinnacle	✓			DPM	✓	
PlanUNC		✓	EGSnrc	EGSnrc/MCNP		
PRIMO		✓	PENELOPE	PENELOPE	✓	
RTGrid	✓		BEAMnrc	EGSnrc		Globus
SMCP	✓		Model	EGSnrc/VMC++		Eclipse
VIMC		✓	BEAMnrc	DOSXYZnrc/VMC++		Eclipse
XiO	✓		Model	XVMC	C	

* QA = quality assurance; C = commercial product

part is handled by XVMC (Fippel 1999). This model has been validated experimentally (Fix 2013).

5.5 SUMMARY OF MONTE CARLO METHOD

Computational time is the key limitation in applying Monte Carlo methods clinically and its mitigation is where most problems can occur. Some of these currently necessary shortcuts are: condensed history, variance reduction techniques, and de-noising techniques for voxels with low dose and high uncertainty levels. Any shortcuts must be implemented carefully with verification of the expected range of clinical techniques. Comparison with experimental measurements, analytic data, other trusted algorithms, and validating against long test simulations that do not use shortcuts is paramount.

Accurate modelling of geometry, materials in the radiation's path, and proper scoring play an important role. This is apparent in the modelling of LINAC heads where precise details of the component dimensions and locations, material composition (i.e. cross section data) and parameters of the incident electron beam striking the target are required for an accurate model of the radiation source(s). For the patient portion, accurate interpretation of CT data in terms of mass density, electron density, and atomic composition is required for proper radiation transport in tissue and possible foreign elements such as prosthetics. Simulation of boundary-crossing for differing tissue types requires acceptable voxel sizes. Voxel size must also be optimized to balance spatial resolution and acceptability of noise in calculated dose, that in turn sets the computational time. Slow computational times can first and foremost be alleviated with faster hardware (see Chapter 7).

The gold standard for validating dose calculation algorithms used for clinical radiation therapy is currently the Monte Carlo method and it has been so for many decades. There are many references in the literature that have benchmarked Monte Carlo results by comparing to measured data. The shortcomings of other calculation techniques, such as in regions of electronic disequilibrium arising from small fields and/or inhomogeneities, are not an issue for a properly implemented Monte Carlo run. Over the last decade, there are also many papers comparing various algorithms under these difficult conditions, including convolution/superposition algorithms, deterministic Boltzmann algorithms to measurements and Monte Carlo results (as the gold standard). Tissue inhomogeneities, small fields and higher energy beams emphasize electronic disequilibrium and this is the basis of critical comparison in many of these publications, such as in Knoos et al. (Knoos et al. 1995), Vandestaeten et al. (Vanderstraeten et al. 2006), and Ojala et al. (Ojala et al. 2014).

In the race to achieve accuracy across a wide range of clinical conditions including radiation transport through accelerator components and the patient, Monte Carlo methods appear to be the frontrunner. However, significant progress has recently been made in applying deterministic Boltzmann methods to modelling dose transport through the patient more efficiently, as will be described in the next chapter.

BIBLIOGRAPHY

Andreo, P. (1991). Monte Carlo techniques in medical radiation physics. *Physics in Medicine and Biology 36*(7), 861–920.

Andreo, P., D. Burns, A. Nahum, J. Seuntjens, and F. Attix (2017). The Monte Carlo Simulation of the Transport of Radiation Through Matter. In Andreo, P. and Burns, D.T. and Nahum, A.E. and Seuntjens, J. and Attix, F.H. (Ed.), *Fundamentals of Ionizing Radiation Dosimetry*, Chapter 8, pp. 349–396. Weinheim, Germany: Wiley-VCH.

Berger, M. (1963). Monte Carlo calculation of the penetration and diffusion of fast charged particles. *Methods in Computational Physics 1*(7), 135–215.

Bielajew, A. (2013). History of Monte Carlo. In Seco, J. and Verhaegen, F. (Ed.), *Monte Carlo Techniques in Radiation Therapy*, Chapter 1, pp. 3–16. Boca Raton, Florida: CRC Press, Taylor & Francis Group.

Bielajew, A. F. and D. Rogers (1986). PRESTA: The parameter reduced electron-step transport algorithm for electron Monte Carlo transport. *Nuclear Instruments and Methods in Physics Research B*(18), 1–6.

Bohr, N. (1948). The Penetration of Atomic Particles Through Matter. *Mat. Fys. Medd. Dan. Vid. Selsk. 18*(8).

Borgers, C. (1998). Complexity of Monte Carlo and deterministic dose calculation methods. *Physics in Medicine and Biology 43*(3), 517–528.

Box, G. and M. E. Muller (1958). A Note on the Generation of Random Normal Deviates. *The Annals of Mathematical Statistics 29*(2), 610–611.

Briesmeister, J. (2000). MCNPTM A General Monte Carlo N-Particle Transport Code. Manual LA13709M, Los Alamos National Laboratory, Los Alamos, New Mexico.

Brualla, L., M. Rodriguez, and A. Lallena (2017). Monte Carlo systems used for treatment planning and dose verification. *Strahlentherapie und Onkologie 193*(4), 243–259.

Butcher, J. (1961). Random Sampling From the Normal Distribution. *The Computer Journal 3*(4), 251–253.

Butler, J. (1956). Machine sampling for given probability distributions. In H. Meyer (Ed.), *Symposium on Monte Carlo Methods*, pp. 249–264. New York, New York: John Wiley and Sons.

Chatfield, C. (1975). *Statistics for Technology*. London, England: Chapman and Hall.

Chetty, I., B. Curran, J. Cygler, J. J. Demarco, G. Ezzell, B. A. Faddegon, I. Kawrakow, P. Keall, H. Liu, C. Ma, D. Rogers, J. Seuntjens, D. Sheikh-Bagheri, and J. Siebers (2007). Report of the AAPM Task No. 105: Issues associated with clinical implementation of Monte Carlo-based photon and electron external beam treatment planning. *Medical Physics 34*(12), 4818–4853.

Feller, W. (1967). *An Introduction to Probability Theory and its Applications, Volume I* (3rd ed.). New York, New York: Wiley.

Fippel, M. (1999). Fast Monte Carlo dose calculation for photon beams based on the VMC electron algorithm. *Medical Physics 26*(8), 1466–1475.

Fippel, M. (2006). Monte Carlo Dose Calculation for Treatment Planning. In Schlegel, W. and Bortfeld, T. and Grosu, A.L. (Ed.), *New Technologies in Radiation Oncology*, Chapter 16, pp. 197–206. Berlin, Germany: Springer-Verlag.

Fippel, M. (2013a). Basics of Monte Carlo Simulations. In Seco, J. and Verhaegen, F. (Ed.), *Monte Carlo Techniques in Radiation Therapy*, Chapter 2, pp. 17–28. Boca Raton, Florida: CRC Press, Taylor & Francis Group.

Fippel, M. (2013b). Variance Reduction Techniques. In Seco, J. and Verhaegen, F. (Ed.), *Monte Carlo Techniques in Radiation Therapy*, Chapter 3, pp. 29–39. Boca Raton, Florida: CRC Press, Taylor & Francis Group.

Fippel, M., F. Haryanto, O. Dohm, F. Nusslin, and S. Kriesen (2003). A virtual photon energy fluence model for Monte Carlo dose calculation. *Medical Physics 30*(3), 301–311.

Fix, M. (2013). Photons: Clinical Considerations and Applications. In Seco, J. and Verhaegen, F. (Ed.), *Monte Carlo Techniques in Radiation Therapy*, Chapter 12, pp. 167–184. Boca Raton, Florida: CRC Press, Taylor & Francis Group.

Gill, R. (2011). The Monty Hall problem is not a probability puzzle: It is a challenge in mathematical modelling. *Statistica Neerlandica 65*(1), 58–71.

Hammersley, J. and D. Handscomb (1964). *Monte Carlo Methods*. London, England: Methuen and Company.

James, F. (1980). Monte Carlo theory and practice. *Reports on Progress in Physics 43*(9), 1145–1189.

Jenkins, T., W. Nelson, and A. Rindi (1988). *Monte Carlo Transport of Electrons and Photons*. New York, New York: Plenum Press.

Kahn, H. (1956). Use of different Monte Carlo sampling techniques. In H. Meyer (Ed.), *Symposium on Monte Carlo Methods*, pp. 146–190. New York, New York: John Wiley and Sons.

Kawrakow, I. (2000). Accurate condensed history Monte Carlo simulation of electron transport, I, EGSnrc, the new EGS4 version. *Medical Physics 27*(5), 485–498.

Kawrakow, I. (2005). On the efficiency of photon beam treatment head simulations. *Medical Physics 32*(7), 2320–2326.

Kawrakow, I., D. Rogers, and B. Walters (2004). Large efficiency improvements in BEAMnrc using directional bremsstrahlung splitting. *Medical Physics 31*(10), 2883–2898.

Knoos, T., A. Ahnesjo, P. Nilsson, and L. Weber (1995). Limitations of a pencil beam approach to photon dose calculations in lung tissue. *Physics in Medicine and Biology 40*(9), 1411.

Landau, L. (1944). On the energy loss of fast particles by ionization. *Journal of Physics (USSR) 8*, 201–205.

Luscher, M. (1994). A portable high-quality random number generator for lattice field theory simulations. *Computer Physics Communications 79*(1), 100–110.

McGrath, E. and D. Irving (1975). Techniques For Efficient Monte Carlo Simulation: Random Number Generation For selected Probability Distributionst. Report ORNL-RSIC-38, Vol II, Oak Ridge National Laboratory, Oak Ridge, Tennessee.

Metropolis, N. (1987). The beginning of the Monte Carlo Method. *Los Alamos Science (Special Issue)*, 125–130.

Muller, M. (1959). A Comparison of Methods for Generating Normal Deviates on Digital Computers. *Journal of the ACM 6*(3), 376–383.

Nelson, W., H. Hirayama, and D. Rogers (1985). The EGS4 Code System. Manual SLAC-265, Stanford Linear Accelerator Center, Stanford, California.

Niederreiter, H. (1992). *Random Number Generation and Quasi-Monte Carlo Methods*. SIAM, Philadelphia, Pennsylvania.

Ojala, J., M. Kapanen, S. Hyodynmaa, T. Wigren, and M. Pitkanen (2014). Performance of dose calculation algorithms from three generations in lung SBRT: Comparison with full Monte Carlo-based dose distributions. *Journal of Applied Clinical Medical Physics 15*(2), 4–18.

Popescu, I. A., C. P. Shaw, S. Zavgorodni, and W. Beckham (2005). Absolute dose calculations for Monte Carlo simulations of radiotherapy beams. *Physics in Medicine and Biology 50*(14), 3375–3392.

Raeside, D. E. (1976). Monte Carlo Principles and Applications. *Physics in Medicine and Biology 21*(2), 181–197.

Rogers, D. (2006). Fifty years of Monte Carlo simulations for medical physics. *Physics in Medicine and Biology 51*(13), R287–R301.

Rogers, D., B. Faddegon, G. Ding, C. Ma, J. We, and M. T.R. (1995). BEAM: A Monte Carlo code to simulate radiotherapy treatment units. *Medical Physics 22*(5), 503–524.

Rogers, D., B. Walters, and I. Kawrakow (2011). BEAMnrc Users Manual. Technical Report PIRS-0509, NRC Institute for National Measurement Standards; National Research Council Canada, Ottawa, Canada.

Salvat, H., J. Fernndez-Varea, and J. Sempau (2008). PENELOPE-2008: A Code System for Monte Carlo Simulation of Electron and Photon Transport. Manual 6416, Nuclear Energy Agency, Spain.

Sawchuk, S., J. McLellan, L. Papiez, G. Sandison, and J. Battista (1992). Alternative scoring and bremsstrahung suppression techniques for use with EGS4 Monte Carlo code. *Medical Physics 19*(Abstract No. 792).

Schlegel, W., T. Bortfeld, A. Grosu, T. Pan, and D. Luo (2006). New Technologies in Radiation Oncology. *Medical Radiology/Radiation Oncology 16*(5), 464.

Sechopoulos, I., D. W. O. Rogers, M. Bazalova-Carter, W. E. Bolch, E. C. Heath, M. F. McNitt-Gray, J. Sempau, and J. F. Williamson (2018). RECORDS: Reporting of montE CarlO RaDiation transport Studies: Report of the AAPM Research Committee Task Group 268. *Medical Physics 45*(1), e1–e5.

Sempau, J., A. Sanchez-Reyes, F. Salvat, H. Oulad ben Tahar, S. Jiang, and J. Fernandez (2001). Monte Carlo simulation of electron beams from an accelerator head using PENELOPE. *Physics in Medicine and Biology 46*(4), 1163–1186.

Sempau, J., S. J. Wilderman, and A. Bielajew (2000). DPM, a fast, accurate Monte Carlo code optimized for photon and electron radiotherapy treatment planning dose calculations. *Physics in Medicine and Biology 45*(8), 2263–2291.

Sheikh-Bagheri, D., I. Kawrakow, B. Walters, and D. Rogers (2006). Monte Carlo Simulations: Efficiency Improvement Techniques and Statistical Considerations. In B. Curren, J. Balter, and I. Chetty (Eds.), *Integrating New Technologies into the Clinic: Monte Carlo and Image-Guided Radiation Therapy, Proceedings of 2006 AAPM Summer School*, pp. 71–91. Madison, Wisconsin: Medical Physics Publishing.

Sikora, M. and M. Alber (2009). A virtual source model of electron contamination of a therapeutic photon beam. *Physics in Medicine and Biology 54* (24), 7329–7344.

Smilowitz, J., I. Das, V. Feygelman, B. Fraass, S. Kry, I. Marshall, D. Mihailidis, Z. Ouhib, T. Ritter, M. Snyder, and L. Fairobent (2015). AAPM Medical Physics Practice Guide-line 5.a.: Commissioning and QA of Treatment Planning Dose Calculations - Megavoltage Photon and Electron Beams. *Journal of Applied Clinical Medical Physics 16*(5), 14–34.

Thomas, G. and R. Finney (1979). *Calculus and Analytic Geometry* (5th ed.). Addison-Wesley, Menlo Park , California.

Vanderstraeten, B., N. Reynaert, L. Paelinck, I. Madani, C. De Wagter, W. De Gersem, W. De Neve, and H. Thierens (2006). Accuracy of patient dose calculation for lung IMRT: A comparison of Monte Carlo, convolution/superposition, and pencil beam computations. *Medical Physics 33*(9), 3149–3158.

Vassiliev, O. (2016). *Monte Carlo Methods for Radiation Transport*. New York, New York: Springer.

Vavilov, P. (1957). Ionization losses of high-energy heavy particles. *Soviet Physics JETP 5*(749), 751.

Verhaegen, F. (2013). Monte Carlo Modeling of External Photon Beams in Radiotherapy. In Seco, J. and Verhaegen, F. (Ed.), *Monte Carlo Techniques in Radiation Therapy*, Chapter 5, pp. 63–86. Boca Raton, Florida: CRC Press, Taylor & Francis Group.

Vos Savant, M. (1990). "Ask Marilyn". In *Parade Magazine*, Volume 16. New York, New York: Parade Publications Incorporated.

Ludwig Boltzmann (centre) and his colleagues in 1887. Boltzmann was an
Austrian physicist who established the foundation of statistical mechanics with
emphasis on the kinetic theory of gas molecules. In Chapter 6, the same principles
are applied to the transport of megavoltage x-rays and secondary electrons.
Photo courtesy of Universitat Graz.

Deterministic Radiation Transport Methods

George Hajdok

London Health Sciences Centre and University of Western Ontario

CONTENTS

6.1 INTRODUCTION

O VER the past decade, photon-based dose calculation algorithms formulated using convolution-superposition methods (e.g. collapsed cone, Philips Healthcare, Madison, WI) have dominated external beam radiation therapy treatment planning. Alternative algorithms based on Monte Carlo simulation (e.g. EGSnrc, National Research Council of Canada, Ottawa, ON) have also matured over the same time period, and often labelled as the "gold standard," but they are not yet widely adopted in commercial planning systems. Instead, Monte Carlo methods have primarily been relegated as an off-line tool for clinical validation (commissioning and quality assurance) of simpler and faster algorithms. In recent years, dose algorithms based on *deterministic radiation transport* methods have been introduced and have gained popularity in clinical practice (e.g. Acuros XB®, Varian Medical Systems, Palo Alto, CA), where they offer competitive dosimetric accuracy compared to their Monte Carlo counterpart (Bush et al. 2011), but with the added benefit of significantly better computational efficiency.

The theory of radiation transport deals with the mathematical description of the transport of radiation particles (or quanta) in matter, taking into account all possible charged and neutral particle interactions, as well as any nuclear decay processes (NCRP 1991; McDermott 2016; Vassiliev 2016; Nahum et al. 2017). More specifically, the theory in question can be expressed using the linear *Boltzmann* transport equation, an integro-differential equation, which can be solved using *deterministic* numerical techniques for the *phase space* number distribution of a given type of radiation particle, or related radiometric quantity. In simpler terms, the solution represents a *snap shot* of the macroscopic state of particles in terms of multiple physical variables. The Boltzmann equation can also be re-cast into an integral form, which can then be *stochastically* solved using traditional Monte Carlo methods. Thus, both deterministic and Monte Carlo approaches fundamentally solve the *same* equation, but go about it using different numerical techniques.

Although introduced recently to the field of radiotherapy dose calculations, transport theory has been applied to many areas of physics and engineering, due to the fact that particle transport processes arise in a wide variety of physical phenomena. The roots of transport theory can be traced back more than a century, when first used by Boltzmann to formulate the kinetic theory of gases (Boltzmann 1872). Subsequently, the theory was applied to derive the radiative transfer equation commonly used in astrophysics (Chandrasekhar 1960), in order to study the spectrum of radiation emerging from astronomical objects. More importantly, transport theory has been utilized to study neutron diffusion in nuclear reactor design and shielding calculations (Duderstadt and Martin 1979; Lewis and Miller 1984), where significant insight and progress have led to the development of novel numerical techniques and tools for solving the linear Boltzmann transport equation (Martin and Duderstadt 1977; Martin et al. 1981; Wareing et al. 2001; Adams and Larsen 2002). Such techniques have been adapted and made available to dose calculation algorithms for radiotherapy treatment planning (Williams et al. 2003; Boman et al. 2005; Hensel et al. 2006; Gifford et al. 2006; Vassiliev et al. 2008, 2010; Das 2015; Bouchard and Bielajew 2015; St Aubin et al. 2015, 2016).

Why are deterministic radiation transport methods difficult to learn? First, one needs to fundamentally understand the theory of the transport formalism itself, which requires a solid background in vector calculus, partial differential equations, and particle (atomic + nuclear) physics. Second, and potentially more challenging, one needs to comprehend how to practically implement and solve the set of coupled differential equations, which requires a thorough knowledge of linear algebra, various types of numerical methods, and computer programming and analysis languages. Despite the fact that medical physicists possess most of the aforementioned educational pre-requisites, application of the deterministic radiation transport method to radiotherapy dose calculation algorithms remains a foreign concept. Part of the difficulty relates to the fact that the method is not yet universally incorporated into medical physics curricula, mainly due to the time required to properly teach both the theory and practical implementation. More importantly, the deterministic radiation transport method is *not* physically intuitive when compared to convolution or Monte Carlo-based dose calculation algorithms. It is much simpler to visualize the spread of particle energy, and hence absorbed dose, using convolution kernels; whereas the solution of the linear Boltzmann transport equation, in the form of either a particle number or fluence density distribution, is only the first step towards calculating a dose distribution.

The primary goal of this chapter is to alleviate the anxiety associated with the deterministic radiation transport method as applied to radiotherapy dose calculations. The hope is for the reader is to no longer view the method, as implemented in commercial treatment planning software, as simply a "black box". The intent here is to enable him or her to understand the basic assumptions, operation, requirements and limitations of the method, in order to competently implement it within a clinical setting.

To achieve the above goal, the following chapter is partitioned into three main sections. First, a brief and concise review of the mathematical elements needed to properly interpret radiation transport theory is presented. Next, both the time-dependent and time-independent forms of the linear Boltzmann transport equation are theoretically derived from first principles (particle balance and conservation), with an accompanying physical explanation and interpretation of the equation. In addition, special cases of the Boltzmann transport equation will be presented, as well as an integral formulation, which connects it to the Monte Carlo method. Next, practical techniques that prepare the linear Boltzmann transport equation for solution, such as series expansion methods, which expand and simplify the functional forms of the phase space distribution function and cross section data; and discretization methods (e.g. multi-group method, discrete ordinates method and finite element method), which discretize the energy, angular, and spatial variables, respectively, are described. Finally, the ingredients and solution strategy on how to solve the discretized linear Boltzmann transport equation are outlined, as well as the practical assumptions used in commercial radiotherapy algorithms. A discussion related to dose calculation accuracy and computational efficiency, as compared to convolution and Monte Carlo methods, is provided.

The author wishes to disclose that he is not a pioneer nor *world expert* in the field of deterministic radiation transport methods. Nonetheless, this chapter is based on a thorough review of the literature, and the material presented represents a concise, fresh and practical synopsis suitable for students, teachers, or clinical medical physicists who wish to tackle a difficult subject in more depth. An attempt has been made to provide a *one-stop shopping* experience that focusses on a didactic approach, and to minimize time for the reader to cross-reference and decipher various publications. However, recommended supplementary textbooks and journal articles are cited in the chapter bibliography.

6.2 MATHEMATICAL PREAMBLE

Before formally deriving the Boltzmann transport equation, the fundamental radiometric quantities used in the equation must be defined, as well as the relevant variables and coordinate systems they depend on. It must be pointed out that notation will vary amongst the various publications on radiation transport theory, as applied to nuclear reactor and radiotherapy applications. As such, in the following sections, a "best-of-both worlds" approach, at the discretion of the author, is used to amalgamate the notation from both applications.

6.2.1 Phase Space Variables

The phase space of a particle describes its dependency on one or more physical variables. In general, a radiation particle can be characterized by a position in space ($\vec{r}$), velocity ($\vec{v}$), and time (t). Both position and velocity are vectors (i.e. consist of a magnitude and direction), each with three coordinate components. Thus, seven independent variables are required to describe a distribution of particles: three spatial coordinates, three velocity coordinates specifying the particle speed and direction-of-travel, and one temporal coordinate.

Closely related to particle velocity are: the particle momentum $(\vec{p})$, a vector quantity; and kinetic (E_K) or total (E) energy, both scalar quantities. Note that any one of these variables can be expressed in terms of the other. Therefore, they may be used inter-changeably depending on the nature of the problem to be solved. Table 6.1 illustrates their inter-relationship in *magnitude* for both classical and relativistic particles.

Table 6.1: Relationship between particle velocity, momentum, and energy. The variable m_0 represents particle rest mass.

Quantity	Classical Particle	Relativistic Particle	Photon
momentum	$\|\vec{p}\| = m\,\|\vec{v}\|$	$\|\vec{p}\| = \gamma\,m\,\|\vec{v}\|$	$E = \|\vec{p}\|\,c$
energy	$E_K = 1/2\,m_0\,\|\vec{v}\|^2$	$E = (\gamma - 1)\,m_0\,c^2$	

In the above table, the Lorentz factor (γ) accounts for the spatio-temporal transformation of a particle as it approaches the speed of light (c), and is given by

$$\gamma(v) = \frac{1}{\sqrt{1 - \beta^2}} \qquad \text{where} \qquad \beta(v) = \frac{v}{c} \tag{6.1}$$

Note that both γ and β depend on the particle speed $(v = |\vec{v}|)$. When the particle speed is much less than the speed of light $(v \ll c)$, then the quantity $\beta \to 0$, and $\gamma \to 1$. Therefore, such a particle can be treated using classical Galilean transformations.

In terms of the conservation of momentum and energy:

- in *classical mechanics* where $v \ll c$, the momentum and *kinetic* energy given as $p = m\,v$ and $E_K = \frac{1}{2}\,m\,v^2$, respectively, are conserved in all collisions.

- in *relativistic mechanics* where $v \approx c$, the momentum $p = \gamma m_0 v$ and *total* energy $E = m_0\,c^2 + E_K$ are conserved in all collisions

In summary, care must be taken to use the appropriate phase space variables, and consider the importance of classical or relativistic effects, for any given radiation transport problem.

6.2.2 Coordinate Systems

The Boltzmann transport equation is typically derived and solved with respect to a Cartesian recti-linear coordinate system (see Figure 6.1), although spherical coordinate variables are often mixed in with the notation. Under certain situations, a cylindrical-only or spherical-only curvi-linear coordinate system (not shown) may also be used to fully exploit the symmetry of a given problem.

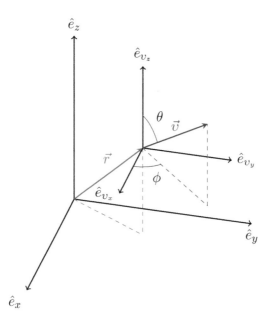

Figure 6.1: Diagram of the 3D Cartesian coordinate system typically used in radiation transport theory. The position ($\vec{r}$) of a particle (denoted in red) is specified with respect to lab coordinate system ($\hat{e}_x, \hat{e}_y, \hat{e}_z$), while the velocity ($\vec{v}$) of a particle (denoted in blue) is specified with a local (particle) coordinate system ($\hat{e}_{v_x}, \hat{e}_{v_y}, \hat{e}_{v_z}$).

In a Cartesian coordinate system, the particle position vector ($\vec{r}$) can be decomposed into the following components:

$$\vec{r} = x\,\hat{e}_x + y\,\hat{e}_y + z\,\hat{e}_z \tag{6.2}$$

where $r = |\vec{r}| = \sqrt{(x^2 + y^2 + z^2)}$ denotes the Euclidean distance between the position of a particle in space and the origin, and ($\hat{e}_x, \hat{e}_y, \hat{e}_z$) represent unit vectors along the (x, y, z) axes, respectively.

Similarly, the particle velocity vector ($\vec{v}$) can be expressed as

$$\vec{v} = v_x\,\hat{e}_{v_x} + v_y\,\hat{e}_{v_y} + v_z\,\hat{e}_{v_z} \tag{6.3}$$

where the quantity $v = |\vec{v}| = \sqrt{(v_x^2 + v_y^2 + v_z^2)}$ represents the speed of the particle, and $(\hat{e}_{v_x}, \hat{e}_{v_y}, \hat{e}_{v_z})$ denotes unit vectors along the (v_x, v_y, v_z) axes, respectively.

As an important aside, note that both Cartesian coordinate systems defined above, the lab reference frame (x, y, z) and particle reference frame (v_x, v_y, v_z), are assumed to be *co-linear* with respect to one another. Therefore, the unit vectors in the lab and particle reference frames are equivalent, $(\hat{e}_x, \hat{e}_y, \hat{e}_z) \equiv (\hat{e}_{v_x}, \hat{e}_{v_y}, \hat{e}_{v_z})$, and can be used inter-changeably. Therefore, in the rest of the chapter, the unit vectors $(\hat{e}_x, \hat{e}_y, \hat{e}_z)$ will be used to define the direction of the particle. In addition, the lab reference frame (x, y, z) is assumed to be stationary.

Instead of explicitly using Cartesian unit vectors $(\hat{e}_x, \hat{e}_y, \hat{e}_z)$ for velocity (or momentum), the direction-of-flight of a particle is traditionally specified using the unit vector[1] $(\hat{\Omega})$ as follows:

$$\vec{v} = |\vec{v}|\,\hat{\Omega} \quad \text{or} \quad \vec{p} = |\vec{p}|\,\hat{\Omega} \qquad \Rightarrow \qquad \hat{\Omega} = \frac{\vec{v}}{|\vec{v}|} = \frac{\vec{p}}{|\vec{p}|} \tag{6.4}$$

where it can formally be defined in terms of three *direction cosines* (η, ξ, μ) using

$$\hat{\Omega} = \eta\,\hat{e}_x + \xi\,\hat{e}_y + \mu\,\hat{e}_z \tag{6.5}$$

subject to the following constraint

$$\eta^2 + \xi^2 + \mu^2 = 1 \tag{6.6}$$

which implies that each value is bounded within the range of 0 and 1.

Direction cosines mathematically represent the projections of a vector onto each of the principal coordinate axes $(\hat{e}_x, \hat{e}_y, \hat{e}_z)$. Moreover, they can also be directly related to two spherical angles: the polar angle (θ), and azimuthal angle (ϕ), which span the domains $0 \le \theta \le \pi$ and $0 \le \phi \le 2\pi$, respectively. In terms of these angles, the direction cosines can be calculated as

$$\hat{\Omega} \equiv (\eta, \xi, \mu) \equiv (\sin\theta\cos\phi, \sin\theta\sin\phi, \cos\theta) \tag{6.7}$$

Note that here the *z-axis* is chosen as the principal axis to define the polar angle θ. Other publications may use the *x-axis* as the reference axis, so care must be taken.

Based on the trigonometric identity $\sin^2\theta + \cos^2\theta = 1$, the first two direction cosines can be re-written in terms of μ as

$$\eta = \sqrt{1 - \mu^2}\,\cos\phi \qquad \text{and} \qquad \xi = \sqrt{1 - \mu^2}\,\sin\phi \tag{6.8}$$

---

[1] Do not confuse $\hat{\Omega}$ with $d\Omega$. The former is a unit vector, while the latter represents a differential element of solid angle.

Using Equations 6.7 and 6.8, the components of the particle velocity vector can be related by

$$v_x = v \sin \theta \cos \phi = v\sqrt{1 - \mu^2} \cos \phi = v\,\eta$$
$$v_y = v \sin \theta \sin \phi = v\sqrt{1 - \mu^2} \sin \phi = v\,\xi \tag{6.9}$$
$$v_z = v \cos \theta = v\mu$$

where the velocity polar and azimuthal angles are related to their components by

$$\theta = \arccos\left(v_z / \sqrt{v_x^2 + v_y^2 + v_z^2}\right) \tag{6.10}$$
$$\phi = \arctan\left(v_y / v_x\right)$$

Expressions for the particle position and velocity vectors in both cylindrical and spherical coordinates can be found in the following references (Lewis and Miller 1984; Duderstadt and Martin 1979).

6.2.3   Radiometry and Dosimetry

In order to use deterministic Boltzmann transport methods for radiotherapy dose calculations, an important distinction must be made between radiometric and dosimetric quantities. Radiometry deals with quantities (e.g. fluence and energy fluence) that characterize the *flow* of radiation (particle or energy), while dosimetry involves quantities (e.g. kinetic energy released per unit mass and absorbed dose) that describe the *conversion* or *deposition* of energy in matter through various types of particle interactions. These two classes of radiation quantities are invariably linked in that dosimetric quantities cannot be determined without radiometric ones. Moreover, the solution of the Boltzmann transport equation is a radiometric quantity, and therefore an *additional* calculation is needed to obtain the desired dosimetric result. Further information on the following quantities can be found in ICRU Report 85 (ICRU 2011) and NCRP Report 108 (NCRP 1991).

Radiometric Quantities

As originally proposed by (Rossi and Roesch 1962; Roesch and Attix 1968; Roesch 1968), the idea of a *radiation field*[2] can be used to characterize a variety of radiometric quantities. In general physics terminology, such a field can be simply defined as any physical quantity that depends on position and time. The two main types of fields are: *scalar fields*, whereby a single number (a scalar) is assigned to each spatial and temporal coordinate; and *vector fields*, in which the vector components are specified at each point in space and time. The concept of a vector field is particularly important in radiation transport because its magnitude is a measure

---

[2]Note that the term "radiation field" here should <u>not</u> be confused with the *geometric field* commonly used in clinical radiation therapy physics to describe a treatment beam field.

of how many radiation particles are flowing at any given moment in phase space, while its direction indicates the local flight path of that flow.

Since radiation transport theory deals with localized particles (charged or uncharged), it differs from a continuum-based theory, such as electro-magnetism or fluid dynamics, in classical physics. In the latter, a continuous description is used to describe the macroscopic fields, whereas in the former, the random nature of particle interactions requires a field based on probability density or distribution functions. Furthermore, the exact number of particles in a region of space and time cannot be specified, but only the *mean* or *expected* number of particles can be stated. Nonetheless, a macroscopic approach and interpretation can still be used to derive the radiation transport equation.

The most fundamental radiometric quantity used in radiation transport theory is the *phase-space density distribution*. All other scalar and vector-based radiometric quantities, as shown below, can be derived from it. As previously shown in Table 6.1, the particle velocity is related to both particle momentum and energy. Therefore, these phase-space variables can be used inter-changeably in the distribution functions introduced below. As a replacement for particle velocity ($\vec{v}$), particle direction ($\hat{\Omega}$) and energy (E) will be used instead. Figure 6.2 summarizes the interrelationship between the various radiometric quantities[3] listed below.

Particle Number Density

The radiometric quantity, $n(\vec{r}, \hat{\Omega}, E, t)$, is the scalar *particle number density distribution*, in units of $[\text{cm}^{-3}\ \text{sr}^{-1}\ \text{MeV}^{-1}]$.

Multiplication of the quantity $n(\vec{r}, \hat{\Omega}, E, t)$ with the differential element $dV\,d\Omega\,dE$ denotes the *expected* number of particles in a differential volume (dV) element about $\vec{r}$, travelling in a cone of direction ($d\Omega$) about $\hat{\Omega}$, with energy in the interval between E and $E + dE$, at time t. Note that $dV\,d\Omega\,dE$ is a cell in a six-dimensional phase space.

The *particle number density*, in units of $[\text{cm}^{-3}]$, is obtained by integrating over both direction and energy of the particles as follows:

$$N(\vec{r}, t) = \int dE \int d\Omega\, n(\vec{r}, \hat{\Omega}, E, t) \tag{6.11}$$

The quantity, $N(\vec{r}, t)\,dV$, represents the *total* number of particles in dV at time t.

---

[3] Be aware that notation for radiometric quantities will vary in the literature. In this chapter, we adopt notation similar to the work by Duderstadt and Martin 1979, whereby: a *lower case* symbol is used to denote a distribution [e.g. $n(\vec{r}, \hat{\Omega}, E, t)$], while an *upper case* is used to represent a non-distribution [e.g. $N(\vec{r}, t)$] function.

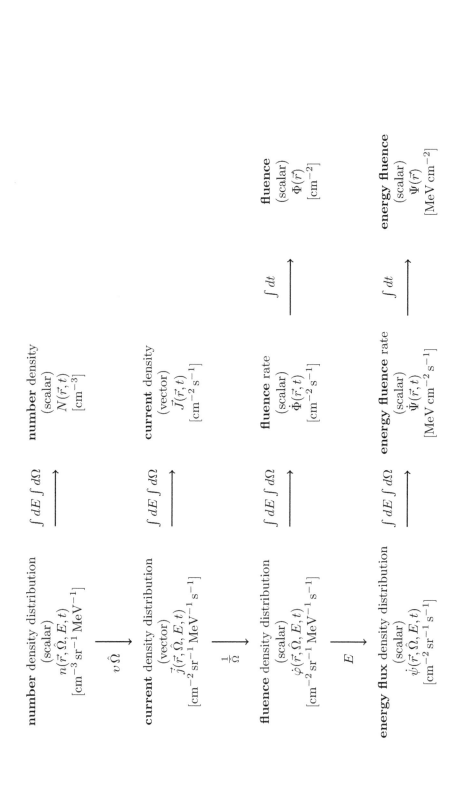

Figure 6.2: Relationship between various radiometric quantities used in the Boltzmann transport equation formalism. Shown for each quantity are its name, type, symbol, and unit. Lower case variables are used to distinguish between a distribution (left column) and non-distribution (right column).

In ICRU Report 85, the notations, $n_{\Omega,E}$ and n, are used for particle number density distribution and particle number density, respectively.

The particle number density distribution and integrated particle number density functions above can both be combined to give the probability density or distribution function, $p(\vec{r}, \hat{\Omega}, E, t)$, as follows:

$$p(\vec{r}, \hat{\Omega}, E, t) = \frac{n(\vec{r}, \hat{\Omega}, E, t)}{N(\vec{r}, t)} \tag{6.12}$$

which represents the *probability* of finding a particle within a differential element of phase space.

Particle Current Density

The radiometric quantity, $\vec{j}(\vec{r}, \hat{\Omega}, E, t)$, is the vector *particle current density distribution*, in units of $[\text{cm}^{-2}\ \text{sr}^{-1}\ \text{MeV}^{-1}\ \text{s}^{-1}]$.

The quantity, $\vec{j}(\vec{r}, \hat{\Omega}, E, t) \cdot \mathrm{d}\vec{A}\, \mathrm{d}\Omega\, \mathrm{d}E\, \mathrm{d}t$, represents the *net* number of particles (outgoing minus incoming) that perpendicularly cross a surface area element $\mathrm{d}\vec{A}$, moving in direction around $\hat{\Omega}$ within the solid angle between $\hat{\Omega}$ and $\hat{\Omega} + \mathrm{d}\Omega$, with energy in the interval between E and $E + \mathrm{d}E$, at time t. Note that $\mathrm{d}\vec{A} = \mathrm{d}A\,\hat{e}_n$, where $\hat{e}_n$ is a unit vector normal (*outward*) to the surface area element.

The *particle current density* is obtained by integrating over both direction and energy of the particles as given by

$$\vec{J}(\vec{r}, t) = \int \mathrm{d}E \int \mathrm{d}\Omega\, \vec{j}(\vec{r}, \hat{\Omega}, E, t) \tag{6.13}$$

In ICRU Report 85, the notations, $\dot{\vec{\Phi}}_{\Omega,E}$ and $\dot{\vec{\Phi}}$, known as the vector particle radiance and vector fluence rate, are used for particle current density distribution and particle current density, respectively.

Figure 6.3 depicts particles travelling in direction $\hat{\Omega}$ across a planar and spherical surface element area $\mathrm{d}A$ with unit normal vector $\hat{e}_n$. Note that any surface area element has two normal vectors, one in the positive direction and the other in the negative direction. The choice of $\hat{e}_n$ dictates the direction-of-flow through $\mathrm{d}A$, and the dot product between $\hat{\Omega}$ and $\hat{e}_n$ can be used to isolate the direction of flow across $\mathrm{d}A$. For example, $\hat{\Omega} \cdot \hat{e}_n > 0$ gives the hemisphere of directions for positive flow through $\mathrm{d}A$, whereas $\hat{\Omega} \cdot \hat{e}_n < 0$ indicates negative flow.

The particle velocity ($\vec{v}$) can be used to relate the particle number and current density distributions as follows:

$$\vec{j}(\vec{r}, \hat{\Omega}, E, t) = \vec{v}\, n(\vec{r}, \hat{\Omega}, E, t) \tag{6.14}$$

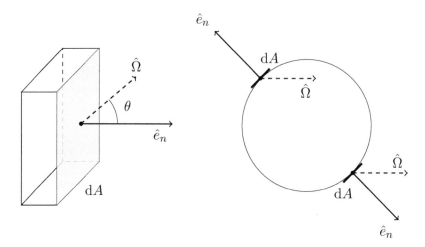

Figure 6.3: The flow of particles across planar and spherical surfaces. The dot product of the particle direction $\hat{\Omega}$ with the surface normal direction $\hat{e}_n$ can be used to describe whether a particle is entering or exiting the volume.

which can be simplified further as

$$\vec{j}(\vec{r}, \hat{\Omega}, E, t) = \hat{\Omega} \left[v \, n(\vec{r}, \hat{\Omega}, E, t) \right] \tag{6.15}$$

where the bracketed term represents the next radiometric quantity.

Particle Fluence (or Flux) Density

The radiometric quantity, $\dot{\varphi}(\vec{r}, \hat{\Omega}, E, t)$, is the scalar *particle fluence*[4] *density distribution*, in units of $[\text{cm}^{-2} \text{ sr}^{-1} \text{ MeV}^{-1} \text{ s}^{-1}]$. Based on Equation 6.15, it can be seen that the fluence density distribution is simply the magnitude of the current density distribution, and therefore

$$\dot{\varphi}(\vec{r}, \hat{\Omega}, E, t) = v \, n(\vec{r}, \hat{\Omega}, E, t) \tag{6.16}$$

Upon integration over all angles and energies gives the *fluence density* or *fluence rate*,

$$\dot{\Phi}(\vec{r}, t) = \int dE \int d\Omega \, \dot{\varphi}(\vec{r}, \hat{\Omega}, E, t) \tag{6.17}$$

Similar to the case of the particle density, the quantity, $\dot{\Phi}(\vec{r}, t) \, dV \, dt$, represents the *total* fluence of particles in dV and time dt located at position $\vec{r}$. Historically, the total fluence has always been defined within a spherical differential element. However, Papiez and Battista have shown that the differential volume can be of

---

[4]In the field of medical physics, the term fluence is preferred over flux.

any general shape (Papiez and Battista 1994).

In ICRU Report 85, the notations, $\dot{\Phi}_{\Omega,E}$ and $\dot{\Phi}$, known as the particle radiance distribution and fluence rate, are used for particle fluence density distribution and fluence rate, respectively.

Although $\dot{\Phi}$ and $\vec{J}$ have the same units, they have significantly different meanings. The product $\Phi \, d\vec{A}$ describes the *total* rate of particles going through $d\vec{A}$, while the dot product $\vec{J} \cdot d\vec{A}$ represents the *net* flow rate of particles passing through $d\vec{A}$.

Particle Energy Fluence (or Flux) Density

The radiometric quantity, $\dot{\psi}(\vec{r}, \hat{\Omega}, E, t)$, is the scalar *particle energy fluence density distribution*[5], in units of $[\text{cm}^{-2} \ \text{sr}^{-1} \ \text{s}^{-1}]$.

Upon integration over all angles and energies gives the *energy fluence density* or *energy fluence rate*,

$$\dot{\Psi}(\vec{r}, t) = \int dE \int d\Omega \ \dot{\psi}(\vec{r}, \hat{\Omega}, E, t) \tag{6.18}$$

In ICRU Report 85, the notations, $\dot{\Psi}_{\Omega,E}$ and $\dot{\Psi}$, known as the particle energy radiance distribution and energy fluence rate, are used for particle energy fluence density and energy fluence rate, respectively.

Dosimetric Quantities

The effects of radiation on matter depend both on the radiation field, as characterized by radiometric quantities, and on interactions between the particles of radiation and the absorbing medium, as described by the macroscopic cross sections for various types of particle interactions. Dosimetric quantities are simply products of these radiometric quantities and all relevant interaction coefficients.

Indirectly ionizing particles, such as photons, interact with matter in a series of processes in which particle energy is converted, and finally, deposited in matter. The term conversion of energy refers to the transfer of incident particle energy to either secondary *charged particles only*, as characterized by the quantity KERMA (Kinetic Energy Released per unit MAss); or both secondary *charged and uncharged particles*, as described by the quantity TERMA (Total Energy Released per unit MAss). Once charged particles have been liberated, they progressively lose energy to the medium through ionization and excitation events. The concept of absorbed dose, energy imparted per unit mass of a region of interest within the medium, is used to quantify the amount of energy deposited in the medium along charged

---

[5]In the nuclear engineering community, a lower case psi, ψ, is typically used to represent the particle fluence (or angular flux) density distribution

particle tracks.

Absorbed Dose

In photon-based dose calculation algorithms, the Boltzmann transport equation must be solved for the *charged particle* fluence density distribution, $\varphi^{e^-/e^+}(\vec{r}, \hat{\Omega}, E)$. Recall that photon energy is deposited in a medium by charged particles (electrons or positrons) only. Therefore, the solution of the Boltzmann transport equation is used to calculate the absorbed dose, $D(\vec{r})$, in a medium by the following equation:

$$D(\vec{r}) = \int_E dE \int_{4\pi} d\Omega \, \frac{S_c}{\rho}(\vec{r}, E) \, \varphi^{e^-/e^+}(\vec{r}, \hat{\Omega}, E) \qquad (6.19)$$

where S_c/ρ represents the mass collisional stopping power for either electrons or positrons set in motion from particle interactions.

6.3 THE BOLTZMANN TRANSPORT EQUATION IN THEORY

In the following section, the time-dependent, differential form of the linear Boltzmann transport equation will be derived intuitively, using concepts from fluid mechanics, and rigorously, using formal mathematics as well as principles from radiation physics. In addition, the integral form of the linear Boltzmann transport equation will then be derived from the differential form, which can then be *indirectly* solved using Monte Carlo methods (Chapter 5). The integral form will also be connected (under simplifying assumptions) to the convolution formalism used in standard dose algorithms (Chapter 4). Finally, the coupled photon-electron Boltzmann transport equation, as applied to external beam radiotherapy, is assembled and discussed, along with the underlying physics.

> **Principle #9: "The Boltzmann transport equation describes an *accounting* principle for tracking how particle fluence is distributed within phase space."**

6.3.1 Balance Equation

A balance equation is a conservation statement for some physical quantity subject to the absence or presence of various physical processes. Within the context of radiation transport theory, a balance equation can be formulated for the number of particles in a region of space (with differential volume dV and surface area dA) by applying *local* particle conservation.

The simplest form of a particle transport balance equation can be written as follows:

$$\begin{bmatrix} \text{number of particles} \\ \text{that flow \underline{in}} \\ \text{volume } \mathrm{d}V \text{ through} \\ \text{surface } \mathrm{d}A \end{bmatrix} = \begin{bmatrix} \text{number of particles} \\ \text{that flow \underline{out} of} \\ \text{volume } \mathrm{d}V \text{ through} \\ \text{surface } \mathrm{d}A \end{bmatrix} \tag{6.20}$$

whereby subtraction of the right hand side by the left hand side yields the *net* number of particles. Such an equation describes the *free flow* of particles, commonly referred to as particle *streaming* in the literature.

A more general form of the particle transport balance equation includes "sources" and "sinks" as given by

$$\begin{bmatrix} \text{net number of} \\ \text{particles from} \\ \text{volume } \mathrm{d}V \end{bmatrix} = \begin{bmatrix} \text{number of particles} \\ \text{produced by a source} \\ \text{within volume } \mathrm{d}V \end{bmatrix} + \begin{bmatrix} \text{number of particles} \\ \text{\underline{removed} from} \\ \text{volume } \mathrm{d}V \end{bmatrix} \tag{6.21}$$

whereby "sources" refer to physical processes (e.g. radioactive decay) that produce particles, and "sinks" represent processes (e.g. absorption and/or scattering events) that remove or transform particles from one type to another.

6.3.2 Continuity Equation

Before formally deriving the linear Boltzmann transport equation for radiation particles, further insight into the equation can be drawn by noting parallels with fluid mechanics and electro-magnetism.

The objective here is to calculate the net rate of change of particle flow through a volume element ($\mathrm{d}V$) with dimensions ($\mathrm{d}x, \mathrm{d}y, \mathrm{d}z$) centred at point $\vec{r}$. Here, the flow is quantified by the product of the particle number density and velocity. Similarly, in fluid mechanics, the flow is determined using the fluid density and velocity, and in electromagnetism, the flow is calculated using the charge density and velocity.

Assume that the surfaces of the volume are parallel with the xy, xz, and yz planes (see Figure 6.4). The particle flow rate into or out of each of the six surfaces of the volume element can be approximated using a Taylor series expansion about the centre point of the volume, whereby particle number density and the velocity components are denoted as n and (v_x, v_y, v_z), respectively.

For example, the particle flow rate at the left face of the volume element is given by

$$nv_y|_{\text{left}} = \left[nv_y - \frac{\partial(nv_y)}{\partial y}\frac{\mathrm{d}y}{2} + \frac{1}{2!}\frac{\partial^2(nv_y)}{\partial y^2}\left(\frac{\mathrm{d}y}{2}\right)^2 + \dots \right] \mathrm{d}x\,\mathrm{d}z \tag{6.22}$$

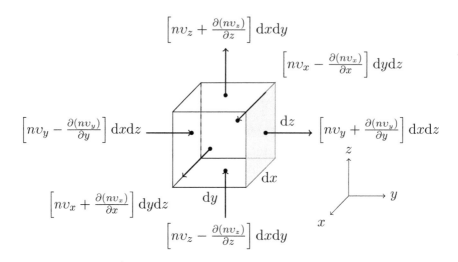

Figure 6.4: Inward and outward flow of particles for a volume element in the $\pm x, \pm y, \pm z$ directions. For example, the Taylor series expansion of the quantities $\pm n v_y$ must be multiplied by $\mathrm{d}x\,\mathrm{d}z$ to represent the total flux through the bounding surfaces at $y \pm \mathrm{d}y/2$. Also shown are the flow components in the x and z directions.

To first order in $\mathrm{d}y$, the density of particles entering the volume element per unit time through face xz located at $y - \mathrm{d}y/2$ is given by

$$n v_y|_{\text{left}} \approx - \left[n v_y - \frac{\partial(n v_y)}{\partial y} \frac{\mathrm{d}y}{2} \right] \mathrm{d}x\,\mathrm{d}z \tag{6.23}$$

Note that only the y-component of the velocity (v_y) is relevant here. The other components of $\vec{v}$ do not contribute to flow through face xz of the volume element. In addition, the minus sign corresponds to *outward* flow along the $-\hat{e}_y$ direction.

Similarly, the outward particle flow density through volume element face xz located at $y + \mathrm{d}y/2$ is given by

$$n v_y|_{\text{right}} \approx + \left[n v_y + \frac{\partial(n v_y)}{\partial y} \frac{\mathrm{d}y}{2} \right] \mathrm{d}x\,\mathrm{d}z \tag{6.24}$$

Combining Equations 6.23 and 6.24 yields an expression for both xz faces

$$\left\{ - \left[n v_y - \frac{\partial(n v_y)}{\partial y} \frac{\mathrm{d}y}{2} \right] + \left[n v_y + \frac{\partial(n v_y)}{\partial y} \frac{\mathrm{d}y}{2} \right] \right\} \mathrm{d}x\,\mathrm{d}z = \left[\frac{\partial(n v_y)}{\partial y} \right] \mathrm{d}x\,\mathrm{d}y\,\mathrm{d}z \tag{6.25}$$

Finally, adding the corresponding contributions from each of the other volume element faces yields

$$\text{net flow rate out} = \left[\frac{\partial}{\partial x}(n v_x) + \frac{\partial}{\partial y}(n v_y) + \frac{\partial}{\partial z}(n v_z) \right] \mathrm{d}x\,\mathrm{d}y\,\mathrm{d}z$$

$$= \left[\vec{\nabla} \cdot (n\vec{v}) \right] \mathrm{d}x\,\mathrm{d}y\,\mathrm{d}z \tag{6.26}$$

and the divergence of the vector $n\vec{v}$ represents the net outflow of particles per unit volume, per unit time.

If the physical problem being examined does not involve the production or removal of particles from within the volume, then Equation 6.26 can be re-arranged and the particle current density $\vec{j} = n\vec{v}$ substituted as follows:

$$\frac{\partial n}{\partial t} + \vec{\nabla} \cdot \vec{j} = 0 \qquad (6.27)$$

which is simply the famous *continuity* equation in physics as applied to particle conservation, and states that the net outflow of particles from a volume element results in a smaller density of particles inside the volume. The continuity equation also applies in electromagnetism (charge conservation), fluid dynamics (mass conservation), thermodynamics (energy conservation), and quantum mechanics (probability conservation).

When a vector field represents the flow of some quantity distributed in space, the divergence of that quantity provides information on the accumulation or depletion of that quantity at the point where the divergence is evaluated. When a vector quantity is non-divergent within a spatial region, it can be interpreted as describing a steady-state "particle-conserving" flow within that region.

6.3.3 Time-Dependent Formulation

The time-dependent Boltzmann transport equation is derived below from both a Eulerian and Langrangian framework (Duderstadt and Martin 1979; Lewis and Miller 1984; Vassiliev 2016), followed by a description and inclusion of *sources* and *sinks*, as stated in the second balance equation (Equation 6.21), in order to form the complete Boltzmann transport equation.

Eulerian Form

Consider once again an arbitrary volume V enclosed by a surface ∂V. The *net* change per unit time in the total number of particles of a given type in V within the phase space element $d\Omega\, dE$ is given by the balance of particles entering and leaving the element $d\Omega\, dE$ through surface patch dA as given by

$$\frac{\partial}{\partial t} \int_V n(\vec{r}, \hat{\Omega}, E, t)\, dV = -\oint_{\partial V} \vec{j}(\vec{r}, \hat{\Omega}, E, t) \cdot \hat{e}_n\, dA \qquad (6.28)$$

where the term on the right hand side corresponds to particles leaving the volume (minus sign) in the direction $\hat{e}_n$.

Using the divergence theorem (or Gauss' theorem) from vector calculus, which states that the flux of a vector function through a closed surface is equal to the

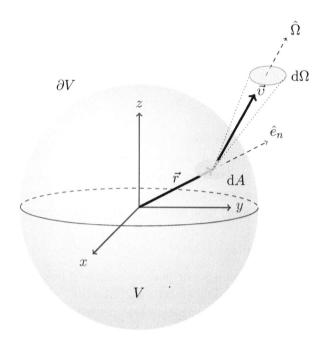

Figure 6.5: Phase space geometry used to describe the flow of particles through an arbitrary volume V enclosed by a surface boundary ∂V. Particle flow through a differential volume element dV within V is represented by the velocity direction $\hat{\Omega}$ and *outward* unit normal vector $\hat{e}_n$ associated with a small patch dA on the surface boundary.

integral of the divergence of that function over the volume enclosed by that surface, then the surface integral in Equation 6.28 can be converted into a volume integral as follows:

$$\oint_{\partial V} \vec{j} \cdot \hat{e}_n \, \mathrm{d}A = \int_V \vec{\nabla} \cdot \vec{j} \, \mathrm{d}V = \int_V \vec{\nabla} \cdot \vec{v} n \, \mathrm{d}V = \int_V \vec{v} \cdot \vec{\nabla} n \, \mathrm{d}V \qquad (6.29)$$

Inserting Equation 6.29 into Equation 6.28 gives

$$\int_V \left[\frac{\partial n}{\partial t} + \vec{v} \cdot \vec{\nabla} n \right] \mathrm{d}V = 0 \qquad (6.30)$$

Since Equation 6.30 is valid for any volume V, then it can be satisfied for all V, if and only if the integrand in square brackets is identically zero, which yields

$$\frac{\partial n}{\partial t} + \vec{v} \cdot \vec{\nabla} n = 0 \qquad (6.31)$$

which is the radiation transport equation for the particle (phase space) number density distribution $n(\vec{r}, \hat{\Omega}, E, t)$ in the absence of any particle losses or gains.

Note that Equation 6.31 is a continuity equation for particle conservation, and is equivalent to Equation 6.27 less formally derived in the previous section. Only here the derivation is more rigorous through the use of Gauss' theorem, whereas previously, an approximation (Taylor series expansion) was used to *intuitively* assemble the equation.

Lagrangian Form

As shown above, the Boltzmann transport equation represents an expression for the temporal rate of change of the particle number density distribution. In differential calculus, the total time derivative of a function with multiple independent variables can be calculated using the chain rule of differentiation. For example, the total time derivative of the multi-variate function, $n(\vec{r}, \vec{v}, t)$, or n for short, is given by

$$\frac{\mathrm{d}n}{\mathrm{d}t} = \frac{\partial n}{\partial t} + \frac{\partial n}{\partial x}\frac{\mathrm{d}x}{\mathrm{d}t} + \frac{\partial n}{\partial y}\frac{\mathrm{d}y}{\mathrm{d}t} + \frac{\partial n}{\partial z}\frac{\mathrm{d}z}{\mathrm{d}t} + \frac{\partial n}{\partial v_x}\frac{\mathrm{d}v_x}{\mathrm{d}t} + \frac{\partial n}{\partial v_y}\frac{\mathrm{d}v_y}{\mathrm{d}t} + \frac{\partial n}{\partial v_z}\frac{\mathrm{d}v_z}{\mathrm{d}t} \qquad (6.32)$$

where the components of the vectors $\vec{r}$ and $\vec{v}$ have been used in the differentiation. In fluid mechanics, such a breakdown of the total time derivative of n is also known as the *substantial* derivative.

Note that the phase space variables have been temporarily switched from $(\vec{r}, \hat{\Omega}, E, t)$ to $(\vec{r}, \vec{v}, t)$. The result above says that some of the change in n is due to the partial derivative of n with respect to t; but also, some of that change is due to the partial derivatives of n with respect to the components of $\vec{r}$ and $\vec{v}$.

Equation 6.32 can be simplified further by using vector notation for $\vec{r}$ and $\vec{v}$ in the derivative in t; and gradient operators for position and velocity, ∇_r and ∇_v, in the partial derivatives with respect to the components of $\vec{r}$ and $\vec{v}$

$$\frac{\mathrm{d}n}{\mathrm{d}t} = \frac{\partial n}{\partial t} + \frac{\mathrm{d}\vec{r}}{\mathrm{d}t} \cdot \vec{\nabla}_r n + \frac{\mathrm{d}\vec{v}}{\mathrm{d}t} \cdot \vec{\nabla}_v n \tag{6.33}$$

where $\vec{\nabla}_r$ and $\vec{\nabla}_v$ are given by

$$\vec{\nabla}_r = \hat{e}_x \frac{\partial}{\partial x} + \hat{e}_y \frac{\partial}{\partial y} + \hat{e}_z \frac{\partial}{\partial z} \quad \text{and} \quad \vec{\nabla}_v = \hat{e}_x \frac{\partial}{\partial v_x} + \hat{e}_y \frac{\partial}{\partial v_y} + \hat{e}_z \frac{\partial}{\partial v_z} \tag{6.34}$$

and note that once again the Cartesian axes for the position and velocity spaces are assumed to be co-linear.

Recognizing that the time derivatives of the position and velocity vectors are simply the velocity, $\vec{v} = d\vec{r}/dt$, and acceleration, $\vec{a} = d\vec{v}/dt$, respectively, then Equation 6.33 becomes

$$\frac{\mathrm{d}n}{\mathrm{d}t} = \frac{\partial n}{\partial t} + \vec{v} \cdot \vec{\nabla}_r n + \vec{a} \cdot \vec{\nabla}_v n \tag{6.35}$$

Therefore, within the Lagrangian framework, the time derivative of the particle number density distribution has produced an additional term, not present in the Eulerian framework. Based on Newtons second law of motion, the acceleration $\vec{a}$ can be recast in terms of a force $\vec{F}$ and mass m as $\vec{a} = \vec{F}/m$. Equation 6.35 simplifies to

$$\frac{\mathrm{d}n}{\mathrm{d}t} = \frac{\partial n}{\partial t} + \vec{v} \cdot \vec{\nabla}_r n + \frac{\vec{F}}{m} \cdot \vec{\nabla}_v n \tag{6.36}$$

The third term on the right hand side represents the effect of *external* forces on the particle number (phase space) density distribution, which serve to alter both the magnitude and direction of the particles. This term should be included if external forces (e.g. electric and magnetic fields) are known to be present in the given problem. For example, in MR-guided radiotherapy, magnetic fields can alter the trajectory of charged particles, and hence perturb the dose distribution within a patient. Further discussion of the effects of these force fields on the Boltzmann transport equation can be found in these references (Fan et al. 2013; Bouchard and Bielajew 2015; St Aubin et al. 2015, 2016). For the time being, external forces will be assumed to be absent in the following derivation.

General Form

Using the form of the Boltzmann transport equation previously derived from the Eulerian and/or Langrangian framework, and including the *sources* and *sinks* terms as specified in Equation 6.21 gives

$$\frac{\mathrm{d}n}{\mathrm{d}t} = \left[\frac{\partial n}{\partial t}\right]_{\text{stream}} + \left[\frac{\partial n}{\partial t}\right]_{\text{source}} + \left[\frac{\partial n}{\partial t}\right]_{\text{abs}} + \left[\frac{\partial n}{\partial t}\right]_{\text{scat}} \tag{6.37}$$

whereby *stream* denotes streaming, and *sinks* have been broken down further into absorptive and scattering physical processes.

Streaming

The streaming term in the Boltzmann transport equation has already been derived in the context of a continuity equation as discussed and summarized in Equation 6.27, and is presented again here,

$$\left[\frac{\partial n}{\partial t}\right]_{\text{stream}} = -\vec{v}\cdot\nabla n(\vec{r},\hat{\Omega},E,t) = -v\,\hat{\Omega}\cdot\nabla n(\vec{r},\hat{\Omega},E,t) \tag{6.38}$$

Sources

Similar to how the particle number or fluence density distribution are defined, the radiometric quantity $\dot{s}(\vec{r},\hat{\Omega},E,t)$, is the scalar *particle source density distribution*, in units of $[\text{cm}^{-2}\ \text{sr}^{-1}\ \text{MeV}^{-1}\ \text{s}^{-1}]$. The product of this quantity with the phase space cell $dV\,d\Omega\,dE\,dt$ represents the number of particles *emitted* from a differential volume element dV located at position $\vec{r}$, travelling in a direction $d\Omega$ about $\hat{\Omega}$, with energies dE about E, and during time t.

The notion of what constitutes a *source* of particles can often lead to confusion, and it becomes important to distinguish between the different types and how they are specified in the literature. Sources can be generically broken down as follows:

$$\left[\frac{\partial n}{\partial t}\right]_{\text{source}} = \dot{s}_{\text{int}}(\vec{r},\hat{\Omega},E,t) + \dot{s}_{\text{scat}}(\vec{r},\hat{\Omega},E,t) + \dot{s}_{\text{ext}}(\vec{r},\hat{\Omega},E,t) \tag{6.39}$$

whereby the superscripts of each term on the right hand side refer to: internal, scattering, and external sources, respectively.

Internal sources apply to particles that originate from *within* the phase space element $dV\,d\Omega\,dE\,dt$ of interest. Any radioactive decay mechanism that transforms energy from the medium and emits or releases it in the form of one or more types of particles falls under this category.

Scattering sources refer to secondary particles created from physical interaction processes that take place *within* the phase space cell, which have altered the energy and direction of primary particles from another phase space cell. Further discussion of scattering sources is provided below.

External sources[6] represent particles known *a priori* that originate *outside* the medium of interest or computational domain. Since these particles do not originate

---

[6]In the nuclear engineering literature, *internal* sources are sometimes referred to and symbolically represented as *external* sources. The reader should be careful on how these sources are defined.

from any particular phase space element, then by definition, should be specified as a separate boundary condition for the *unknown* number or fluence density distribution to be solved from the Boltzmann transport equation. However, under certain conditions, external sources can be stated as an independent source, similar to internal and scatter sources. Further elaboration of external sources is found in Section 6.3.5.

Thus, the source term in the Boltzmann transport equation is meant to represent known particles that *drive* the calculation of the unknown particle number or fluence density distribution. As such, the second term in Equation 6.37 can be simply stated as

$$\left[\frac{\partial n}{\partial t}\right]_{\text{source}} = \dot{s}(\vec{r}, \hat{\Omega}, E, t) \tag{6.40}$$

and depending on the application, the particular type of source will be explicitly identified when required.

Sinks (Absorption)

Absorption refers to physical processes that completely remove radiation particles through interactions with the medium. In the case of photons, examples that fit into the category include the photoelectric effect, pair production, and photonuclear disintegration.

Thus, the third term in Equation 6.37 can be written as

$$\left[\frac{\partial n}{\partial t}\right]_{\text{abs}} = -\upsilon \, \Sigma_a(\vec{r}, E) \, n(\vec{r}, \hat{\Omega}, E, t) \tag{6.41}$$

where $\Sigma_a(\vec{r}, E)$ is the *macroscopic* linear absorption coefficient[7], defined by

$$\Sigma_a(\vec{r}, E) = \rho_a(\vec{r}) \, \sigma_a(\vec{r}, E) \tag{6.42}$$

and $\rho_a(\vec{r})$ is the number of absorbing atoms per unit volume (atomic density), and σ_a is the *microscopic* absorption coefficient. Since ρ_a has units of reciprocal volume and σ_a an area, then Σ_a has units of reciprocal length as expected.

Sinks (Scattering)

Scattering has *two* effects on the particle number density distribution. Consider particles in a small volume element dV travelling in a small solid angle $d\Omega$ and within an energy range between E and $E + dE$. The mean number of photons in

---

[7]An upper case sigma Σ has been chosen to represent the macroscopic attenuation coefficient, instead of the more traditional lower case mu μ, which has already been reserved for the direction cosine of the particle z coordinate axis.

the (phase space) group is $n(\vec{r}, \hat{\Omega}, E, t)\, dV d\Omega dE$. Scattering processes that occur in the volume element can either *increase* or *decrease* the number of photons in the group. The decrease comes about because particles within the group can change direction, lose energy or both as a result of scattering. On the other hand, particles *not* in the group under consideration can scatter *into* the angular range $d\Omega$ and energy band dE.

Scattering *out* of the group is described by exactly the same mathematics as in the absorption case; as far as removal from the group is concerned, where there is *no* distinction between full absorption and out-scattering. Thus, by analogy to Equation 6.41

$$\left[\frac{\partial n}{\partial t}\right]_{\text{scat,out}} = -v\,\Sigma_s(\vec{r}, E)\, n(\vec{r}, \hat{\Omega}, E, t) \tag{6.43}$$

where $\Sigma_s(\vec{r}, E)$ is the linear scattering coefficient, as defined by

$$\Sigma_s(\vec{r}, E) = \rho_s(\vec{r})\, \sigma_s(\vec{r}, E) \tag{6.44}$$

where $\rho_s(\vec{r})$ is the number of scatterers per unit volume (electron density). As in the absorption term, Σ_s has units of reciprocal length, and can depend on position and particle energy.

Scattering *into* the phase space group under consideration is more complicated. It involves integrals over direction and energy since particles with *any* direction or energy can, in principle, scatter into the group. On the other hand, no integral over position or time is required since scattering processes generally occur at a definite location and time. Thus, the goal here is to look for an integral transform that connects $n(\vec{r}, \hat{\Omega}, E, t)$ to $n(\vec{r}, \hat{\Omega}', E', t)$ for all other combinations of $\hat{\Omega}'$ and E'.

The general form of the term that describes scattering *into* the group of interest can be represented as

$$\left[\frac{\partial n}{\partial t}\right]_{\text{scat,in}} = +v \int dE' \int d\Omega'\, \Sigma_s(\vec{r}, \hat{\Omega}', E', \hat{\Omega}, E)\, n(\vec{r}, \hat{\Omega}', E', t) \tag{6.45}$$

where the kernel $\Sigma_s(\vec{r}, \hat{\Omega}', E', \hat{\Omega}, E)$ is defined as the probability per unit pathlength that a particle with phase space coordinate[8] $(\hat{\Omega}', E')$ will, as a result of an interaction at $\vec{r}$, produce secondary particles that scatter into phase space coordinate $(\hat{\Omega}, E)$.

The *macroscopic* kernel in Equation 6.45 can be related to the more familiar *microscopic* differential scattering cross section as discussed in Chapter 2. Therefore,

---

[8]Note that *primed* phase space coordinates are used to signify particles outside the phase space group of interest. In contrast, medical physicists often use *primed* notation to represent physical variables following a particle interaction.

by arguments similar to those used in Equations 6.41 and 6.43 (via dimensional analysis), the relationship for the kernel is given by

$$\Sigma_s(\vec{r}, \hat{\Omega}', E', \hat{\Omega}, E) = n_s(\vec{r}, E) \frac{\mathrm{d}^2\sigma_s}{\mathrm{d}\Omega \mathrm{d}E}(\vec{r}, \hat{\Omega}', E', \hat{\Omega}, E) \tag{6.46}$$

where $n_s(\vec{r}, E)$ is the number of scattering centres per unit volume and $\mathrm{d}^2\sigma_s/\mathrm{d}\Omega\mathrm{d}E$ represents the *doubly* differential scattering cross section, and has units of squared length per energy per steradian.

Based on expressions for scattering *into* and *out* of the phase space group developed above, a total expression for the scattering component of Equation 6.37 is given by

$$\left[\frac{\partial n}{\partial t}\right]_{\mathrm{scat}} = \left[\frac{\partial n}{\partial t}\right]_{\mathrm{scat,in}} + \left[\frac{\partial n}{\partial t}\right]_{\mathrm{scat,out}}$$

$$= +v \int \mathrm{d}E' \int \mathrm{d}\Omega' \, \Sigma_s(\vec{r}, \hat{\Omega}', E', \hat{\Omega}, E) \, n(\vec{r}, \hat{\Omega}', E', t)$$

$$- v \, \Sigma_s(\vec{r}, E) \, n(\vec{r}, \hat{\Omega}, E, t) \tag{6.47}$$

Final Form

Combining the individual terms from Equations 6.38, 6.40, 6.41, and 6.47, and using the relationship between the particle number and fluence density distribution (Equation 6.16), then the final form of the *time-dependent* Boltzmann transport equation is given by

$$\frac{1}{v}\frac{\partial \dot{\varphi}}{\partial t}(\vec{r}, \hat{\Omega}, E, t) + \hat{\Omega} \cdot \vec{\nabla}\dot{\varphi}(\vec{r}, \hat{\Omega}, E, t) + \Sigma_t(\vec{r}, E) \, \dot{\varphi}(\vec{r}, \hat{\Omega}, E, t)$$

$$= \dot{s}(\vec{r}, \hat{\Omega}, E, t) + \int \mathrm{d}E' \int \mathrm{d}\Omega' \, \Sigma_s(\vec{r}, \hat{\Omega}', E', \hat{\Omega}, E) \, \dot{\varphi}(\vec{r}, \hat{\Omega}', E', t)$$

$$\tag{6.48}$$

where Σ_t is the total linear attenuation coefficient, as given by

$$\Sigma_t = \Sigma_a + \Sigma_s = n_a\sigma_a + n_s\sigma_s \tag{6.49}$$

In summary, the Boltzmann transport equation can be regarded as a balance equation, which expresses particle number conservation within a given phase space element. Mathematically, Equation 6.48 represents a first order *linear*, integro-differential equation, whereby the linearity property arises from the fact that particles do not interfere with one another, and the medium of interest does not change.

6.3.4 Time-Independent Formulation

The transport of particles (e.g. high-energy photons and electrons) typically encountered in radiotherapy dose calculation problems can generally be described using a *stationary* or *steady-state* Boltzmann transport equation. The rationale for a time-independent formulation can be justified by noting that such particles generally move at a speed on the order of the speed of light, and a steady state can be achieved in far less time than the duration in which the incident beam of particles is typically "on".

Therefore, if a source of radiation emits these particles at a constant rate over a given time interval $[0, t]$, then

$$\int_0^T \frac{\partial n}{\partial t'} \, \mathrm{d}t' = \dot{n}(T) - \dot{n}(0) = 0 \tag{6.50}$$

Furthermore, in the steady-state approximation, the time dependence of the particle fluence density and source distribution functions can be simplified as follows:

$$\varphi(\vec{r}, \hat{\Omega}, E) = \int_0^T \dot{\varphi}(\vec{r}, \hat{\Omega}, E, t) \, \mathrm{d}t' \tag{6.51}$$

and

$$s(\vec{r}, \hat{\Omega}, E) = \int_0^T \dot{s}(\vec{r}, \hat{\Omega}, E, t) \, \mathrm{d}t' \tag{6.52}$$

where the notation (dot over variable) for *rate* is removed as the result of time integration.

Thus, the linear *time-independent* Boltzmann transport equation can be simplified to

$$\hat{\Omega} \cdot \vec{\nabla} \, \varphi(\vec{r}, \hat{\Omega}, E) + \Sigma_t(\vec{r}, E) \, \varphi(\vec{r}, \hat{\Omega}, E)$$
$$= s(\vec{r}, \hat{\Omega}, E) + \int dE' \int d\Omega' \, \Sigma_s(\vec{r}, \hat{\Omega}', \hat{\Omega}, E', E) \, \varphi(\vec{r}, \hat{\Omega}', E') \tag{6.53}$$

In order to solve either Equations 6.48 or 6.53 for the particle fluence density distribution, boundary conditions must be specified, as will be discussed in the next section.

6.3.5 Boundary Conditions

In general, to uniquely solve any differential equation, boundary and/or initial conditions are required. With respect to the linear Boltzmann transport equation, *boundary* sources need to be specified. These sources represent particles that can either enter the (computational) volume V through an *outer* boundary ∂V, or originate within the phase space element dV. Also, in radiation transport problems, it is customary to assume that the system volume is *convex*, which means that particles that "leak" out of the volume cannot re-enter through the boundary again.

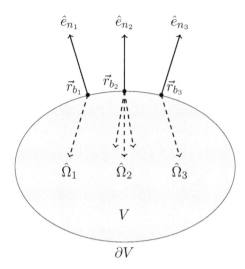

Figure 6.6: Schematic highlighting how boundary conditions are defined for Boltzmann transport problems. The computational domain V has a boundary ∂V at which particles can enter $\hat{\Omega} \cdot \hat{e}_n < 0$ (shown) or leave $\hat{\Omega} \cdot \hat{e}_n > 0$ (not shown) the volume. For external particle sources, the particle fluence density needs to be defined at the points (e.g. $\vec{r} \equiv \{\vec{r}_{b_1}, \vec{r}_{b_2}, \vec{r}_{b_3}\}$) on the computational boundary.

As mentioned previously, *external* sources, independent of the particle fluence within the system volume, can be described using a *boundary* particle fluence density distribution, $\varphi^b(\vec{r}, \hat{\Omega}, E)$, which must be specified for: (i) all points on the outer boundary of the system ($\vec{r} \epsilon \partial V$), (ii) all particle directions pointing into the system ($\hat{\Omega} \cdot \hat{e}_n < 0$, where $\hat{e}_n$ is the outer unit normal vector at $\vec{r} \epsilon \partial V$), and (iii) all particle energies E. These conditions can be summarized as follows:

$$\varphi(\vec{r}, \hat{\Omega}, E) = \varphi^b(\vec{r}, \hat{\Omega}, E), \qquad \vec{r} \epsilon \partial V, \ \hat{\Omega} \cdot \hat{e}_n < 0, \ 0 < E < \infty \qquad (6.54)$$

In the special case where $\varphi^b(\vec{r}, \hat{\Omega}, E) = 0$, then ∂V is called a *vacuum* boundary.

Note that particle boundary sources are *independent* of the particle fluence density distribution, $\varphi(\vec{r}, \hat{\Omega}, E)$, within V. Generally, radiation transport problems are driven by these *known* external sources, which generate the original particles in the problem. After particles are introduced into V by these boundary sources, they then transport within V based on the physical interaction processes interactions that take place within the system (see Figure 6.6).

6.3.6 Adjoint Formulation

In the following, we introduce the concept of an *adjoint* operator and function, and use it to define the adjoint Boltzmann transport equation. Before discussing the adjoint, the mathematics of *inner products* and *operators* need to be introduced.

Inner Product

The inner product, also known as the scalar product, of two *arbitrary* (real) functions f and g with phase space variables, $(\vec{r}, \hat{\Omega}, E)$, can be defined as

$$\langle f, g \rangle = \iiint f(\vec{r}, \hat{\Omega}, E)\, g(\vec{r}, \hat{\Omega}, E)\, dV\, d\Omega\, dE \tag{6.55}$$

whereby integration is performed over all phase space variables.

The inner product between the functions f and g is a *scalar* quantity that provides a measure of similarity between two functions (i.e. a way to compare to functions). An inner product $\langle f, g \rangle = 0$ means f and g are orthogonal, if $\langle f, g \rangle =$ small number signifies f and g are very different, and if $\langle f, g \rangle =$ big number then f and g are very similar.

Some useful properties of the inner product, as defined in Equation 6.55 include: symmetry, linearity, and homogeneity; and they all can be summarized by

$$\langle f, g \rangle = \langle g, f \rangle \tag{6.56}$$

$$\langle \alpha f + \beta g, h \rangle = \alpha \langle f, h \rangle + \beta \langle g, h \rangle \tag{6.57}$$

where α, β are real scalars. Note that these properties of the inner product can be modified if the functions of interest are not real (i.e. complex).

Operators

An operator *maps* the elements of a (vector) space of functions to produce other elements of the same (vector) space. The most common (linear) operators include the differential and integral operators.

A linear, homogeneous equation can be represented as

$$\mathcal{L}[f(x)] = g(x) \tag{6.58}$$

where $\mathcal{L}$ denotes a linear operator, while $f(x)$ and $g(x)$ represent the unknown function and known driving function, respectively.

Then, the linear operator $\mathcal{L}$ possesses the following properties:

$$\mathcal{L}[f_1(x) + f_2(x)] = \mathcal{L}[f_1(x)] + \mathcal{L}[f_2(x)] \tag{6.59}$$

$$\mathcal{L}[\alpha f(x)] = \alpha \mathcal{L}[f(x)] \tag{6.60}$$

$$\mathcal{L}_1 \mathcal{L}_2 [f(x)] = \mathcal{L}_2 \mathcal{L}_1 [f(x)] \tag{6.61}$$

The linear time-independent Boltzmann transport equation, Equation 6.53, can be expressed in the form of Equation 6.58 as follows:

$$\mathcal{L}\varphi = s \tag{6.62}$$

where the generic operator $\mathcal{L}$ can be decomposed into two other (physically meaningful) linear operators $\mathcal{T}$ and $\mathcal{K}$ by

$$\mathcal{L} = \mathcal{T} - \mathcal{K} \tag{6.63}$$

where $\mathcal{T}$ denotes the *transport* operator

$$\mathcal{T} = \hat{\Omega} \cdot \vec{\nabla} + \Sigma_t \tag{6.64}$$

and $\mathcal{K}$ represents the *scattering* operator

$$\mathcal{K} = \int dE' \int d\Omega' \, \Sigma_s(\vec{r}, \hat{\Omega}', E' \rightarrow \hat{\Omega}, E) \, \varphi(\vec{r}, \hat{\Omega}', E') \tag{6.65}$$

Thus, the linear Boltzmann transport equation can be summarized in a compact form, making it amenable to mathematical manipulation; it can be interpreted as a simple algebraic equation of the form $Ax = b$, which can be solved with standard algebraic techniques.

Adjoint

Given an arbitrary, linear operator $\mathcal{L}$, which may be differential, integral, or integro-differential; two real functions f and g; and definition of the inner product, then the *adjoint* of the operator, denoted by $\mathcal{L}^\dagger$, can be defined by the following inner product identity

$$\langle \mathcal{L}f, g \rangle = \langle f, \mathcal{L}^\dagger g \rangle \tag{6.66}$$

Equation 6.66 represents the formal definition of an adjoint operator. When $\mathcal{L}^\dagger$ is applied to the left member of *any* inner product, then the same result is produced when $\mathcal{L}$ is used on the right member of the same inner product.

Without proof, the adjoint of the linear differential ($\mathcal{D}$) and integral ($\mathcal{I}$) operators, acting on arbitrary functions f and g, are given by

$$\mathcal{D}f = \frac{df}{dx} \quad \rightarrow \quad \mathcal{D}^\dagger g = -\frac{dg}{dx} \tag{6.67}$$

and

$$\mathcal{I}f = \int k(x' \rightarrow x) \, f(x') \, dx' \quad \rightarrow \quad \mathcal{I}^\dagger g = \int k(x \rightarrow x') \, g(x') \, dx' \tag{6.68}$$

where $k(x' \to x)$ represents an arbitrary kernel function.

Thus, the adjoint of the derivative operator simply negates the derivative operation, while the adjoint of an integral operator has a transposed kernel.

The relationships in Equations 6.67 and 6.68 can now be applied to the operators, Equations 6.63 and 6.65, of the (non-adjoint) Boltzmann transport equation, to give

$$\mathcal{L}^\dagger = -\hat{\Omega} \cdot \vec{\nabla} + \Sigma_t - \mathcal{K}^\dagger \tag{6.69}$$

and

$$\mathcal{K}^\dagger = \int dE' \int d\Omega' \, \Sigma_s(\vec{r}, \hat{\Omega}, E \to \hat{\Omega}', E') \tag{6.70}$$

Therefore, Equations 6.69 and 6.70 can now be used to construct the *adjoint* Boltzmann transport equation as given by

$$-\hat{\Omega} \cdot \vec{\nabla} \, \varphi^\dagger(\vec{r}, \hat{\Omega}, E) + \Sigma_t(\vec{r}, E) \, \varphi^\dagger(\vec{r}, \hat{\Omega}, E)$$
$$= s^\dagger(\vec{r}, \hat{\Omega}, E) + \int dE' \int d\Omega' \, \Sigma_s(\vec{r}, \hat{\Omega}', \hat{\Omega}, E', E) \, \varphi^\dagger(\vec{r}, \hat{\Omega}', E') \tag{6.71}$$

or compactly in operator notation as

$$\mathcal{L}^\dagger \varphi^\dagger = s^\dagger \tag{6.72}$$

where $\varphi^\dagger(\vec{r}, \hat{\Omega}', E')$ is the *adjoint* particle fluence density distribution, and $s^\dagger(\vec{r}, \hat{\Omega}, E)$ is an *adjoint* source density distribution.

In essence, the adjoint Boltzmann transport equation represents a *backward* transport process, which describes the flow of pseudo-particles (adjoint fluence density distribution) that can be created, destroyed, and re-distributed in phase space. In contrast to the standard (forward) Boltzmann transport equation, these adjoint particles are: created according to an adjoint source at some final (instead of an initial) point in time, stream or travel in *reverse* directions, and can *gain* energy in scattering collisions.

6.3.7 Integral Formulation

The linear Boltzmann transport equation can also be re-cast into an integral form through integration of the time-independent (steady-state) integro-differential form (Equation 6.53). As seen in Figure 6.7, the integration is performed along a *straight* line from $\vec{r}$ to a different point $\vec{r}_\ell$ (by "looking back" in the opposite direction of particle flow $\hat{\Omega}$).

The equation for vector $\vec{r}_\ell$ in terms of the vector $\vec{r}$ is given by

$$\vec{r}_\ell = \vec{r} - \ell\,\hat{\Omega} \tag{6.73}$$

where $\ell = |\vec{r} - \vec{r}_\ell|$ represents the distance between the two vectors.

Based on Equation 6.73, the streaming operator simply represents the directional derivative along the direction of particle travel, and the linear Boltzmann transport equation can be re-written as

$$-\frac{\mathrm{d}\varphi}{\mathrm{d}\ell}(\vec{r} - \ell\,\hat{\Omega}, \hat{\Omega}, E) + \Sigma_t(\vec{r} - \ell\,\hat{\Omega}, E)\,\varphi(\vec{r} - \ell\,\hat{\Omega}, \hat{\Omega}, E) = s_{\mathrm{eff}}(\vec{r} - \ell\,\hat{\Omega}, \hat{\Omega}, E) \tag{6.74}$$

where $s_{\mathrm{eff}}(\vec{r} - \ell\,\hat{\Omega}, \hat{\Omega}, E)$ is an effective source term that combines both the scattering source and internal and/or external sources, as given by

$$s_{\mathrm{eff}}(\vec{r} - \ell\,\hat{\Omega}, \hat{\Omega}, E) = s(\vec{r} - \ell\,\hat{\Omega}, \hat{\Omega}, E)$$
$$+ \int dE' \int d\Omega'\, \Sigma_s(\vec{r} - \ell\,\hat{\Omega}, \hat{\Omega}', \hat{\Omega}, E', E)\,\varphi(\vec{r} - \ell\,\hat{\Omega}, \hat{\Omega}', E') \tag{6.75}$$

The derivative of ℓ can be removed by using the following exponential factor

$$\exp\left[-\int_0^\ell \Sigma_t(\vec{r} - \ell'\,\hat{\Omega}, \hat{\Omega}, E)\,\mathrm{d}\ell'\right] = \exp\left[-\tau(\vec{r}, \vec{r} - \ell\,\hat{\Omega}, \hat{\Omega}, E)\right] \tag{6.76}$$

where $\tau(\vec{r}, \vec{r} - \ell\,\hat{\Omega}, \hat{\Omega}, E)$ is known as the *optical (or radiological) pathlength*, which physically represents the line integral of the total macroscopic cross section along the line of particle travel between $\vec{r}$ and $\vec{r}_\ell$.

By multiplying the exponential factor with the particle fluence density distribution, and taking the derivative over the path-length ℓ, the following useful property is obtained:

$$\frac{\mathrm{d}}{\mathrm{d}\ell}\left\{\exp\left[-\tau(\vec{r}, \vec{r} - \ell\,\hat{\Omega}, \hat{\Omega}, E)\right]\varphi(\vec{r} - \ell\,\hat{\Omega}, \hat{\Omega}, E)\right\}$$
$$= \exp\left[-\tau(\vec{r}, \vec{r} - \ell\,\hat{\Omega}, \hat{\Omega}, E)\right] \times \left\{\frac{\mathrm{d}}{\mathrm{d}\ell}\varphi(\vec{r} - \ell\,\hat{\Omega}, \hat{\Omega}, E)\right.$$
$$\left. - \Sigma_t(\vec{r} - \ell\,\hat{\Omega}, \hat{\Omega}, E)\,\varphi(\vec{r} - \ell\,\hat{\Omega}, \hat{\Omega}, E)\right\} \tag{6.77}$$

Therefore, substituting Equation 6.74 into 6.77 leads to

$$-\frac{\mathrm{d}}{\mathrm{d}\ell}\left\{\exp\left[-\tau(\vec{r}, \vec{r} - \ell\,\hat{\Omega}, \hat{\Omega}, E)\right]\varphi(\vec{r} - \ell\,\hat{\Omega}, \hat{\Omega}, E)\right\}$$
$$= \exp\left[-\tau(\vec{r}, \vec{r} - \ell\,\hat{\Omega}, \hat{\Omega}, E)\right] s_{\mathrm{eff}}(\vec{r} - \ell\,\hat{\Omega}, \hat{\Omega}, E) \tag{6.78}$$

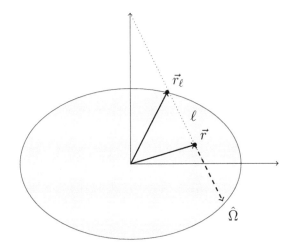

Figure 6.7: Schematic showing the geometry for transforming the linear Boltzmann transport equation from differential form to integral form. A position vector $\vec{r}_\ell$ is defined (on the computational boundary) by *looking back* in the opposite direction of particle motion, denoted by $\hat{\Omega}$.

Finally, by integrating both sides of Equation 6.78 between 0 and ℓ gives

$$-\int_0^\ell d\ell' \frac{d}{d\ell'} \left\{ \exp\left[-\tau(\vec{r}, \vec{r} - \ell'\,\hat{\Omega}, \hat{\Omega}, E)\right] \varphi(\vec{r} - \ell'\,\hat{\Omega}, \hat{\Omega}, E) \right\}$$

$$= \int_0^\ell d\ell' \exp\left[-\tau(\vec{r}, \vec{r} - \ell'\,\hat{\Omega}, \hat{\Omega}, E)\right] s_{\text{eff}}(\vec{r} - \ell'\,\hat{\Omega}, \hat{\Omega}, E) \quad (6.79)$$

which leads to the following solution:

$$\varphi(\vec{r}, \hat{\Omega}, E) = \varphi(\vec{r} - \ell\,\hat{\Omega}, \hat{\Omega}, E) \exp\left[-\tau(\vec{r}, \vec{r} - \ell\,\hat{\Omega}, \hat{\Omega}, E)\right]$$

$$+ \int_0^\ell d\ell' s_{\text{eff}}(\vec{r} - \ell'\,\hat{\Omega}, \hat{\Omega}, E) \exp\left[-\tau(\vec{r}, \vec{r} - \ell'\,\hat{\Omega}, \hat{\Omega}, E)\right] \quad (6.80)$$

The integral Boltzmann transport equation physically represents contributions to the fluence density distribution from particles generated by a source (internal/external/scattering) along *intermediate* positions $\vec{r} - \ell\hat{\Omega}$, which are attenuated exponentially according to the macroscopic cross section between their point of origin and $\vec{r}$. Note that in general, Equation 6.80 does *not* yield a closed analytic solution, since the *unknown* particle fluence density distribution also appears in the effective scattering source term (see Equation 6.75).

6.3.8 Connection with Convolution and Monte Carlo Methods

In the following, an attempt is made to *loosely* connect the mathematics associated with the linear Boltzmann transport equation and the main photon-based dose calculation algorithms described in this book. First, the concept of an *observable* is introduced, which in turn can be calculated from either a *detector* response function or *importance* (or influence) function. Then, the notion of a *Green's* function will be applied to the linear Boltzmann transport equation, which will be used to show the different pathways toward solution of the absorbed dose between the deterministic, Monte Carlo, and convolution-based algorithms.

Observable

An observable represents a physical quantity that can be calculated and validated through measurement. Within the context of radiation transport, the inner product concept can be used to calculate an observable $\mathcal{O}$ as follows:

$$\mathcal{O} = \langle f, \varphi \rangle = \iiint f(\vec{r}, \hat{\Omega}, E) \, \varphi(\vec{r}, \hat{\Omega}, E) \, dV \, d\Omega \, dE \tag{6.81}$$

where $f(\vec{r}, \hat{\Omega}, E)$ represents a (real) function that *acts* on the particle fluence density distribution to give an observable physical quantity. In external beam radiation therapy, the main observable or quantity of interest is the absorbed dose. Below, two types of functions will be described that can be used to determine the absorbed dose based on the definition provided in Equation 6.81.

Detector Response Function

As mentioned previously, the calculation of absorbed dose is a two-step process. Knowing the radiation field as characterized by the particle fluence density distribution, $\varphi(\vec{r}, \hat{\Omega}, E)$, in combination with a detector response function, the observable expression in Equation 6.81 becomes

$$\mathcal{O} = \langle \mathcal{D}, \varphi \rangle = \iiint \mathcal{D}(\vec{r}, \hat{\Omega}, E) \, \varphi(\vec{r}, \hat{\Omega}, E) \, dV \, d\Omega \, dE \tag{6.82}$$

According to Equation 6.82, the detector response function represents the contribution to an observable or detector reading from a unit path length of particle with phase space coordinates $(\vec{r}, \hat{\Omega}, E)$.

Comparing the observable expression in Equation 6.82 with the definition of absorbed dose in Equation 6.19, a connection can be made that the detector response function is simply the mass-collisional stopping power, S_c/ρ. Therefore, if the absorbed dose, $D(\vec{r})$, is the observable quantity of interest, then

$$\mathcal{O}(\vec{r}) \equiv D(\vec{r}) = \iint \frac{S_c}{\rho}(\vec{r}, E) \, \varphi(\vec{r}, \hat{\Omega}, E) \, d\Omega \, dE \tag{6.83}$$

where integration has been performed over each volume element.

Importance (or Influence) Function

Alternatively, based on the *adjoint* formulation of the linear Boltzmann transport equation discussed earlier, observable quantities can also be represented by the following:

$$\mathcal{O} = \langle \varphi^{\dagger}, s \rangle = \iiint \varphi^{\dagger}(\vec{r}, \hat{\Omega}, E) \, s(\vec{r}, \hat{\Omega}, E) \, \mathrm{d}V \, \mathrm{d}\Omega \, \mathrm{d}E \qquad (6.84)$$

where $s(\vec{r}, \hat{\Omega}, E)$ is a known internal/external source function, and $\varphi^{\dagger}(\vec{r}, \hat{\Omega}, E)$ is the adjoint particle fluence density distribution, and represents the average contribution to an observable from a particle originating at phase space element $(\vec{r}, \hat{\Omega}, E)$.

Assume for the moment a point mono-energetic (E_0), mono-directional $(\hat{\Omega}_0)$ source that locally injects one particle into the volume of interest in phase space by

$$s(\vec{r}, \hat{\Omega}, E) = \delta(\vec{r} - \vec{r}_0) \, \delta(\hat{\Omega} - \hat{\Omega}_0) \, \delta(E - E_0) \qquad (6.85)$$

where $\vec{r}_0$ is the spatial location of the injected particle, and $\delta(...)$ is the Kronecker delta impulse function.

Inserting Equation 6.85 into 6.84, the observable quantity evaluates to

$$\mathcal{O} = \varphi^{\dagger}(\vec{r}_0, \hat{\Omega}_0, E_0) \qquad (6.86)$$

Then, equating Equation 6.86 with 6.83 provides an alternate interpretation of the absorbed dose as given by

$$\mathcal{O}(\vec{r}) \equiv D(\vec{r}) = \varphi^{\dagger}(\vec{r}|\vec{r}_0, \hat{\Omega}_0, E_0)$$
$$= \iint \frac{S_c}{\rho}(\vec{r}, E) \, \varphi(\vec{r}, \hat{\Omega}, E|\vec{r}_0, \hat{\Omega}_0, E_0) \, \mathrm{d}\Omega \, \mathrm{d}E \qquad (6.87)$$

whereby $\varphi(\vec{r}, \hat{\Omega}, E|\vec{r}_0, \hat{\Omega}_0, E_0)$ is the solution of the (forward) transport equation corresponding to a point source; and $\varphi^{\dagger}(\vec{r}|\vec{r}_0, \hat{\Omega}_0, E_0)$ is the absorbed dose at point $\vec{r}$ in the volume element due to a point source.

Moreover, the adjoint particle fluence density distribution is a direct measure of the *importance* (or influence) of a locally injected particle to the absorbed dose. For this reason, the adjoint fluence is commonly referred to as the importance function, and directly relates the absorbed dose in a receptor and a distribution of sources.

Therefore, having defined how to calculate an observable quantity (e.g. absorbed dose) from either the detector response or importance functions, these two concepts can be shown to be related as follows:

$$\mathcal{O} = \langle \mathcal{D}, \varphi \rangle = \langle \varphi^{\dagger}, s \rangle \qquad (6.88)$$

The equality of the two inner products is a statement of *reciprocity* between the source and detection of particles at a given point. As previously introduced in Chapters 2 and 3, reciprocity is a fundamental concept that source and detector can be interchanged without changing the response of the detector.

Green's Function

Below, a particularly interesting case of the generalized reciprocity relationship given in Equation 6.88 is considered.

In the case of localized forward and adjoint sources in phase space, the resulting solutions of the forward and adjoint transport equations with homogeneous boundary conditions are simply the forward and adjoint Green's functions, with the following reciprocity relationship

$$\langle \mathcal{D}, \mathcal{G} \rangle = \langle \mathcal{G}, s \rangle \tag{6.89}$$

Since the Boltzmann transport equation is linear, then the solution can be expressed in terms of a Green's function, $G(\vec{r}, \hat{\Omega}, E | \vec{r}', \hat{\Omega}', E')$, as given by

$$\varphi(\vec{r}, \hat{\Omega}, E) = \int_V \mathrm{d}r' \int_{4\pi} \mathrm{d}\Omega' \int_E \mathrm{d}E' \, G(\vec{r}, \hat{\Omega}, E | \vec{r}', \hat{\Omega}', E') \, s(\vec{r}', \hat{\Omega}', E') \tag{6.90}$$

Here, the Green's function is an operator that maps the particle fluence from one point in phase space $(\vec{r}', \hat{\Omega}', E')$ to another point in phase space $(\vec{r}, \hat{\Omega}, E)$.

Based on Equation 6.90, the corresponding equation for an observable has the following form:

$$\mathcal{O} = \langle \mathcal{D}, \mathcal{G} \rangle = \int \mathrm{d}r \int \mathrm{d}\Omega \int \mathrm{d}E \int \mathrm{d}r' \int \mathrm{d}\Omega' \int \mathrm{d}E' \, \mathcal{D}(\vec{r}, \hat{\Omega}, E)$$
$$\times G(\vec{r}, \hat{\Omega}, E | \vec{r}', \hat{\Omega}', E') \, s(\vec{r}', \hat{\Omega}', E') \tag{6.91}$$

Under the assumptions of a homogeneous, semi-infinite medium, in which particle scattering with the medium has azimuthal symmetry, then Equation 6.90 can be written as

$$\varphi(\vec{r}, \hat{\Omega}, E) = \int \mathrm{d}r' \int \mathrm{d}\Omega' \int \mathrm{d}E' \, G(|\vec{r} - \vec{r}'|, \hat{\Omega} \cdot \hat{\Omega}', E') \, s(\vec{r}', \hat{\Omega}', E') \tag{6.92}$$

where the Green's function depends on the distance between the source point $\vec{r}'$ and $\vec{r}$, the polar angle between the particle directions, and the energy. Thus, the Green's function under these assumptions is invariant to translations and rotations.

Therefore, using Equations 6.91 and 6.92, the absorbed dose, $D(\vec{r})$, can be calculated by integrating over $\hat{\Omega}$ and E with a response function $\mathcal{D}(\vec{r}, \hat{\Omega}, E)$ as shown by

$$D(\vec{r}) = \int \mathrm{d}r' \int \mathrm{d}\Omega' \int \mathrm{d}E' \, k(|\vec{r} - \vec{r}'|, \hat{\Omega}', E') \, s(\vec{r}', \hat{\Omega}', E') \tag{6.93}$$

where the kernel $k(|\vec{r} - \vec{r}'|, \hat{\Omega}', E')$ is given by

$$k(|\vec{r} - \vec{r}'|, \hat{\Omega}', E') = \int d\Omega \int dE\, \mathcal{D}(\vec{r}, \hat{\Omega}, E)\, G(|\vec{r} - \vec{r}'|, \hat{\Omega} \cdot \hat{\Omega}', E') \tag{6.94}$$

and can be interpreted as a functional relationship between particle fluence at phase space location $(\vec{r}', \hat{\Omega}', E')$ to an observable (dose) calculated at $\vec{r}$. Equation 6.93 also represents a way to estimate a tally in a Monte Carlo simulation of particle transport. Note that special precautions must be made to use the methodology depicted in Equation 6.93 for the case of a heterogeneous medium.

Finally, Equation 6.94 can also be interpreted as a photon *dose spread array*, discussed and shown previously in Chapters 3 and 4, as calculated from Monte Carlo simulations, and used as a kernel in convolution-based dose calculation algorithms.

6.3.9 Theoretical Approximations

Various approximations can be made to the linear Boltzmann transport equation, Equation 6.53, to simplify the mathematics, and to gain insight for any given problem.

Fluence (Unscattered or Uncollided)

In the absence of particle *in-scattering*, the linear Boltzmann transport equation simplifies to

$$\hat{\Omega} \cdot \vec{\nabla}\, \varphi^{(0)}(\vec{r}, \hat{\Omega}, E) + \Sigma_t(\vec{r}, E)\, \varphi^{(0)}(\vec{r}, \hat{\Omega}, E) = s(\vec{r}, \hat{\Omega}, E) \tag{6.95}$$

The solution to Equation 6.95 can be recognized as the solution to the integral form of the linear Boltzmann transport equation, Equation 6.80, as given by

$$\varphi^{(0)}(\vec{r}, \hat{\Omega}, E) = \varphi^{(0)}(\vec{r} - \ell\hat{\Omega}, \hat{\Omega}, E)\, \exp\left[-\tau(\vec{r}, \vec{r} - \ell''\hat{\Omega}, \hat{\Omega}, E)\right]$$
$$+ \int_0^\ell d\ell'\, s(\vec{r} - \ell'\hat{\Omega}, \hat{\Omega}, E)\, \exp\left[-\tau(\vec{r}, \vec{r} - \ell''\hat{\Omega}, \hat{\Omega}, E)\right] \tag{6.96}$$

except that the effective source term, $s_{eff}(\vec{r} - \ell'\hat{\Omega}, \hat{\Omega}, E) = s(\vec{r} - \ell'\hat{\Omega}, \hat{\Omega}, E)$, simplies to just the internal/external source density distribution.

If $\vec{r}_l = \vec{r} - \ell\hat{\Omega}$ is located on the boundary of the (computational) volume of interest, then the non-re-entrant boundary condition, $\varphi^{(0)}(\vec{r} - \ell\hat{\Omega}, \hat{\Omega}, E) = 0$, can be applied to Equation 6.95, which gives

$$\varphi^{(0)}(\vec{r}, \hat{\Omega}, E) = \int_0^\ell d\ell'\, s(\vec{r} - \ell'\hat{\Omega}, \hat{\Omega}, E)\, \exp\left[-\tau(\vec{r}, \vec{r} - \ell''\hat{\Omega}, \hat{\Omega}, E)\right] \tag{6.97}$$

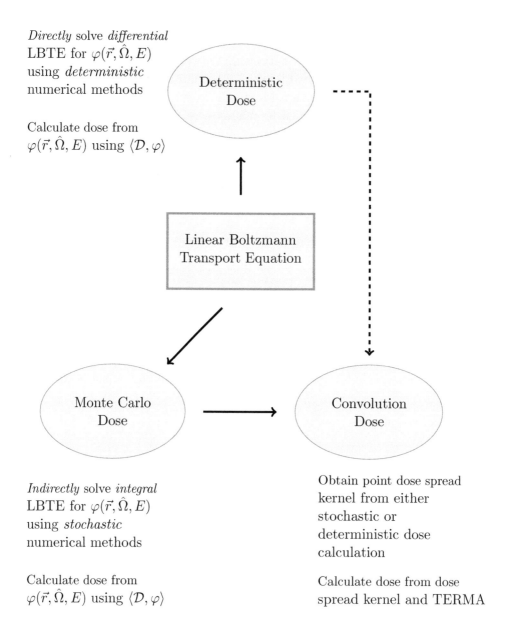

Figure 6.8: Flowchart highlighting the relationship between the three main dose algorithms presented in the book and the governing equation for particle transport, as described by the linear Boltzmann transport equation (LBTE). The dotted arrow represents a pathway "less travelled".

Equation 6.97 can be transformed from a line integral equation to volume integral equation using the following relationship

$$\int_0^\infty f(\vec{r}) - \ell\hat{\Omega})\mathrm{d}\ell = \int_{V'} f(\vec{r}_\ell)\,\delta^{(2)}\left(\hat{\Omega} - \frac{\vec{r} - \vec{r}_\ell}{|\vec{r} - \vec{r}_\ell|^2}\right)\frac{\mathrm{d}^3\vec{r}_l}{|\vec{r} - \vec{r}_\ell|^2} \qquad (6.98)$$

which yields

$$\varphi^{(0)}(\vec{r}, \hat{\Omega}, E) = \int_{V'} s(\vec{r}_l, \hat{\Omega}, E)\,\exp\left[-\tau(\vec{r}, \vec{r}_l, \hat{\Omega}, E)\right]$$
$$\times\,\delta^{(2)}\left(\hat{\Omega} - \frac{\vec{r} - \vec{r}_\ell}{|\vec{r} - \vec{r}_\ell|^2}\right)\frac{\mathrm{d}^3\vec{r}_l}{|\vec{r} - \vec{r}_\ell|^2} \qquad (6.99)$$

where $\delta^{(2)}(...)$ represents a two-dimensional ring Kronecker delta function.

Equation 6.99 can be applied to the case of an external ideal point source, as defined by

$$s(\vec{r}, \hat{\Omega}, E) = s_0\,\delta(\vec{r} - \vec{r}_0)\,\delta(\hat{\Omega} - \hat{\Omega}_0)\,\delta(E - E_0) \qquad (6.100)$$

and can be inserted into the equation to give

$$\varphi^{(0)}(\vec{r}, \hat{\Omega}, E) = \frac{s_0}{|\vec{r} - \vec{r}_0|^2}\,\delta\left(\hat{\Omega} - \frac{\vec{r} - \vec{r}_0}{|\vec{r} - \vec{r}_0|}\right)\exp\left[-\tau(\vec{r}_0, \vec{r}_0 + |\vec{r} - \vec{r}_0|\hat{\Omega}, E)\right] \qquad (6.101)$$

which represents the *uncollided* fluence density distribution at position $\vec{r}$ in direction $\hat{\Omega}$ due to a point source at position $\vec{r}_0$ that emits particles in direction $\hat{\Omega}_0$.

In the case of an *isotropic* point source, Equation 6.101 can be integrated over the variables $\vec{r}_0$, $\hat{\Omega}_0$, and $\hat{\Omega}$ to yield the following familiar result:

$$\varphi^{(0)}(r) = \frac{s_0}{4\pi r^2}\,\exp\left(-\Sigma_t\,r\right) \qquad (6.102)$$

which has the expected exponential attenuation term due to particle interaction within the medium, as well as the inverse square fall-off due to geometric spreading of the particles.

Geometry (Semi-infinite Medium)

In a semi-infinite planar geometry, whereby the medium is infinite in the (x,y) plane, all position dependent functions in the linear Boltzmann transport equation reduce their spatial dimensionality from three ($\vec{r} \equiv \{x, y, z\}$) down to one ($\vec{r} \equiv \{z\}$). Therefore, the total number of phase space variables is effectively reduced from six to four.

Therefore, the streaming operator simplifies to

$$\left[\hat{\Omega} \cdot \vec{\nabla}\right]\varphi(z, \hat{\Omega}, E) = \mu\frac{\partial}{\partial z}\varphi(z, \hat{\Omega}, E) \qquad (6.103)$$

where $\mu = \Omega_z = \cos\theta$, and θ is the polar angle between $\hat{\Omega}$ and z-axis.

Scattering (Azimuthal Symmetry)

Since most particle interactions possess azimuthal symmetry, the polar angle between the incident and final particles in the interaction can be used to simplify the argument of the differential cross section (via a dot product) as follows:

$$\sigma(\hat{\Omega}' \cdot \hat{\Omega}, E', E) \equiv \frac{\mathrm{d}^2\sigma}{\mathrm{d}\Omega \mathrm{d}E}(\hat{\Omega}' \cdot \hat{\Omega}, E', E) \tag{6.104}$$

Also, in deterministic radiation transport codes, the doubly differential cross sections are generally assumed to be separable in angle and energy as shown by

$$\sigma(\hat{\Omega}' \cdot \hat{\Omega}, E', E) \equiv \frac{\mathrm{d}\sigma}{\mathrm{d}E}(E', E) \times \frac{\mathrm{d}f}{\mathrm{d}\mu}(\mu) \tag{6.105}$$

where $f(\mu)$ is a normalized angular distribution. For interactions in which the angle and energy are kinematically *linked*, the singly differential cross section can still be used by enforcing the angle-energy relationship through a Dirac delta function.

6.3.10 Neutral and Charged Particle Formulations

Until now, a generalized treatment of the linear Boltzmann transport equation has been presented without specific details regarding the types of particles involved. For any given particle, the form of the transport equation will differ based on the types of collisions or interactions they undergo with the medium. In the context of external beam (photon-based) radiation therapy, three types of particles need to be considered: photons (x-rays), electrons, and positrons. The Boltzmann transport equation associated with each of these particles will be presented, along with a brief description of their relevant interactions, and will end with a summary of the coupled-nature of these equations.

Figure 6.9 shows an *interaction matrix* to help with the construction of the linear Boltzmann transport equation for each type of particle, particularly with the coupled terms in the equation(s). Based on the matrix, each equation is expected to have three scattering source terms, in addition to a boundary (external) source term.

<u>Photon Formulation</u>

The linear time-independent, Boltzmann transport equation for photons can be stated, based on Equation 6.53, with appropriate subscripts and/or superscripts (γ)

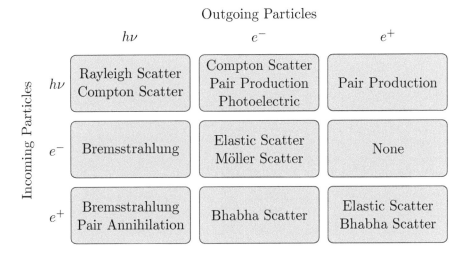

Figure 6.9: Interaction matrix used to sort the scattering sources in the coupled Boltzmann transport equation(s) for photons, electrons, and positrons. For each given incoming particle, the row indicates which particle interactions can occur, and the column shows what particles are produced for the given interaction. Note that *no* electron interaction produces positrons.

on the functions and variables, as given by

$$\hat{\Omega}_\gamma \cdot \vec{\nabla} \varphi^\gamma(\vec{r}, \hat{\Omega}_\gamma, E_\gamma) + \Sigma_t^\gamma(\vec{r}, E_\gamma)\, \varphi^\gamma(\vec{r}, \hat{\Omega}_\gamma, E_\gamma)$$
$$= s^{\gamma\gamma}(\vec{r}, \hat{\Omega}_\gamma, E_\gamma) + s^{e^-\gamma}(\vec{r}, \hat{\Omega}_\gamma, E_\gamma) + s^{e^+\gamma}(\vec{r}, \hat{\Omega}_\gamma, E_\gamma) + s^\gamma(\vec{r}, \hat{\Omega}_\gamma, E_\gamma)$$

$$(6.106)$$

where $s^{\gamma\gamma}(\vec{r}, \hat{\Omega}_\gamma, E_\gamma)$ is the *photon-photon* scattering source, and is given by

$$s^{\gamma\gamma}(\vec{r}, \hat{\Omega}_\gamma, E_\gamma) = \rho_{a,e}(\vec{r}) \int dE'_\gamma \int d\Omega'_\gamma\, \sigma_s^{\gamma\gamma}(\hat{\Omega}'_\gamma, \hat{\Omega}_\gamma, E'_\gamma, E_\gamma)$$
$$\times\, \varphi^\gamma(\vec{r}, \hat{\Omega}_\gamma, E_\gamma)$$

$$(6.107)$$

where $\sigma_s^{\gamma\gamma}(\hat{\Omega}'_\gamma, \hat{\Omega}_\gamma, E'_\gamma, E_\gamma)$ is the doubly differential cross section for photon interactions that create secondary photons, and $s^{e^-\gamma}(\vec{r}, \hat{\Omega}_\gamma, E_\gamma)$ is the *electron-photon* scattering source, and is given by

$$s^{e^-\gamma}(\vec{r}, \hat{\Omega}_\gamma, E_\gamma) = \rho_{a,e}(\vec{r}) \int dE'_e \int d\Omega'_e\, \sigma_s^{e^-\gamma}(\hat{\Omega}'_e, \hat{\Omega}_\gamma, E'_e, E_\gamma)$$
$$\times\, \varphi^{e^-}(\vec{r}, \hat{\Omega}'_e, E'_e)$$

$$(6.108)$$

where $\sigma_s^{e^-\gamma}(\hat{\Omega}'_e, \hat{\Omega}_\gamma, E'_e, E_\gamma)$ is the doubly differential cross section for electron interactions that create secondary photons, and $s^{e^+\gamma}(\vec{r}, \hat{\Omega}_\gamma, E_\gamma)$ is the *positron-photon* scattering source, and is given by

$$s^{e^+\gamma}(\vec{r}, \hat{\Omega}_\gamma, E_\gamma) = \rho_{a,e}(\vec{r}) \int dE'_p \int d\Omega'_p \, \sigma_s^{e^+\gamma}(\hat{\Omega}'_p, \hat{\Omega}_\gamma, E'_p, E_\gamma)$$
$$\times \, \varphi^{e^+}(\vec{r}, \hat{\Omega}'_p, E'_p) \tag{6.109}$$

where $\sigma_s^{e^+\gamma}(\hat{\Omega}'_p, \hat{\Omega}_\gamma, E'_p, E_\gamma)$ is the doubly differential cross section for positron interactions that create secondary photons, and $s^\gamma(\vec{r}, \hat{\Omega}_\gamma, E_\gamma)$ is the *external* photon source for photons that originate from outside the computational volume.

Based on Figure 6.9, the possible photon interactions include: Rayleigh (coherent) scatter, Compton (incoherent) scatter, photoelectric absorption, pair production, and triplet production. Therefore, the removal term in Equation 6.106 accounts for these interactions through the macroscopic attenuation coefficient, $\Sigma_t^\gamma(\vec{r}, E_\gamma)$, as given by

$$\Sigma_t^\gamma = \Sigma_{coh}^\gamma + \Sigma_{inc}^\gamma + \Sigma_{pe}^\gamma + \Sigma_{pp}^\gamma + \Sigma_{tp}^\gamma \tag{6.110}$$

where the independent variables have been suppressed for brevity.

Similarly, for the photon-photon, $s^{\gamma\gamma}(\vec{r}, \hat{\Omega}_\gamma, E_\gamma)$, and electron or positron-photon, $s^{e^-/e^+\gamma}(\vec{r}, \hat{\Omega}_\gamma, E_\gamma)$, scattering sources, the relevant scattering interactions are: Rayleigh (coherent) scatter, Compton (incoherent) scatter, photoelectric absorption for the former; and bremsstrahlung and pair annihilation (positron only) for the latter. Therefore, their corresponding doubly differential scattering cross sections are given by

$$\sigma_s^{\gamma\gamma} = \sigma_{coh}^{\gamma\gamma} + \sigma_{inc}^{\gamma\gamma} \tag{6.111}$$

and

$$\sigma_s^{e^-/e^+\gamma} = \sigma_{brem}^{e^-/e^+\gamma} + \sigma_{pa}^{e^+\gamma} \tag{6.112}$$

Theoretical expressions for these photon cross sections can be found in Chapter 2 and the following reference (Podgorsak 2010).

Electron Formulation

Similar to the photon case, the linear time-independent, Boltzmann transport equation for electrons can be stated, based on Equation 6.53, with appropriate subscripts/superscripts (e^-) on the functions and variables, as given by

$$\hat{\Omega}_e \cdot \vec{\nabla}\varphi^{e^-}(\vec{r}, \hat{\Omega}_e, E_e) + \Sigma_t^{e^-}(\vec{r}, E_e)\,\varphi^{e^-}(\vec{r}, \hat{\Omega}_e, E_e)$$
$$= s^{e^-e^-}(\vec{r}, \hat{\Omega}_e, E_e) + s^{e^+e^-}(\vec{r}, \hat{\Omega}_e, E_e) + s^{\gamma e^-}(\vec{r}, \hat{\Omega}_e, E_e) + s^{e^-}(\vec{r}, \hat{\Omega}_e, E_e)$$
$$\tag{6.113}$$

where $s^{e^-e^-}(\vec{r},\hat{\Omega}_e,E_e)$ is the *electron-electron* scattering source, and is given by

$$s^{e^-e^-}(\vec{r},\hat{\Omega}_e,E_e) = \rho_{a,e}(\vec{r}) \int dE'_e \int d\Omega'_e \, \sigma_s^{e^-e^-}(\hat{\Omega}'_e,\hat{\Omega}_e,E'_e,E_e)$$
$$\times \, \varphi^{e^-}(\vec{r},\hat{\Omega}'_e,E'_e) \tag{6.114}$$

where $\sigma_s^{e^-e^-}(\hat{\Omega}'_e,\hat{\Omega}_e,E'_e,E_e)$ is the doubly differential cross section for electron interactions that create secondary electrons or delta rays, and $s^{e^+e^-}(\vec{r},\hat{\Omega}_e,E_e)$ is the *positron-electron* scattering source, and is given by

$$s^{e^+e^-}(\vec{r},\hat{\Omega}_e,E_e) = \rho_{a,e}(\vec{r}) \int dE'_p \int d\Omega'_p \, \sigma_s^{e^+e^-}(\hat{\Omega}'_p,\hat{\Omega}_e,E'_p,E_e)$$
$$\times \, \varphi^{e^+}(\vec{r},\hat{\Omega}'_p,E'_p) \tag{6.115}$$

where $\sigma_s^{e^+e^-}(\hat{\Omega}'_p,\hat{\Omega}_e,E'_p,E_e)$ is the doubly differential cross section for positron interactions that create secondary electrons or delta rays, and $s^{\gamma e^-}(\vec{r},\hat{\Omega}_e,E_e)$ is the *photon-electron* scattering source, and is given by

$$s^{\gamma e^-}(\vec{r},\hat{\Omega}_e,E_e) = \rho_{a,e}(\vec{r}) \int dE'_\gamma \int d\Omega'_\gamma \, \sigma_s^{\gamma e^-}(\hat{\Omega}'_\gamma,\hat{\Omega}_e,E'_\gamma,E_e)$$
$$\times \, \varphi^{\gamma}(\vec{r},\hat{\Omega}'_\gamma,E'_\gamma) \tag{6.116}$$

where $\sigma_s^{\gamma e^-}(\hat{\Omega}'_\gamma,\hat{\Omega}_e,E'_\gamma,E_e)$ is the doubly differential cross section for photon interactions that create secondary electrons, and $s^{e^-}(\vec{r},\hat{\Omega}_e,E_e)$ is the *external* electron source for electrons that originate from outside the computational volume.

Based on Figure 6.9, the possible electron interactions include: Rutherford (elastic) scatter, Möller (inelastic) scatter, and bremsstrahlung. Therefore, the removal term in Equation 6.113 will account for these interactions through the macroscopic attenuation coefficient, $\Sigma_t^{e^-}(\vec{r},E_e)$, as given by

$$\Sigma_t^{e^-} = \Sigma_{el}^{e^-} + \Sigma_{inel}^{e^-} + \Sigma_{brem}^{e^-} \tag{6.117}$$

where the dependent variables have been suppressed for brevity.

Similarly, for the electron or positron-electron, $s^{e^-e^-}(\vec{r},\hat{\Omega}_e,E_e)$ or $s^{e^+e^-}(\vec{r},\hat{\Omega}_e,E_e)$, and photon-electron, $s^{\gamma e^-}(\vec{r},\hat{\Omega}_e,E_e)$, scattering sources, the relevant scattering interactions are: Rutherford (elastic) scatter, Möller or Bhahba (inelastic) scatter, and bremsstrahlung for the former; and Compton (incoherent) scatter, photoelectric absorption, pair and triplet production for the latter. Therefore, their corresponding doubly differential scattering cross sections are given by

$$\sigma_s^{e^-/e^+e^-} = \sigma_{el}^{e^-/e^+e^-} + \sigma_{inel}^{e^-/e^+e^-} + \sigma_{brem}^{e^-/e^+e^-} \tag{6.118}$$

and

$$\sigma_s^{\gamma e^-} = \sigma_{inc}^{\gamma e^-} + \sigma_{pp}^{\gamma e^-} + \sigma_{tp}^{\gamma e^-} \tag{6.119}$$

Theoretical expressions for these electron cross sections can be found in Chapter 2 and the following reference (Podgorsak 2010).

Positron Formulation

Since the positron is the anti-particle to the electron, due to symmetry, the linear time-independent, Boltzmann transport equation for positrons can be stated, based on Equation 6.53, with appropriate subscripts and/or superscripts (e^+) on the functions and variables, as given by

$$\hat{\Omega}_p \cdot \vec{\nabla} \varphi^{e^+}(\vec{r}, \hat{\Omega}_p, E_p) + \Sigma_t^{e^+}(\vec{r}, E_p) \, \varphi^{e^+}(\vec{r}, \hat{\Omega}_p, E_p)$$
$$= s^{e^+ e^+}(\vec{r}, \hat{\Omega}_p, E_p) + s^{\gamma e^+}(\vec{r}, \hat{\Omega}_p, E_p) + s^{e^+}(\vec{r}, \hat{\Omega}_p, E_p) \tag{6.120}$$

where $s^{e^+ e^+}(\vec{r}, \hat{\Omega}_p, E_p)$ is the *positron-positron* scattering source, and is given by

$$s^{e^+ e^+}(\vec{r}, \hat{\Omega}_p, E_p) = \rho_{a,e}(\vec{r}) \int dE_p' \int d\Omega_p' \, \sigma_s^{e^+ e^+}(\hat{\Omega}_p', \hat{\Omega}_p, E_p', E_p)$$
$$\times \varphi^{e^+}(\vec{r}, \hat{\Omega}_p', E_p') \tag{6.121}$$

where $\sigma_s^{e^+ e^+}(\hat{\Omega}_p', \hat{\Omega}_p, E_p', E_p)$ is the doubly differential cross section for positron interactions that create secondary positrons, and $s^{\gamma e^+}(\vec{r}, \hat{\Omega}_p, E_p)$ is the *photon-positron* scattering source, and is given by

$$s^{\gamma e^+}(\vec{r}, \hat{\Omega}_p, E_p) = \rho_{a,e}(\vec{r}) \int dE_\gamma' \int d\Omega_\gamma' \, \sigma_s^{\gamma e^+}(\hat{\Omega}_\gamma', \hat{\Omega}_p, E_\gamma', E_p)$$
$$\times \varphi^\gamma(\vec{r}, \hat{\Omega}_\gamma', E_\gamma') \tag{6.122}$$

where $\sigma_s^{\gamma e^+}(\hat{\Omega}_\gamma', \hat{\Omega}_p, E_\gamma', E_p)$ is the doubly differential cross section for photon interactions that create secondary positrons, and $s^{e^+}(\vec{r}, \hat{\Omega}_p, E_p)$ is the *external* positron source for positrons that originate from outside the computational volume.

Based on Figure 6.9, the possible positron interactions include: Rutherford (elastic) scatter, Bhabha (inelastic) scatter, and bremsstrahlung, and pair annihilation. Therefore, the removal term in Equation 6.120 will account for these interactions through the macroscopic attenuation coefficient, $\Sigma_t^{e^+}(\vec{r}, E_p)$, as given by

$$\Sigma_t^{e^+} = \Sigma_{el}^{e^+} + \Sigma_{inel}^{e^+} + \Sigma_{brem}^{e^+} + \Sigma_{pa}^{e^+} \tag{6.123}$$

where the dependent variables have suppressed for brevity.

Note that <u>no</u> electron interaction creates secondary positrons, so an electron-positron scattering source, $s^{e^-e^+}(\vec{r}, \hat{\Omega}_p, E_p)$, is not included in the positron Boltzmann transport equation. Similarly, for the positron-positron, $s^{e^+e^+}(\vec{r}, \hat{\Omega}_p, E_p)$, and photon-positron, $s^{\gamma e^+}(\vec{r}, \hat{\Omega}_p, E_p)$, scattering sources, the relevant scattering interactions are: Rutherford (elastic) scatter, Möller or Bhabha (inelastic) scatter, and bremsstrahlung for the former; and pair and triplet production for the latter. Therefore, their corresponding doubly differential scattering cross sections are given by

$$\sigma_s^{e^+e^+} = \sigma_{el}^{e^+e^+} + \sigma_{inel}^{e^+e^+} + \sigma_{brem}^{e^+e^+} \tag{6.124}$$

and

$$\sigma_s^{\gamma e^+} = \sigma_{pp}^{\gamma e^+} + \sigma_{tp}^{\gamma e^+} \tag{6.125}$$

Theoretical expressions for these positron cross sections can be found in the following reference (Podgorsak 2010).

Coupled Photon, Electron and Positron Formulation

In summary, the coupled photon, electron and positron linear Boltzmann transport equations are re-listed together here

$$\hat{\Omega}_\gamma \cdot \vec{\nabla} \varphi^\gamma(\vec{r}, \hat{\Omega}_\gamma, E_\gamma) + \Sigma_t^\gamma(\vec{r}, E_\gamma)\, \varphi^\gamma(\vec{r}, \hat{\Omega}_\gamma, E_\gamma)$$
$$= s^{\gamma\gamma}(\vec{r}, \hat{\Omega}_\gamma, E_\gamma) + s^{e^-\gamma}(\vec{r}, \hat{\Omega}_\gamma, E_\gamma) + s^{e^+\gamma}(\vec{r}, \hat{\Omega}_\gamma, E_\gamma) + s^\gamma(\vec{r}, \hat{\Omega}_\gamma, E_\gamma)$$

$$\hat{\Omega}_e \cdot \vec{\nabla} \varphi^{e^-}(\vec{r}, \hat{\Omega}_e, E_e) + \Sigma_t^{e^-}(\vec{r}, E_e)\, \varphi^{e^-}(\vec{r}, \hat{\Omega}_e, E_e)$$
$$= s^{e^-e^-}(\vec{r}, \hat{\Omega}_e, E_e) + s^{e^+e^-}(\vec{r}, \hat{\Omega}_e, E_e) + s^{\gamma e^-}(\vec{r}, \hat{\Omega}_e, E_e) + s^{e^-}(\vec{r}, \hat{\Omega}_e, E_e)$$

$$\hat{\Omega}_p \cdot \vec{\nabla} \varphi^{e^+}(\vec{r}, \hat{\Omega}_p, E_p) + \Sigma_t^{e^+}(\vec{r}, E_p)\, \varphi^{e^+}(\vec{r}, \hat{\Omega}_p, E_p)$$
$$= s^{e^+e^+}(\vec{r}, \hat{\Omega}_p, E_p) + s^{\gamma e^+}(\vec{r}, \hat{\Omega}_p, E_p) + s^{e^+}(\vec{r}, \hat{\Omega}_p, E_p)$$

These integro-differential equations are *coupled* due to the presence of the various scattering sources present on the right hand side of each equation. Each contains a particle fluence density distribution from one of the other equations. In principle, one equation cannot be analytically solved without the other equations.

Within the context of photon-based dose calculations, the goal is to solve the coupled linear Boltzmann transport equations above for both the electron and positron fluence distribution functions, $\varphi^{e^-}(\vec{r}, \hat{\Omega}_e, E_e)$ and $\varphi^{e^+}(\vec{r}, \hat{\Omega}_p, E_p)$.

Solutions for $\varphi^{e^-}(\vec{r}, \hat{\Omega}_e, E_e)$ and $\varphi^{e^+}(\vec{r}, \hat{\Omega}_p, E_p)$ can in turn be used to determine the absorbed dose within the computational volume of interest as follows:

$$D(\vec{r}) = \int_E \mathrm{d}E_e \int_{4\pi} \mathrm{d}\Omega \frac{S_c^{e^-}}{\rho}(\vec{r}, E_e)\, \varphi^{e^-}(\vec{r}, \hat{\Omega}_e, E_e)$$

$$+ \int_E \mathrm{d}E_p \int_{4\pi} \mathrm{d}\Omega \frac{S_c^{e^+}}{\rho}(\vec{r}, E_p)\, \varphi^{e^+}(\vec{r}, \hat{\Omega}_p, E_p) \quad (6.126)$$

The next section details how the *theoretical* development of the coupled linear Boltzmann transport equations outlined above can be *practically* implemented and solved within the constraints of computational resources and time.

6.4   THE BOLTZMANN TRANSPORT EQUATION IN PRACTICE

As mentioned previously, novel deterministic numerical techniques and tools for solving the linear Boltzmann transport equation have been adapted from the nuclear engineering community, and made available to dose calculation algorithms for radiotherapy treatment planning. In recent years, one such dose algorithm known as Acuros XB® (Varian Medical Systems, Palo Alto, CA), or AXB for short, has been made commercially available. AXB was specifically developed for the clinical radiotherapy community in order to offer a computationally efficient alternative to Monte Carlo-based dose algorithms. The deterministic prototype solver that forms the backbone of the AXB algorithm was originally developed from research conducted in the X-division of the Los Alamos National Laboratory. A spin-off company (Transpire Inc, Gig Harbor, WA) commercialized the solver, also known as Attila®, and licensed the technology as a general purpose radiation transport software product. AXB represents a ground up re-write of the Attila solver, and has been optimized for external beam photon dose calculations.

In the following section, a general strategy to obtain a practical solution to the coupled photon-electron linear Boltzmann transport equation is introduced. The ultimate goal of such a strategy is to use deterministic numerical methods to convert a governing equation (i.e. Boltzmann transport equation) into matrix form that can then be solved algebraically. The strategy, as presented below, has been broken down in terms of: (1) *series expansions* of the particle fluence density distribution and differential scattering cross section using spherical harmonics and Legendre polynomials, respectively, (2) *phase space discretization* of the energy, anglular, and spatial variables using the multi-group method, discrete ordinates method, and finite element method, respectively, and (3) *algebraic solution* techniques, such as energy and source iteration plus synthetic acceleration, that provide efficient ways to systematically solve a large number of simultaneous partial differential equations. Once the strategy has been described, the underlying *approximations* used in practically implementing the linear Boltzmann transport equation are discussed, then followed by an outline of the details of a practical solution for external beam radiotherapy dose calculation algorithms. Finally, a *comparison* of 3D dose distributions calculated with commercial deterministic, Monte Carlo, and convolution-based algorithms will be reviewed.

6.4.1 Series Expansions

One method to simplify and approximate an arbitrary, unknown function is to represent it in terms of basis functions with unknown expansion coefficients. For example, in Fourier analysis, any function can be expanded in terms of a series of appropriately weighted cosine and sine functions. If a set of basis functions is linearly independent, then the bases are *orthogonal*. The orthogonality property of basis functions provides a way to determine the unknown expansion coefficients above by simplifying integrals involving the unknown function. The series expansions for the particle fluence density distribution in terms of spherical harmonics, and particle differential scattering cross section using Legendre polynomials is presented below.

6.4.1.1 Transformation of Fluence

Spherical harmonics (Arfken et al. 2012) represent the angular decomposition of an unknown function using a set of basis functions on the unit sphere. Using spherical coordinate representation, they can formally be defined by

$$Y_l^m(\hat{\Omega}) \equiv Y_l^m(\theta, \phi) = \sqrt{\frac{(2l+1)}{4\pi} \frac{(l-m)!}{(l+m)!}} \, P_l^m(\cos\theta) \, \exp(im\phi) \qquad (6.127)$$

where $P_l^m(\cos\theta)$ represents the associated Legendre polynomials, and l and m are the *order* and *degree* indices, respectively. These indices satisfy the constraint that $l \in \mathbb{N}$ and $-l \le m \le l$, which limits the number $(2l+1)$ of basis functions for each order l.

The first few (real) spherical harmonics are given by

$$l = 0: \quad Y_0^0(\theta, \phi) = \sqrt{\frac{1}{4\pi}}$$

$$Y_1^{-1}(\theta, \phi) = \sqrt{\frac{3}{4\pi}} \sin\theta \sin\phi$$

$$l = 1: \quad Y_1^0(\theta, \phi) = \sqrt{\frac{3}{4\pi}} \cos\theta$$

$$Y_1^{+1}(\theta, \phi) = \sqrt{\frac{3}{4\pi}} \sin\theta \cos\phi$$

The orthogonality condition of the spherical harmonics for basis functions of different l and m is given by

$$\int d\Omega \, Y_l^{m*}(\hat{\Omega}) \, Y_{l'}^{m'}(\hat{\Omega}) = \delta_{ll'} \, \delta_{mm'} \qquad (6.128)$$

where * denotes the complex conjugate, and $\delta_{ll'}$ represents the Kronecker delta function. Equation 6.128 suggests that the inner product of any two distinct spherical harmonic basis functions is zero.

Since the spherical harmonics define a complete basis over the unit sphere, then any (real-valued) spherical function can be expanded as a linear combination of the basis functions. Therefore, in terms of the particle fluence density distribution,

$$\varphi(\vec{r}, \hat{\Omega}, E) = \sum_{l=0}^{\infty} \sum_{m=-l}^{l} \varphi_l^m(\vec{r}, E) \, Y_l^m(\hat{\Omega}) \tag{6.129}$$

Equation 6.129 is *exact* provided that the sum over l goes to infinity. In practice, a limit is specified, which serves to approximate the particle fluence density distribution, as given by

$$\varphi(\vec{r}, \hat{\Omega}, E) \approx \sum_{l=0}^{N_L} \sum_{m=-l}^{l} \varphi_l^m(\vec{r}, E) \, Y_l^m(\hat{\Omega}) \tag{6.130}$$

where N_L is the total number of harmonics in the series expansion.

By applying the orthogonality property in Equation 6.128, the coefficients $\varphi_l^m(\vec{r}, E)$ of the particle fluence density distribution can be computed by projecting $\varphi(\vec{r}, \hat{\Omega}, E)$ onto each spherical harmonic basis function $Y_l^m(\hat{\Omega})$ as follows:

$$\varphi_l^m(\vec{r}) = \int d\Omega \, Y_l^{m*}(\hat{\Omega}) \, \varphi(\vec{r}, \hat{\Omega}, E) \tag{6.131}$$

6.4.1.2 Transformation of Cross Sections

Similar to the transformation of the particle fluence density distribution, the differential scattering cross sections can also be simplified using spherical harmonics. However, due to the azimuthal symmetry involved with high-energy photon and electron scattering interactions, the particle direction following the interaction generally depends only on polar angle only ($\hat{\Omega}' \cdot \hat{\Omega} = \cos\theta$). Therefore, the much simpler *Legendre* series (Arfken et al. 2012) can be used instead of the more general spherical harmonics.

The Legendre series represents an expansion in terms of polynomial functions, and can formally be expressed using Roderigues' formula as follows:

$$P_l(x) = \frac{1}{2^l l!} \frac{d^l}{dx^l} (x^2 - 1)^l \tag{6.132}$$

where once again l represents the *order* of expansion.

The first few Legendre polynomials for $x = \cos\theta$ are given by

$$P_0(\cos\theta) = 1 \qquad P_2(\cos\theta) = \tfrac{1}{2}(3\cos^2\theta - 1)$$
$$P_1(\cos\theta) = \cos\theta \qquad P_3(\cos\theta) = \tfrac{1}{2}(5\cos^3\theta - 3\cos\theta)$$

Similar to the spherical harmonics, the Legendre polynomials also possess an orthogonality condition for basis functions of different l as shown by

$$\int_{-1}^{1} d(\cos\theta)\, P_l(\cos\theta)\, P_{l'}(\cos\theta) = \frac{2}{2l+1}\,\delta_{ll'} \tag{6.133}$$

Another useful property is the addition theorem, which relates the Legendre polynomials to the spherical harmonics as given by

$$P_{l'}(\cos\theta_s) = \frac{4\pi}{2l'+1} \sum_{m'=-l'}^{l'} Y_l^{m'}(\hat{\Omega})\, Y_{l'}^{m'*}(\hat{\Omega}) \tag{6.134}$$

Since the Legendre polynomials define a complete basis, then any (real-valued) one-dimensional function can be expanded as a linear combination of its basis functions. Therefore, in terms of the differential scattering cross section with azimuthal symmetry,

$$\Sigma_s(\vec{r}, \cos\theta_s, E', E) = \sum_{l=0}^{\infty} \Sigma_{s,l}(\vec{r}, E', E)\, P_l(\cos\theta_s) \tag{6.135}$$

where $P_l(\cos\theta_s)$ denotes a Legendre polynomial.

In practice, Equation 6.135 is approximated based on a limit on the order, as given by

$$\Sigma_s(\vec{r}, \cos\theta_s, E', E) \approx \sum_{l=0}^{N_L} \Sigma_{s,l}(\vec{r}, E', E)\, P_l(\cos\theta_s) \tag{6.136}$$

where N_L is the total number of polynomials in the series expansion.

By applying the orthogonality property in Equation 6.133, the coefficients $\Sigma_{s,l}(\vec{r}, E', E)$ of the particle differential scattering cross section can be determined by

$$\Sigma_{s,l}(\vec{r}, E', E) = \frac{2l+1}{2} \int_{-1}^{1} d(\cos\theta)\, P_l(\cos\theta)\, \Sigma_s(\vec{r}, \cos\theta_s) \tag{6.137}$$

Now, combining the series expansions for the particle fluence density distribution (Equation 6.130) and differential scattering cross section (Equation 6.136), and inserting them into the scattering source of the Boltzmann transport equation (Equation 6.53) gives

$$\sum_{l=0}^{N_L} \sum_{m=-l}^{l} \sum_{l'=0}^{N_L} \varphi_l^m\, \Sigma_{s,l'}(\vec{r}, E', E) \int d\Omega'\, Y_l^m(\hat{\Omega}')\, P_{l'}(\cos\theta_s) \tag{6.138}$$

By applying the addition theorem in Equation 6.134, the Legendre polynomials in the integral of Equation 6.138 can be replaced to simplify the integral as follows:

$$\int d\Omega' \, Y_l^m(\hat{\Omega}') \, P_{l'}(\cos\theta_s)$$

$$= \frac{4\pi}{2l'+1} \sum_{m'=-l'}^{l'} \int d\Omega' \, Y_l^m(\hat{\Omega}') \, Y_{l'}^{m'}(\hat{\Omega}) \, Y_{l'}^{m'*}(\hat{\Omega}')$$

$$= \frac{4\pi}{2l'+1} \, Y_{l'}^m(\hat{\Omega}) \, \delta_{ll'} \tag{6.139}$$

where the integral over $d\Omega'$ has been eliminated by the orthogonality of spherical harmonics as defined in Equation 6.128.

Finally, by using Equation 6.139, the sum over l' in Equation 6.138 can be removed, and the form of the linear Boltzmann transport equation using series expansions becomes

$$\left[\hat{\Omega} \cdot \vec{\nabla} + \Sigma_t(\vec{r}, E) \right] \varphi(\vec{r}, \hat{\Omega}, E)$$

$$= s(\vec{r}, \hat{\Omega}, E) + \sum_{l=0}^{N_L} \frac{4\pi}{2l+1} \Sigma_{s,l}(\vec{r}, E', E) \sum_{m=-l}^{l} \varphi_l^m(\vec{r}, E') \, Y_l^m(\hat{\Omega}) \tag{6.140}$$

Thus, the angular integral in the scattering source term has been replaced with a sum of spherical harmonic expansions. The coefficients $\Sigma_{s,l}(E', E)$ of the doubly differential scattering cross-section can be pre-tabulated as a three dimensional matrix in terms of spherical harmonic order l, and incoming E' and outgoing E particle energies, respectively. The unknown to be calculated would then be the coefficients $\varphi_l^m(\vec{r}, E')$ of the particle fluence density distribution.

6.4.2 Phase Space Discretization

In order to practically implement the discontinuous form of the linear (time-independent) Boltzmann transport equation (Equation 6.53), proper discretization of all phase space variables must be carried out. In this section, various discretization methods for energy, angular, and spatial variables are discussed in detail.

6.4.2.1 Energy Discretization

The *multi-group* approximation (Duderstadt and Martin 1979; Lewis and Miller 1984) is the most common method used to discretize the energy-dependent terms of the linear Boltzmann transport equation. The mathematical form of the multi-group transport equation is closely related to the original transport equation, except that the energy variable is discrete rather than continuous. Essentially, energy-dependent quantities are averaged over discrete energy groups, while integrals over the continuous energy variable are replaced by discrete sums over these same energy groups.

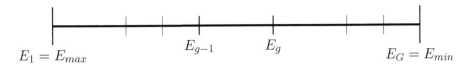

Figure 6.10: Schematic showing the division of the particle energy into N_G discrete groups. The energy index is defined such that a lower value represents a higher particle energy, and a higher index value represents a lower energy.

Below, the multi-group notation is introduced and the associated transport equation is derived.

As shown in Figure 6.10, the basic idea behind the multi-group method is to determine a particle energy range relevant to the problem to be solved, and hence divide it into N_g intervals, whereby E_1 and E_{N_g} represent the maximum and minimum particle energies, respectively. By convention, E_1 would be chosen to correspond to the maximum energy of the source particles, and similarly, E_{N_g} would either be equal to zero or some cut-off energy. The particles in energy *group g* represent those particles within an energy interval (ΔE_g) between E_g and E_{g-1}. Note that these energy intervals need *not* be uniform over the energy range of interest. Care must be taken to ensure that energy-dependent quantities and integrals are smooth within each energy group. In practice, the choice of the total number of energy groups N_G yields a trade-off between dose algorithm accuracy and computation time.

In the multi-group method, the particle fluence density distribution can be expressed as

$$\varphi_g(\vec{r}, \hat{\Omega}) = \int_g dE \, \varphi(\vec{r}, \hat{\Omega}, E) \tag{6.141}$$

where for brevity the following shorthand notation is used

$$\int_g dE = \int_{E_g}^{E_{g-1}} dE \tag{6.142}$$

Using the above definitions, the integral over E' in the collision integral can be replaced with the sum of integrals over all energy groups as given by

$$\int_0^\infty dE' = \sum_{g'=1}^{N_G} \int_{g'} dE' \tag{6.143}$$

and integrating between E_g and E_{g-1} yields

$$\left[\hat{\Omega} \cdot \vec{\nabla} + \Sigma_{t,g}(\vec{r}) \right] \varphi_g(\vec{r}, \hat{\Omega}) = s_g(\vec{r}, \hat{\Omega}) + \sum_{g'=1}^{N_G} \int d\Omega' \, \Sigma_{s,g'g}(\vec{r}, \hat{\Omega}' \cdot \hat{\Omega}) \, \varphi_{g'}(\vec{r}, \hat{\Omega}')$$

$$\tag{6.144}$$

where the following notation is introduced:

$$\varphi_g(\vec{r}, \hat{\Omega}) = \int_{E_g}^{E_{g-1}} dE\, \varphi(\vec{r}, \hat{\Omega}, E) \tag{6.145}$$

$$s_g(\vec{r}, \hat{\Omega}) = \int_{E_g}^{E_{g-1}} dE\, s(\vec{r}, \Omega, E) \tag{6.146}$$

$$\Sigma_{t,g}(\vec{r}) = \frac{1}{\varphi_g(\vec{r}, \hat{\Omega})} \int_{E_g}^{E_{g-1}} dE\, \Sigma_t(\vec{r}, E)\, \varphi(\vec{r}, \Omega, E) \tag{6.147}$$

$$\Sigma_{s,g'g}(\vec{r}, \hat{\Omega}' \cdot \hat{\Omega})$$
$$= \frac{1}{\varphi_{g'}(\vec{r}, \hat{\Omega}')} \int_{E_g}^{E_{g-1}} dE \int_{E_{g'}}^{E_{g'-1}} dE'\, \Sigma_s(\vec{r}, \hat{\Omega}', E', \hat{\Omega}, E)\, \varphi(\vec{r}, \hat{\Omega}', E') \tag{6.148}$$

In the above definitions, there is an implicit assumption that the ratio $\varphi(\vec{r}, \Omega, E)/\varphi_g(\vec{r}, \Omega)$ is independent of both $\vec{r}$ and $\hat{\Omega}$. The assumption is known as the *energy separability* approximation and is given by

$$\varphi(\vec{r}, \hat{\Omega}, E) \approx f(E)\, \varphi_g(\vec{r}, \hat{\Omega}) \quad \text{for} \quad E_g < E \le E_{g-1} \tag{6.149}$$

where $f(E)$ is known as the spectral weighting function.

Proper choice of the form of $f(E)$ has a direct impact on the accuracy of the multi-group energy discretization of the various factors used in the linear Boltzmann transport equation. For example, if these factors are known to be "smooth", then a simple spectral weighting function can be used

$$f(E) = \frac{1}{\Delta E_g}, \quad E_g \le E < E_{g-1}, \tag{6.150}$$

where $\Delta E_g = E_{g-1} - E_g$.

The uncertainty associated with the above form of $f(E)$ is mitigated by using more energy groups, which in turn reduces the widths of each energy bin. Therefore, the group total cross-section, $\Sigma_t(\vec{r})$, and the group scattering cross-section, $\Sigma_{s,g'g}(\vec{r}, \hat{\Omega}' \cdot \hat{\Omega})$ can be pre-tabulated for each material in the problem to be solved.

6.4.2.2 Angular Discretization

The most common method for discretizing the angular variables in the linear Boltzmann transport equation is the *discrete ordinates* method (Duderstadt and Martin 1979; Lewis and Miller 1984; Vassiliev 2016), whereby quadrature sets are used to approximate integrals over the direction variables. In terms of the particle fluence density distribution, $\phi_g(\vec{r}, \hat{\Omega})$, the particle direction unit vector, $\hat{\Omega}$, is discretized

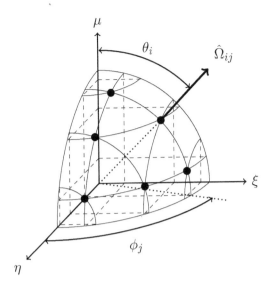

Figure 6.11: Discretization of particle direction (polar and azimuthal angles) using the discrete ordinates method. The integration weights of each point on the spherical grid depend on the type of quadrature used.

in terms of the polar and azimuthal angles as $\hat{\Omega}_{ij} \equiv (\mu_i, \phi_j)$, where $i = 1, ..., N_\mu$ and $j = 1, ..., N_\phi$. These discrete directions are *specifically* chosen such that they optimize numerical integration over the differential solid angle $d\Omega$, as required by the scattering source term in the Boltzmann transport equation.

In numerical analysis, a quadrature set can be used to approximate the definite integral of a function, usually as a *weighted* sum of function values at specific points within the domain of integration. Gaussian quadrature is a common one-dimensional numerical integration technique, whereby an unknown $f(x)$ is decomposed in terms of a weighting function $w(x)$ and an approximate polynomial function $g(x)$ and integrated or summed as given by

$$\int_{-1}^{1} f(x)\, dx = \int_{-1}^{1} w(x)\, g(x)\, dx \approx \sum_{i=1}^{n} w_i\, g(x_i) \tag{6.151}$$

where w_i represent specific weights defined at each x_i over the range of integration $[-1, 1]$.

The expression for the weighting function defines the type of Gaussian quadrature. Two useful types of Gaussian quadrature include: Gauss-Legendre quadrature, $w(x) = 1$, and Gauss-Chebyshev quadrature, $w(x) = 1/\sqrt{1 - x^2}$. In the former, the discrete weights are given by

$$w_i = \frac{2}{[1 - x_i^2\, P_n(x_i)]} \qquad (\text{Gauss} - \text{Legendre}) \tag{6.152}$$

where $P_n(x)$ is the Legendre polynomial function of order n. In the latter,

$$w_j = \frac{\pi}{n} \qquad \text{(Gauss − Chebyshev)} \tag{6.153}$$

where once again n is the number of points within the integration interval.

In applying the Gaussian quadrature methodology described above to the scattering source term in the linear Boltzmann transport equation, the integral over solid angle can be split into two components: one integral over $\mu = \cos\theta$ and one over ϕ' as follows:

$$\int_{4\pi} d\Omega' \, \Sigma_s(\vec{r}, \hat{\Omega}', \hat{\Omega}, E', E) \, \varphi(\vec{r}, \hat{\Omega}', E')$$
$$= \int_{-1}^{1} d\mu \int_{0}^{2\pi} d\phi' \, \Sigma_s(\vec{r}, \mu, \phi', \phi, E', E) \, \varphi(\vec{r}, \mu, \phi', E') \tag{6.154}$$

Since integration using Gaussian quadrature is defined over the interval $[-1, 1]$, then the interval $[0, 2\pi]$ over ϕ' must be changed. Making the substitution $\nu = \cos\phi'$ transforms the ϕ' integral as follows:

$$\int_{-1}^{1} d\mu \int_{0}^{2\pi} d\phi' \, \Sigma_s(\vec{r}, \mu, \phi', \phi, E', E) \, \varphi(\vec{r}, \mu, \phi', E')$$
$$= 2 \int_{-1}^{1} d\mu \int_{-1}^{1} \frac{d\nu}{\sqrt{1 - \nu^2}} \, \Sigma_s(\vec{r}, \mu, \nu, \phi, E', E) \, \varphi(\vec{r}, \mu, \nu, E') \tag{6.155}$$

As seen in Equation 6.155, the two-dimensional integration over solid angle has been replaced with two one-dimensional integrals over μ and ν, which can be solved with Gaussian quadrature using the Gauss-Legendre and Gauss-Chebyshev weight functions, respectively, defined above.

Finally, combining the two 1D quadratures into the 2D integral produces the following result:

$$\int_{-1}^{1} d\mu \int_{-1}^{1} \frac{d\nu}{\sqrt{1 - \nu^2}} \, \Sigma_s(\vec{r}, \mu, \nu, \phi, E', E) \, \varphi(\vec{r}, \mu, \nu, E')$$
$$\approx \sum_{i=1}^{N_\mu} \sum_{j=1}^{N_\phi} w_i \, w_j \, \Sigma_s(\vec{r}, \mu_i, \nu_j, \phi, E', E) \, \varphi(\vec{r}, \mu_i, \nu_j, E')$$
$$\approx \sum_{i=1}^{N_\mu} \sum_{j=1}^{N_\phi} w_i \, w_j \, \Sigma_s(\vec{r}, \hat{\Omega}'_{ij}, \hat{\Omega}_{ij}, E', E) \, \varphi(\vec{r}, \hat{\Omega}'_{ij}, E') \tag{6.156}$$

where the short hand notation $\hat{\Omega}_{ij} \equiv (\mu_i, \nu_j)$ has been used.

Therefore, using the discrete ordinates method, the group Boltzmann transport equation can be simply re-written and solved for each of the discrete directions $\hat{\Omega}_{ij}$, as defined by the Gaussian quadrature of the scattering source term, and given by

$$\left[\hat{\Omega}_{ij} \cdot \vec{\nabla} + \Sigma_{t,g}(\vec{r})\right] \varphi_g(\vec{r}, \hat{\Omega}_{ij})$$

$$= s_g(\vec{r}, \hat{\Omega}_{ij}) + \sum_{l=0}^{N_L} \frac{4\pi}{2l+1} \sum_{g'=1}^{N_g} \Sigma_{s,lg'g}(\vec{r}) \sum_{m=-l}^{l} \varphi_{g'l}^m(\vec{r}) Y_l^m(\hat{\Omega}_{ij}) \quad (6.157)$$

where $i = 1, ..., N_\mu$, $j = 1, ..., N_\phi$, and $g = 1, ..., N_g$.

By applying Gaussian quadrature to the 2D angular integral in the scattering source term, the particle fluence density distribution coefficients, $\varphi_{g'l}^m$, can be determined by

$$\varphi_{g'l}^m(\vec{r}) = \int d\hat{\Omega}\, Y_l^m(\hat{\Omega})\, \varphi_{g'}(\vec{r}, \hat{\Omega}) \approx \sum_{i'=1}^{N_\mu} \sum_{j'=1}^{N_\phi} w_{i'}\, w_{j'}\, Y_l^m(\hat{\Omega}_{i'j'})\, \varphi_{g'}(\vec{r}, \hat{\Omega}_{i'j'}) \quad (6.158)$$

Thus, for each energy group g, there is a system of $N_\mu \times N_\phi$ equations. These equations will be coupled, since the equation for any given discrete direction $\hat{\Omega}_{ij}$ consists of fluences from all other directions. The only remaining variable to be discretized now is the spatial variable.

6.4.2.3 Spatial Discretization

The *finite element* method is a numerical technique capable of solving complex differential equations, which would be too difficult to solve analytically. In the method, the problem geometry is divided into many sub-volumes, called *finite elements*, whereby each sub-volume is represented by a set of *elemental* equations for the physical problem of interest. These elemental equations can then be systematically combined into a *global* system of equations, and then solved algebraically to give a final solution.

In general, the finite element method can be better understood by breaking down the method into the following sub-topics:

1. element mesh generation of problem geometry

2. element shape functions

3. weak formulation of governing equation

4. approximation of weak formulation using Galerkin method

5. elemental equation/matrix formulation

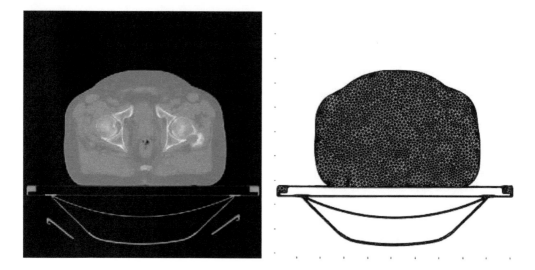

Figure 6.12: CT image (left) of male pelvis anatomy and corresponding computational mesh (right) using tetrahedral finite elements.

6. coordinate transformation from global-to-local coordinates

7. global equation/matrix formulation

These steps will be discussed below within the context of solving the linear Boltzmann transport equation for the particle fluence density distribution, $\varphi(\vec{r}, \hat{\Omega}, E)$. Note that the finite element method is used here to discretize the spatial variable only. However, as an alternative to the multi-group and discrete ordinates methods, as applied to the energy and angular variables, respectively; the finite element approach may also be used to discretize these other variables. Finally, in the foregoing discussion, the mathematical rigor of the finite element method is *relaxed* in order to facilitate an *introductory* understanding of the technique. An in-depth resource on the topic can be found in (Zienkiewicz et al. 2005) for interested readers.

Mesh Generation

The first step in applying the finite element method to a physical problem is to construct a computational mesh of the problem geometry, whereby the total volume V of the geometry is sub-divided into a collection of non-overlapping sub-volumes V_e, or finite elements, given by

$$V = \bigcup_{e=1}^{N_e} V_e^{(e)} \tag{6.159}$$

where N_e denotes the total number of finite elements used to discretize the problem geometry. In the context of radiotherapy dose calculations, the problem geometry

(A)

(B)

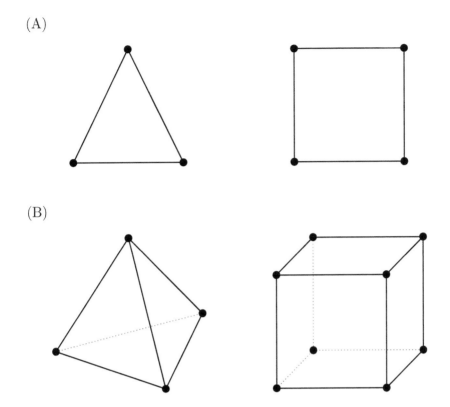

Figure 6.13: Examples of 2D (A) triangular, rectangular, and 3D (B) tetrahedral, hexahedral *linear* finite element geometries.

is usually a 3D CT image or dose grid.

A wide variety of finite element *shapes* can be used to *tesselate* or *tile* the problem geometry. For example, Figure 6.13 shows commonly used element shapes such as triangles or rectangles for 2D problems, and tetrahedrons or hexahedrons in 3D problems. Note that each of these element types can be characterized by the number of *nodes* or *vertices* (corner points of the element), *edges* (lines connecting any two vertices), and *faces* (plane defined by multiple edges) used to represent their geometrical shape. For example, triangles consist of 3 vertices and edges plus 1 face, while tetrahedrons possess 4 vertices and faces plus 6 edges. In addition, the edges or lines connecting individual vertices can be straight or curved. However, more complex shapes come with an associated computational cost, in terms of processing speed or memory.

A collection of finite elements making up the computational grid can be classified as either *structured*, which display regular connectivity between elements, or *unstructured*, which possess non-uniform connectivity. In the latter, variable-sized elements of the same shape may be used to divide the computational grid. For example, smaller elements can be used in areas with a material boundary or high-gradient regions, while larger elements can be reserved for more homogeneous areas.

In practice, an *adaptive* mesh strategy can also be used to solve a physical problem with the finite element method. In such an approach, a *course* mesh of the problem geometry is first used to obtain an initial solution to the governing (partial) differential equation(s). Based on the error of the initial solution, the mesh can be *re-fined* in areas where more accuracy is required, which can then be used to make a subsequent attempt at a solution. This process continues until the solution converges to the desired level of accuracy.

Shape Functions

For a given finite element, the particle fluence density distribution, $\varphi(x, y, z)$, can be approximated by

$$\hat{\varphi}(x, y, z) = \sum_{k=1}^{N_k} \varphi_k \, h_k(x, y, z) = [\varphi]^T [h] \qquad (6.160)$$

where φ_k are coefficients that represent the particle fluence at each vertex of the finite elemet and $h_k(x, y, z)$ is the finite element basis (or shape) function for node k. Basically, Equation 6.160 represents an interpolation formula for the particle fluence.

For each node, the basis function, $h_k(x, y, z)$, must satisfy the following condition:

$$h_k(x_{k'}, y_{k'}, z_{k'}) = \delta_{kk'} \qquad \text{where} \qquad k, k' = 1, 2, ..., N_k \qquad (6.161)$$

where $\delta_{kk'}$ represents the Kronecker delta function, which for any given node, assigns a value of one on that node and zero for all other nodes.

In the case of tetrahedral finite elements, which consist of four nodes, the particle fluence density distribution is given by

$$\varphi_k(x, y, z) = \varphi_1 \, h_1(x, y, z) + \varphi_2 \, h_2(x, y, z) + \varphi_3 \, h_3(x, y, z) + \varphi_4 \, h_4(x, y, z) \quad (6.162)$$

where $h_k(x, y, z)$ denotes the components of the shape function.

For example, a tetrahedral finite element, with a *linear*, piecewise continuous shape function components are defined as

$$h_k(x, y, z) = \frac{a_k + b_k x + c_k y + d_k z}{6V_e} \qquad k = 1, .., 4 \qquad (6.163)$$

whereby V_e represents the volume of the tetrahedral element, and can be calculated

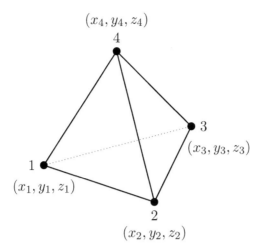

Figure 6.14: Node number assignment and coordinates of a tetrahedral finite element.

from the co-ordinates of each vertex as follows:

$$V_e = \frac{1}{6} \det \begin{vmatrix} 1 & x_1 & y_1 & z_1 \\ 1 & x_2 & y_2 & z_2 \\ 1 & x_3 & y_3 & z_3 \\ 1 & x_4 & y_4 & z_4 \end{vmatrix}$$

The constants for each shape function in Equation 6.163 can be determined by expanding the following determinants into their respective co-factors as

$$a_1 = \det \begin{vmatrix} x_2 & y_2 & z_2 \\ x_3 & y_3 & z_3 \\ x_4 & y_4 & z_4 \end{vmatrix} \qquad b_1 = -\det \begin{vmatrix} 1 & y_2 & z_2 \\ 1 & y_3 & z_3 \\ 1 & y_4 & z_4 \end{vmatrix}$$

$$c_1 = -\det \begin{vmatrix} x_2 & 1 & z_2 \\ x_3 & 1 & z_3 \\ x_4 & 1 & z_4 \end{vmatrix} \qquad d_1 = -\det \begin{vmatrix} x_2 & y_2 & 1 \\ x_3 & y_3 & 1 \\ x_4 & y_4 & 1 \end{vmatrix} \qquad (6.164)$$

whereby solutions to the other constants can be found through cyclic interchange of the subscripts in the order 1,2,3,4.

Note that, as an alternative to Equation 6.163, higher order polynomials may be used to represent shape functions. However, they come with a computational cost.

The components of the shape function must satisfy the following criterion at all points of the domain as given by

$$h_1(x, y, z) + h_2(x, y, z) + h_3(x, y, z) + h_4(x, y, z) = 1 \qquad (6.165)$$

With an approximate expression for the particle fluence density distribution and choice of a finite element shape function, the next step is to derive an equation for the unknown coefficients φ_k in Equation 6.160.

Weak (Integral Law) Formulation

The weak or integral law form of the linear Boltzmann transport equation is obtained by multiplying it by an arbitrary *test* function, $q(\vec{r})$, and integrating the product over the phase space variable of interest $\vec{r}$, as given by

$$\int_V d\vec{r}\, \hat{\Omega} \cdot \vec{\nabla} \varphi(\vec{r}, \hat{\Omega}, E)\, q(\vec{r}) + \int_V d\vec{r}\, \Sigma_t(\vec{r}, E)\, \varphi(\vec{r}, \hat{\Omega}, E)\, q(\vec{r})$$
$$= \int_V d\vec{r} \int dE' \int d\Omega'\, \Sigma_s(\vec{r}, \hat{\Omega}', \hat{\Omega}, E', E)\, \varphi(\vec{r}, \hat{\Omega}', E')\, q(\vec{r}) + \int_V d\vec{r}\, s(\vec{r}, \hat{\Omega}, E)\, q(\vec{r})$$
$$(6.166)$$

Using inner product and operator notation defined previously, Equation 6.166 can be compactly summarized as

$$\langle \hat{\Omega} \cdot \vec{\nabla} \varphi, q \rangle + \langle \Sigma_t \varphi, q \rangle = \langle \mathcal{K} \varphi, q \rangle + \langle s, q \rangle \tag{6.167}$$

In the following discussion, an external source expressed in the form of an incoming particle boundary condition is assumed:

$$\varphi(\vec{r}, \hat{\Omega}, E) = \varphi_b(\vec{r}, \hat{\Omega}, E) \quad \text{for} \quad \forall \vec{r} \in \partial V \text{ such that } \hat{\Omega} \cdot \hat{e}_n < 0 \tag{6.168}$$

Next, the first term of Equation 6.167 can be integrated by parts to give

$$-\langle \varphi, \hat{\Omega} \cdot \vec{\nabla} q \rangle + \langle \varphi, q \rangle_{\partial V_+} - \langle \varphi, q \rangle_{\partial V_-} + \langle \Sigma_t \varphi, q \rangle = \langle \mathcal{K} \varphi, q \rangle + \langle s, q \rangle \tag{6.169}$$

whereby

$$\langle \varphi, q \rangle_{\partial V_\pm} = \int_{\partial V_\pm} dA\, |\hat{\Omega} \cdot \hat{e}_n|\, \varphi(\vec{r_b})\, q(\vec{r_b}) \tag{6.170}$$

The boundary condition in Equation 6.168 can be *explicitly* substituted into Equation 6.169, and moving the known terms to the right hand side gives

$$-\langle \varphi, \hat{\Omega} \cdot \vec{\nabla} q \rangle + \langle \varphi, q \rangle_{\partial V_+} + \langle \Sigma_t \varphi, q \rangle = \langle \mathcal{K} \varphi, q \rangle + \langle s, q \rangle + \langle \varphi_b, q \rangle_{\partial V_-} \tag{6.171}$$

Note that Equation 6.171 represents the weak or integral law form of the linear Boltzmann transport equation, and without proof is equivalent to the original form of the integro-differential equation with boundary conditions. More importantly, it is in a form amenable to approximation by way of the finite element method.

Galerkin Method

Instead of attempting to find a solution to $\varphi(\vec{r}, \hat{\Omega}, E)$ within an infinite-dimensional function space, an approximate solution $\hat{\varphi}(\vec{r}, \hat{\Omega}, E)$ can be sought such that Equation 6.171 is satisfied for all test functions $q^h(\vec{r}, \hat{\Omega}, E)$ within a finite-dimensional subspace, S^h, where h is a parameter that depends on the finite element mesh spacing used in the approximate solution.

The finite subspace is specially constructed with basis functions $q_k^h(\vec{r})$ for $k = 1, 2, .., N_k$, where N_k is the dimension of the function subspace (typcially the number of nodes in the mesh). Note that these basis functions are *local*, in the sense that

$$\int_V d\vec{r}\, q_k(\vec{r})\, q_{k'}(\vec{r}) \tag{6.172}$$

is non-zero only when the nodes k and k' are close together.

By restricting the solution of the integral law within the above subspace, an approximate form of the weak (or integral law) can be stated as: *find* $\hat{\varphi}(\vec{r}) \in S^h$ such that for all $q^h(\vec{r}) \in S^h$,

$$-\langle \varphi, \hat{\Omega} \cdot \vec{\nabla} q \rangle + \langle \varphi, q \rangle_{\partial V_+} + \langle \Sigma_t\, \varphi, q \rangle - \langle \mathcal{K}\, \varphi, q \rangle = \langle s, q \rangle + \langle \varphi_b, q \rangle_{\partial V_-} \tag{6.173}$$

which can be further reduced into the simple form:

$$\Big\langle q(\vec{r}), \mathcal{L}\, \hat{\varphi}(\vec{r}, \hat{\Omega}, E) \Big\rangle - \Big\langle q(\vec{r}), s(\vec{r}, \hat{\Omega}, E) \Big\rangle = 0 \tag{6.174}$$

Since the subspace S^h is finite dimensional and $\hat{\varphi}(\vec{r}) \in S^h$, then $\hat{\varphi}$ can be expanded in terms of the basis functions for S^h,

$$\hat{\varphi}(\vec{r}) = \sum_{k=1}^{N_k} \varphi_k\, h_k(\vec{r}) \tag{6.175}$$

In the Galerkin method, the shape function $h(\vec{r})$ of the finite element is chosen as the test function $q(\vec{r})$. Therefore, Equation 6.174 can be re-written as

$$\Big\langle h_{k'}(\vec{r}), \mathcal{L}\, \hat{\varphi}(\vec{r}, \hat{\Omega}, E) \Big\rangle - \Big\langle h_{k'}(\vec{r}), s(\vec{r}, \hat{\Omega}, E) \Big\rangle = 0 \qquad \text{for } k' = 1, .., N_k \tag{6.176}$$

Substitution of Equation 6.175 into 6.174 gives

$$\sum_{k=1}^{N_k} \Big[\varphi_k \langle h_{k'}(\vec{r}), \mathcal{L}\, h_k \rangle - s_k \langle h_{k'}(\vec{r}), h_k(\vec{r}) \rangle \Big] = 0 \qquad \text{for } k' = 1, .., N_k \tag{6.177}$$

which can be written in matrix form as

$$\sum_k \varphi_k \Big\langle h_{k'}(x), \mathcal{L}[h_k(x)] \Big\rangle = \Big\langle h_{k'}(x), s(x) \Big\rangle \rightarrow [L_{k'k}][\varphi_k] = [s_{k'}]$$

where $[L_{k'k}]$ is the matrix defined by

$$[L_{k'k}] = \begin{bmatrix} \langle h_1, \mathcal{L}[h_1] \rangle & \langle h_1, \mathcal{L}[h_2] \rangle & \cdots & \langle h_1, \mathcal{L}[h_{N_k}] \rangle \\ \langle h_2, \mathcal{L}[h_1] \rangle & \langle h_2, \mathcal{L}[h_2] \rangle & & \\ \vdots & & \ddots & \\ \langle h_{N_{k'}}, \mathcal{L}[h_1] \rangle & & & \langle h_{N_{k'}}, \mathcal{L}[h_{N_k}] \rangle \end{bmatrix}$$

and $[\varphi_k]$ and $[s_{k'}]$ are column vectors

$$[\varphi_k] = \begin{bmatrix} \varphi_1 \\ \varphi_2 \\ \vdots \\ \varphi_{N_k} \end{bmatrix} \quad \text{and} \quad [s_{k'}] = \begin{bmatrix} \langle h_1, s] \rangle \\ \langle h_2, s] \rangle \\ \vdots \\ \langle h_{N_{k'}}, s] \rangle \end{bmatrix}$$

Therefore, the choice of a finite element basis function not only simplified the calculation, but the numerical estimate or result converges to the actual solution as the finite element mesh spacing is refined. Furthermore, Equation 6.177 is a statement of minimizing the residual in the least squares sense.

Elemental Equation/Matrix Formulation

Below, the Galerkin method will be applied to each term of the linear Boltzmann transport equation to give the elemental equations for each finite element for a single type of particle (Martin and Duderstadt 1977; Martin et al. 1981; Wareing et al. 2001; Vassiliev 2016).

Source Term

Similar to the particle fluence density distribution, the particle source density distribution is expanded in terms of basis functions, to give the following:

$$\int_{V_e} d\vec{r} \, h_{k'}(\vec{r}) \, s_g(\vec{r}, \hat{\Omega}_{ij}) = \int_{V_e} d\vec{r} \, h_{k'}(\vec{r}) \sum_{k=1}^{N_k} h_k(\vec{r}) \, s_{gk}(\hat{\Omega}_{ij})$$

$$= \sum_{k=1}^{N_k} s_{gk}(\hat{\Omega}_{ij}) \int_{V_e} d\vec{r} \, h_{k'}(\vec{r}) \, h_k(\vec{r}) \qquad (6.178)$$

whereby $s_{gk}(\hat{\Omega}_{ij})$ denotes the (unknown) expansion coefficients of the particle source.

Removal Term

In the case of the removal term, the material within the finite element is *assumed* to be homogeneous, which removes the position dependence of the total cross, as follows:

$$\int_{V_e} d\vec{r}\, h_{k'}(\vec{r})\, \Sigma_{t,g}(\vec{r})\, \varphi_g(\vec{r}, \hat{\Omega}_{ij}) = \Sigma_{t,g} \int_{V_e} d\vec{r}\, h_{k'}(\vec{r}) \sum_{k=1}^{N_k} h_k(\vec{r})\, \varphi_{gk}(\hat{\Omega}_{ij})$$

$$= \Sigma_{t,g} \sum_{k=1}^{N_k} \varphi_{gk}(\hat{\Omega}_{ij}) \int_{V_e} d\vec{r}\, h_{k'}(\vec{r})\, h_k(\vec{r}) \qquad (6.179)$$

whereby $\varphi_{gk}(\hat{\Omega}_{ij})$ represents the (unknown) expansion coefficients of the particle fluence.

Scattering Term

As in the above cases, the order of integration and summation can be re-arranged, to give

$$\int_{V_e} d\vec{r}\, h_{k'}(\vec{r}) \sum_{l=0}^{N_L} \frac{4\pi}{2l+1} \sum_{g'=1}^{N_G} \Sigma_{s,lg'g} \sum_{m=-l}^{l} Y_l^m(\hat{\Omega}_{ij})\, \varphi_{g'l}^m(\vec{r})$$

$$= \int_{V_e} d\vec{r}\, h_{k'}(\vec{r}) \sum_{l=0}^{N_L} \frac{4\pi}{2l+1} \sum_{g'=1}^{N_G} \Sigma_{s,lg'g} \sum_{m=-l}^{l} Y_l^m(\hat{\Omega}_{ij}) \sum_{k=1}^{N_k} h_k(\vec{r})\, \varphi_{g'lk}^m$$

$$= \sum_{l=0}^{N_L} \frac{4\pi}{2l+1} \sum_{g'=1}^{N_G} \Sigma_{s,lg'g} \sum_{m=-l}^{l} Y_l^m(\hat{\Omega}_{ij}) \sum_{k=1}^{N_k} \varphi_{g'lk}^m \int_{V_e} d\vec{r}\, h_{k'}(\vec{r})\, h_k(\vec{r}) \qquad (6.180)$$

whereby $\varphi_{g'lk}^m$ is the (unknown) expansion coefficients of the differential cross section, and according to Equation 6.158, is given by

$$\varphi_{g'lk}^m = \int d\Omega\, Y_l^{m*}(\hat{\Omega}) \varphi_{g'k}(\hat{\Omega}) \approx \sum_{i'=1}^{N_\mu} \sum_{j'=1}^{N_\phi} w_{i'}\, w_{j'}\, Y_l^{m*}(\hat{\Omega}_{i'j'})\, \varphi_{g'k}(\hat{\Omega}_{i'j'}) \qquad (6.181)$$

where once again $w_{i'}$ and $w_{j'}$ correspond to the quadrature weight functions over polar and azimuthal angles, respectively.

Streaming Term

Due to the presence of the divergence operator, the streaming term is the most difficult part to mathematically manipulate, as shown by

$$\int_{V_e} d\vec{r}\, h_{k'}(\vec{r}) \left[\hat{\Omega}_{ij} \cdot \vec{\nabla} \right] \varphi_g(\vec{r}, \hat{\Omega}_{ij})$$

$$= \int_{V_e} d\vec{r} \left[\hat{\Omega}_{ij} \cdot \vec{\nabla} \right] h_{k'}(\vec{r})\, \varphi_g(\vec{r}, \hat{\Omega}_{ij}) - \int_{V_e} d\vec{r}\, \varphi_g(\vec{r}, \hat{\Omega}_{ij}) \left[\hat{\Omega}_{ij} \cdot \vec{\nabla} \right] h_{k'}(\vec{r})$$

$$(6.182)$$

where integration by parts has been used to separate the expression into two terms.

The first term on the right hand side of Equation 6.182 can be simplified using the divergence theorem to convert it from a volume integral to a surface integral as follows:

$$\int_{V_e} d\vec{r} \left[\hat{\Omega}_{ij} \cdot \vec{\nabla}\right] h_{k'}(\vec{r}) \varphi_g(\vec{r}, \hat{\Omega}_{ij}) = \oint_{\partial V_e} dA \left[\hat{\Omega}_{ij} \cdot \hat{e}_n\right] h_{k'}(\vec{r}) \varphi_g(\vec{r}, \hat{\Omega}_{ij}) \quad (6.183)$$

and

$$\oint_{\partial V_e} dA \left[\hat{\Omega}_{ij} \cdot \hat{e}_n\right] h_{k'}(\vec{r}) \varphi_g(\vec{r}, \hat{\Omega}_{ij}) = \sum_{f=1}^{N_f} \left[\hat{\Omega}_{ij} \cdot \hat{e}_{n,f}\right] \int_{\partial V_e} dA\, h_{k'}(\vec{r}) \varphi_{fg}(\vec{r}, \hat{\Omega}_{ij})$$

$$(6.184)$$

where N_f denotes the total number of faces for the finite element, $\hat{e}_{n,f}$ represents the outward unit normal vector for a given face f, and $\varphi_{fg}(\vec{r}, \hat{\Omega}_{ij})$ is the fluence density distribution on the element boundary.

Next, approximating the particle fluence by a sum gives

$$\sum_{f=1}^{N_f} \left[\hat{\Omega}_{ij} \cdot \hat{e}_{n,f}\right] \int_{\partial V_e} dA\, h_{k'}(\vec{r}) \varphi_{fg}(\vec{r}, \hat{\Omega}_{ij})$$

$$= \sum_{f=1}^{N_f} \left[\hat{\Omega}_{ij} \cdot \hat{e}_{n,f}\right] \int_{\partial V_e} dA\, h_{k'}(\vec{r}) \sum_{k=1}^{N_k} h_k(\vec{r}) \varphi_{fgk}(\hat{\Omega}_{ij})$$

$$= \sum_{k=1}^{N_k} \varphi_{fgk}(\hat{\Omega}_{ij}) \sum_{f=1}^{N_f} \int_{\partial V_e} dA \left[\hat{\Omega}_{ij} \cdot \hat{e}_n\right] h_{k'}(\vec{r}) h_k(\vec{r}) \quad (6.185)$$

where the surface integral is over face f of the product of two basis functions.

The second term on the right hand side of Equation 6.182 is given by

$$\int_{V_e} d\vec{r}\, \varphi_g(\vec{r}, \hat{\Omega}_{ij}) \left[\hat{\Omega}_{ij} \cdot \vec{\nabla}\right] h_{k'}(\vec{r})$$

$$= \int_{V_e} d\vec{r} \sum_{k=1}^{N_k} h_k(\vec{r}) \varphi_{gk}(\hat{\Omega}_{ij}) \left[\hat{\Omega}_{ij} \cdot \vec{\nabla}\right] h_{k'}(\vec{r})$$

$$= \sum_{k=1}^{N_k} \varphi_{gk}(\hat{\Omega}_{ij}) \int_{V_e} d\vec{r}\, h_k(\vec{r}) \left[\hat{\Omega}_{ij} \cdot \vec{\nabla}\right] h_{k'}(\vec{r}) \quad (6.186)$$

Boundary Term

In order to complete the derivation, the element boundary fluences need to be defined. Here, the fluence functions on the upstream side of the boundary are given by

$$\varphi_{fg}(\vec{r}, \hat{\Omega}_{ij}) = \begin{cases} \varphi & \hat{\Omega}_{ij} \cdot \hat{e}_{n,f} > 0 \\ \varphi^{inc} & \hat{\Omega}_{ij} \cdot \hat{e}_{n,f} < 0 \end{cases} \quad (6.187)$$

where φ^{inc} is the incoming fluence density distribution from the upstream element through face f of the element.

Note that integration over finite element faces needs to be done carefully. In the case of the *discontinuous* finite element method, the continuity of the fluence across the surface of the element is *not* required, which means that the fluence may have two different values on each side of the finite element face.

Complete Equation

The complete discretized linear Boltzmann transport equation for a single finite element and type of particle can be expressed in matrix form as

$$\sum_{k=1}^{N_k} \mathcal{T}_{g'kk'}(\hat{\Omega}_{ij}) \, \varphi_{gk}(\hat{\Omega}_{ij}) = s_{gk'}(\hat{\Omega}_{ij})$$

$$+ \sum_{k=1}^{N_k} \sum_{g'=1}^{N_G} \sum_{i'=1}^{N_\mu} \sum_{j'=1}^{N_\phi} \mathcal{K}_{kk'i'j'g'g}(\hat{\Omega}_{i'j'} \cdot \hat{\Omega}_{ij}) \, \varphi_{g'k}(\hat{\Omega}_{i'j'}) \quad (6.188)$$

where the transport matrix is given by:

$$\mathcal{T}_{g'kk'}(\hat{\Omega}_{ij}) = \sum_{f=1}^{N_f} \int_{A_e} dA \left[\hat{\Omega}_{ij} \cdot \hat{e}_{n,f} \right] h_{k'}(\vec{r}) \, h_k(\vec{r})$$

$$- \int_{V_e} d\vec{r} \, h_k(\vec{r}) \left[\hat{\Omega}_{ij} \cdot \vec{\nabla} \right] h_{k'}(\vec{r}) + \Sigma_{t,g} \int_{V_e} d\vec{r} \, h_{k'}(\vec{r}) \, h_k(\vec{r}) \quad (6.189)$$

and the scattering matrix is given by

$$\mathcal{K}_{kk'i'j'g'g}(\hat{\Omega}_{i'j'} \cdot \hat{\Omega}_{ij}) = w_{i'} \, w_{j'} \sum_{l=0}^{L} \frac{4\pi}{2l+1} \Sigma_{s,lg'g}$$

$$\times \sum_{m=-l}^{l} Y_{lm}(\hat{\Omega}_{ij}) \, Y_l^{m*}(\hat{\Omega}_{i'j'}) \int_{V_e} d\vec{r} \, h_{k'}(\vec{r}) \, h_k(\vec{r}) \quad (6.190)$$

for $k' = 1, ..., N_k$, $g = 1, ..., N_g$, $i = 1, ..., N_\mu$, and $j = 1, ..., N_j$.

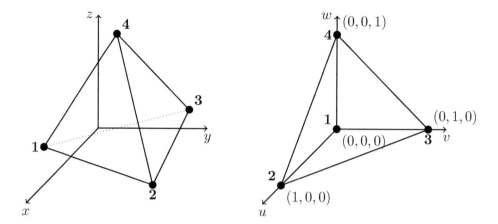

Figure 6.15: Finite element coordinate transformation from global (left) to normalized local (right) coordinates in the case of a tetrahedral finite element.

Equation 6.190 contains three types of integrals in the global spatial coordinate system that need to be evaluated. However, they can be considerably simplified if they are transformed into a local coordinate system for a specific type of finite element.

Global-to-Local Coordinate Transformation

In general, finite elements will be non-orthogonal, and it becomes necessary to transform from global coordinates, which have difficult limits of integration $\vec{r} = (x, y, z)$, to local coordinates $\vec{r}_e = (u, v, w)$, with simple limits of integration.

As seen in Figure 6.15, the normalized local coordinate system is defined such that the physical cell takes on a simple shape and size. For example, a linear tetrahedral element can be defined by four co-ordinates as shown in Table 6.2.

Table 6.2: Normalized local coordinates for each node of a linear tetrahedral finite element.

Node k	u_k	v_k	w_k
1	0	0	0
2	1	0	0
3	0	1	0
4	0	0	1

Therefore, the shape functions for a linear tetrahedral finite element in the local coordinate system take the simple form

$$h_1 = 1 - u - v - w$$
$$h_2 = u$$
$$h_3 = v$$
$$h_4 = w \tag{6.191}$$

Since the shape functions are defined with respect to the local coordinate system, then they can be used to express a global spatial position with respect to a local one as

$$x(\vec{r}_e) = x(u, v, w) = \sum_{k=1}^{N_k} h_k(u, v, w)\, x_k$$

$$y(\vec{r}_e) = y(u, v, w) = \sum_{k=1}^{N_k} h_k(u, v, w)\, y_k \tag{6.192}$$

$$z(\vec{r}_e) = z(u, v, w) = \sum_{k=1}^{N_k} h_k(u, v, w)\, z_k$$

Using the chain rule for partial differentiation, the derivatives of the shape functions with respect to the global coordinate system can be expressed as

$$\frac{\partial h_k}{\partial x} = \frac{\partial h_k}{\partial u}\frac{\partial u}{\partial x} + \frac{\partial h_k}{\partial v}\frac{\partial v}{\partial x} + \frac{\partial h_k}{\partial w}\frac{\partial w}{\partial x}$$

$$\frac{\partial h_k}{\partial y} = \frac{\partial h_k}{\partial u}\frac{\partial u}{\partial y} + \frac{\partial h_k}{\partial v}\frac{\partial v}{\partial y} + \frac{\partial h_k}{\partial w}\frac{\partial w}{\partial y} \tag{6.193}$$

$$\frac{\partial h_k}{\partial z} = \frac{\partial h_k}{\partial u}\frac{\partial u}{\partial z} + \frac{\partial h_k}{\partial v}\frac{\partial v}{\partial z} + \frac{\partial h_k}{\partial w}\frac{\partial w}{\partial z}$$

Equation 6.193 can be written in matrix form as follows:

$$\begin{bmatrix} \dfrac{\partial h_k}{\partial x} \\[2mm] \dfrac{\partial h_k}{\partial y} \\[2mm] \dfrac{\partial h_k}{\partial z} \end{bmatrix} = \begin{bmatrix} \dfrac{\partial u}{\partial x} & \dfrac{\partial v}{\partial x} & \dfrac{\partial w}{\partial x} \\[2mm] \dfrac{\partial u}{\partial y} & \dfrac{\partial v}{\partial y} & \dfrac{\partial w}{\partial y} \\[2mm] \dfrac{\partial u}{\partial z} & \dfrac{\partial v}{\partial z} & \dfrac{\partial w}{\partial z} \end{bmatrix} \begin{bmatrix} \dfrac{\partial h_k}{\partial u} \\[2mm] \dfrac{\partial h_k}{\partial v} \\[2mm] \dfrac{\partial h_k}{\partial w} \end{bmatrix} = \mathcal{J}^{-1} \begin{bmatrix} \dfrac{\partial h_k}{\partial u} \\[2mm] \dfrac{\partial h_k}{\partial v} \\[2mm] \dfrac{\partial h_k}{\partial w} \end{bmatrix}$$

where the matrix on the right hand side is known as the inverse of the Jacobian matrix $\mathcal{J}^{-1}$ used to relate the derivatives between the global and local coordinate systems as follows:

$$\frac{\partial}{\partial \vec{r}} = \mathcal{J}^{-1}(\vec{r}_e)\, \frac{\partial}{\partial \vec{r}_e} \tag{6.194}$$

The Jacobian matrix can be evaluated by calculating the derivatives of Equation 6.192 with respect to the local coordinate system

$$
\mathcal{J}(\vec{r}_e) =
\begin{bmatrix}
\dfrac{\partial}{\partial u} x(\vec{r}_e) & \dfrac{\partial}{\partial u} y(\vec{r}_e) & \dfrac{\partial}{\partial u} z(\vec{r}_e) \\[2mm]
\dfrac{\partial}{\partial v} x(\vec{r}_e) & \dfrac{\partial}{\partial v} y(\vec{r}_e) & \dfrac{\partial}{\partial v} z(\vec{r}_e) \\[2mm]
\dfrac{\partial}{\partial w} x(\vec{r}_e) & \dfrac{\partial}{\partial w} y(\vec{r}_e) & \dfrac{\partial}{\partial w} z(\vec{r}_e)
\end{bmatrix}
$$

In order to evaluate volume integrals in the local coordinate system, the following transformation for the differential volume element must be used

$$
dV = dx\, dy\, dz = \det \mathcal{J}(\vec{r}_e)\, du\, dv\, dw \tag{6.195}
$$

Based on the definition of the Jacobian matrix and differential volume elements in the normalized local coordinate system described above, the following element integral can be re-written as

$$
\int_{V_e} dV\, h_{k'}(x, y, z)\, h_k(x, y, z) = \det(\mathcal{J}) \int_{V_e} dV\, h_{k'}(u, v, w)\, h_k(u, v, w) \tag{6.196}
$$

In the case of a linear tetrahedral element, Equation 6.195 is given by

$$
\det(\mathcal{J}) \int_0^1 \int_0^{1-u} \int_0^{1-u-v} h_{k'}(u, v, w)\, h_k(u, v, w) = \det(\mathcal{J}) \times
\begin{cases}
\frac{1}{60} & \text{for} \quad k = k' \\[2mm]
\frac{1}{120} & \text{for} \quad k \neq k'
\end{cases}
\tag{6.197}
$$

Thus, the advantage of a linear tetrahedron can be seen from the simple result of the integration in Equation 6.197. Using a normalized local coordinate system, these integrals apply to every finite element in the problem domain, and all that is required is the calculation of the determinant of the Jacobian matrix for each element, which can be implemented as a pre-processing step.

As for surface integrals, the differential area element is given by

$$
d\partial V = |\det \mathcal{J}_s(\vec{r}_s)|\, dx_s\, dy_s \tag{6.198}
$$

where u_s and v_s represent the local coordinates for the element face and $\mathcal{J}_s$ is the magnitude of the surface Jacobian as given by

$$
\mathcal{J}_s(\vec{r}_s) =
\begin{bmatrix}
\dfrac{\partial}{\partial u_s} x(\vec{r}_s) \\[2mm]
\dfrac{\partial}{\partial u_s} y(\vec{r}_s) \\[2mm]
\dfrac{\partial}{\partial u_s} z(\vec{r}_s)
\end{bmatrix}
\times
\begin{bmatrix}
\dfrac{\partial}{\partial v_s} x(\vec{r}_s) \\[2mm]
\dfrac{\partial}{\partial v_s} y(\vec{r}_s) \\[2mm]
\dfrac{\partial}{\partial v_s} z(\vec{r}_s)
\end{bmatrix}
$$

Global Matrix Assembly

Before solving the discretized linear Boltzmann transport equation for the particle fluence density distribution, all of the element equations must be assembled into one global equation. The following two pseudo-code algorithms outline how to carry out the operation for the global transport matrix $L^{(g)}$ and global source vector $s^{(g)}$.

In each case, a connectivity matrix $C[N_e, N_k^{(e)}]$ needs to be defined that simply provides the address in the global matrix or vector in which the ith and/or jth entry of the local transport matrix or source vector for element e maps to.

Algorithm 1 Assembly of Global Transport Matrix

1: *Initialize $N_k^{(e)}$ = total number of local nodes for each finite element*

2: *Initialize $N_k^{(g)}$ = total number of global nodes for all finite elements in domain*

3: *Initialize N_e = total number of finite elements in domain*

4: *Initialize $C[N_e, N_k^{(e)}]$ = finite element connectivity array*

5: *Initialize $L^{(e)}[N_k^{(e)}, N_k^{(e)}]$ = element transport matrix*

6: *Initialize $L^{(g)}[N_k^{(g)}, N_k^{(g)}]$ = global transport matrix*

7:

8: **for** $i = 1, N_k^{(g)}$ **do**

9:      **for** $j = 1, N_k^{(g)}$ **do**

10:          $L^{(g)}[i, j] = 0$

11:      **end**

12: **end**

13:

14: **for** $e = 1, N_e$ **do**

15:      *generate $L^{(e)}$*

16:      **for** $i = 1, N_k^{(e)}$ **do**

17:          **for** $j = 1, N_k^{(e)}$ **do**

18:              $L^{(g)}[C[e, i], C[e, j]] = L^{(g)}[C[e, i], C[e, j]] + L^{(e)}[i, j]$

19:          **end**

20:      **end**

21: **end**

22:

23: **return** $L^{(g)}$

Once the global matrix and vector have been defined, they then may be solved by one of many linear algebraic techniques.

Algorithm 2 Assembly of Global Source Matrix

1: *Initialize* $N_k^{(e)}$ = total number of local nodes for each finite element
2: *Initialize* $N_k^{(g)}$ = total number of global nodes for all finite elements in domain
3: *Initialize* N_e = total number of finite elements in domain
4: *Initialize* $C[N_e, N_k^{(e)}]$ = finite element connectivity array
5: *Initialize* $s^{(e)}[N_k^{(e)}]$ = element source vector
6: *Initialize* $s^{(g)}[N_k^{(g)}]$ = global source vector
7:
8: **for** $i = 1, N_k^{(g)}$ **do**
9:     $S^{(g)}[i] = 0$
10: **end**
11:
12: **for** $e = 1, N_e$ **do**
13:     *generate* $s^{(e)}$
14:     **for** $i = 1, N_k^{(e)}$ **do**
15:         $s^{(g)}[C[e, i]] = s^{(g)}[C[e, i]] + S^{(e)}[i]$
16:     **end**
17: **end**
18:
19: **return** $S^{(g)}$

6.4.3 Solution Techniques

In the following section, several iterative strategies are discussed that are typically used in solving the discretized linear Boltzmann transport equation. Below, each strategy is explained conceptually and independently for better clarity. More in-depth versions of these iterative techniques can be found in (Lewis and Miller 1984; Wareing et al. 2001).

6.4.3.1 Energy Iteration

The multi-group version of the linear Boltzmann transport equation is used to demonstrate the energy iteration method

$$\left[\hat{\Omega} \cdot \vec{\nabla} + \Sigma_{t,g}(\vec{r})\right] \varphi_g(\vec{r}, \hat{\Omega}) = \sum_{g'=1}^{N_G} \int d\Omega\, \Sigma_{s,g'g}(\vec{r}, \hat{\Omega}' \cdot \hat{\Omega})\, \varphi_{g'}(\vec{r}, \hat{\Omega}') + s(\vec{r}, \hat{\Omega}) \quad (6.199)$$

where g represents an index for a particular energy group.

The above equation can be simplified using operator notation. The streaming operator for group g is defined by

$$\mathcal{T}_{gg}\, \varphi_g = \left[\hat{\Omega} \cdot \vec{\nabla} + \Sigma_{t,g}(\vec{r})\right] \quad (6.200)$$

and the group-to-group scattering operator as

$$\mathcal{K}_{gg'} \varphi_{g'} = \int d\Omega \, \Sigma_{s,g'g}(\vec{r}, \hat{\Omega}' \cdot \hat{\Omega}) \, \varphi_{g'}(\vec{r}, \hat{\Omega}') \tag{6.201}$$

Therefore, using these two simplifying notations, then the multi-group transport operator is given by

$$\mathcal{L}_{gg'} = \delta_{gg'} \, \mathcal{T}_{gg} - \mathcal{K}_{gg'} \tag{6.202}$$

where $\delta_{gg'}$ is the Kronecker delta function.

Equation 6.202 can now be organized as a coupled set of operator equations as follows:

$$
\begin{bmatrix}
\mathcal{L}_{11} & \mathcal{L}_{12} & \cdots & \mathcal{L}_{1g} & \cdots & \mathcal{L}_{1G} \\
\mathcal{L}_{21} & \mathcal{L}_{22} & & & & \\
\vdots & & \ddots & & & \\
\mathcal{L}_{g1} & & & \mathcal{L}_{gg} & & \\
& & & & \ddots & \\
\mathcal{L}_{G1} & & & & & \mathcal{L}_{GG}
\end{bmatrix}
\begin{bmatrix}
\varphi_1 \\ \varphi_2 \\ \vdots \\ \varphi_g \\ \vdots \\ \varphi_G
\end{bmatrix}
=
\begin{bmatrix}
s_1 \\ s_2 \\ \vdots \\ s_g \\ \vdots \\ s_G
\end{bmatrix}
$$

or, in matrix notation (using boldface variables)

$$\mathcal{L} \, \varphi = \mathbf{s} \tag{6.203}$$

In problems involving photon or electron transport, there is generally no scattering of particles from lower to higher energy groups. Since g increases as energy decreases, then $\mathcal{L}_{gg'} = 0$ for $g' > g$, which turns Equation 6.203 into a *lower diagonal matrix* as given by

$$
\begin{bmatrix}
\mathcal{L}_{11} & 0 & \cdots & 0 & \cdots & 0 \\
\mathcal{L}_{21} & \mathcal{L}_{22} & & & & \\
\vdots & & \ddots & 0 & & \vdots \\
\mathcal{L}_{g1} & & & \mathcal{L}_{gg} & & \\
& & & & \ddots & 0 \\
\mathcal{L}_{G1} & & & & & \mathcal{L}_{GG}
\end{bmatrix}
\begin{bmatrix}
\varphi_1 \\ \varphi_2 \\ \vdots \\ \varphi_g \\ \vdots \\ \varphi_G
\end{bmatrix}
=
\begin{bmatrix}
s_1 \\ s_2 \\ \vdots \\ s_g \\ \vdots \\ s_G
\end{bmatrix}
$$

which allows the matrix to be easily inverted and solved algebraically as

$$\varphi = \mathcal{L}^{-1} \mathbf{s} \tag{6.204}$$

The above matrix equation can be solved successively for energy groups of increasing index as

$$\varphi_1 = \mathcal{L}_{11}^{-1} s_1$$
$$\varphi_2 = \mathcal{L}_{22}^{-1} \left(s_2 - \mathcal{L}_{21}^{-1} \varphi_1 \right)$$
$$\vdots$$
$$\varphi_g = \mathcal{L}_{gg}^{-1} \left(s_g - \sum_{g'<g} \mathcal{L}_{gg'}^{-1} \varphi_{g'} \right) \qquad (6.205)$$

6.4.3.2 Source Iteration

The solution to fixed-source particle transport problems, such as those encountered in external beam radiotherapy, can be written in the following way. Consider splitting the particle fluence density distribution, $\varphi(\vec{r}, \hat{\Omega}, E)$, into *two* components

$$\varphi(\vec{r}, \hat{\Omega}, E) = \varphi^{(0)}(\vec{r}, \hat{\Omega}, E) + \varphi^{(n)}(\vec{r}, \hat{\Omega}, E) \qquad (6.206)$$

where $\varphi^{(0)}(\vec{r}, \hat{\Omega}, E)$ denotes the uncollided or unscattered component, and $\varphi_n(\vec{r}, \hat{\Omega}, E)$ represents the collided or scattered component for $1 < n < \infty$.

Substituting Equation 6.206 into 6.53 allows the formulation of two linear Boltzmann transport equation(s), one for each component of $\varphi(\vec{r}, \hat{\Omega}, E)$, as given by

$$\left[\hat{\Omega} \cdot \vec{\nabla} + \Sigma_t(\vec{r}, E) \right] \varphi^{(0)}(\vec{r}, \hat{\Omega}, E) = s(\vec{r}, \hat{\Omega}, E)$$

$$\left[\hat{\Omega} \cdot \vec{\nabla} + \Sigma_t(\vec{r}, E) \right] \varphi^{(n)}(\vec{r}, \hat{\Omega}, E)$$
$$= \int dE' \int d\Omega' \, \Sigma_s(\vec{r}, \hat{\Omega}', \hat{\Omega}, E', E) \, \varphi^{(n-1)}(\vec{r}, \hat{\Omega}', E') \qquad (6.207)$$

In operator notation, these two equations can be compactly represented as

$$\mathcal{T} \varphi^{(0)} = s$$
$$\mathcal{T} \varphi^{(n)} = \mathcal{K} \varphi^{(n-1)} \qquad (6.208)$$

Thus, for a given problem, if the total number of collided components is assumed to be N, then there would be N (integro-differential) linear Boltzmann transport equations to solve, and the solution for $\varphi(\vec{r}, \hat{\Omega}, E)$ would be approximated by:

$$\varphi(\vec{r}, \hat{\Omega}, E) \approx \sum_{n=0}^{N} \varphi^{(n)}(\vec{r}, \hat{\Omega}, E) \qquad (6.209)$$

Equation 6.209 is known as the *Neumann* series

As mentioned earlier, an analytic solution for the uncollided/unscattered component can be determined through a simple ray-tracing exercise of Equation 6.101 in Section 6.3.9.

6.4.3.3 Synthetic Acceleration

Using synthetic acceleration, a low-order approximation, such as diffusion theory, is used as a mechanism for accelerating the numerical solution of the linear Boltzmann transport equation.

Based on Equation 6.199, the within-group linear Boltzmann transport equation can be written as

$$\mathcal{L}\,\varphi = s \tag{6.210}$$

where

$$\mathcal{L}\,\varphi = \left[\hat{\Omega}\cdot\vec{\nabla} + \Sigma_t - \Sigma_s\int d\Omega\right]\varphi \tag{6.211}$$

and iteration on the scattering source becomes

$$\mathcal{H}^0\,\hat{\varphi}^\ell = \mathcal{H}^1\,\phi^\ell + s \tag{6.212}$$

where ℓ is the iteration index, and for un-accelerated iteration, $\varphi^{\ell+1} = \hat{\varphi}^\ell$, and ϕ^ℓ is the *scalar* fluence.

The streaming-collision operator is defined by

$$\mathcal{H}^0\,\varphi = \left[\hat{\Omega}\cdot\vec{\nabla} + \Sigma_t\right]\varphi \tag{6.213}$$

and the within-group *isotropic* scattering operator by

$$\mathcal{H}^1\,\varphi = \Sigma_s\int d\Omega\,\varphi = \mathcal{H}^1\,\phi \tag{6.214}$$

where the normalization $\int d\Omega = 1$ is used.

To derive the synthetic method, Equation 6.210 must be first integrated over $\hat{\Omega}$, such that

$$\int d\Omega\,\mathcal{H}\,\varphi = S \tag{6.215}$$

Now, the operator on the left hand side can be written as a sum of the low-order diffusion operator

$$\mathcal{H}_L = -\vec{\nabla}D\vec{\nabla} + \Sigma_r \tag{6.216}$$

and a correction $(\mathcal{H} - \mathcal{H}_L)$ given by

$$\int \mathrm{d}\Omega \left[\mathcal{H}_L + (\mathcal{H} - \mathcal{H}_L)\right] \varphi = s \tag{6.217}$$

Since $\mathcal{H}_L$ is independent of $\hat{\Omega}$, then

$$\int \mathrm{d}\Omega \, \mathcal{H}_L \varphi = \mathcal{H}_L \phi \tag{6.218}$$

Then re-arranging terms yields

$$\mathcal{H}_L \phi = s - \int \mathrm{d}\Omega \left(\mathcal{H} - \mathcal{H}_L\right) \varphi \tag{6.219}$$

The above expression suggests an acceleration scheme as follows:

$$\mathcal{H}_L \, \phi^{\ell+1} = s - \int \mathrm{d}\Omega \left(\mathcal{H} - \mathcal{H}_L\right) \hat{\varphi}^{\ell} \tag{6.220}$$

where $\hat{\varphi}^{\ell}$ is determined from Equation 6.212. Noting that $\mathcal{H} = \mathcal{H}^0 - \mathcal{H}^1$ and combining Equations 6.212 and 6.220, the expression may be expressed in a more convenient form as:

$$\mathcal{H}_L \left(\phi^{\ell+1} - \hat{\phi}^{\ell}\right) = \mathcal{H}^1 \left(\hat{\phi}^{\ell} - \phi^{\ell}\right) \tag{6.221}$$

Therefore, knowing ϕ^{ℓ} and $\hat{\phi}^{\ell}$, then the diffusion equation need only be solved, by obtaining $\mathcal{H}_L^{-1}$ to obtain $\varphi^{\ell+1}$.

In the case of isotropic scattering, $\mathcal{H}^1 \phi = \Sigma_s \phi$. Therefore,

$$\varphi^{\ell+1} = \hat{\phi}^{\ell} + \mathcal{H}_L^{-1} \Sigma_s \left(\hat{\phi}^{\ell} - \phi^{\ell}\right) \tag{6.222}$$

Clearly, as $\hat{\phi}^{\ell} \to \phi^{\ell}$, the system converges.

More advanced versions of diffusion synthetic acceleration, typically used in the nuclear engineering and medical physics fields, can be found in the following reference (Wareing et al. 2001).

6.4.4 Practical Approximations

A variety of approximations can be used to considerably reduce the complexity of the coupled photon and electron linear Boltzmann transport equations, in order to help practically solve them within a reasonable amount of computing time and memory resources. A summary of the common physics and phase space discretization approximations is presented below.

Physics Approximations

The basic *physics* approximations used in commercial, grid-based linear Boltzmann transport equation solvers can be summarized as follows:

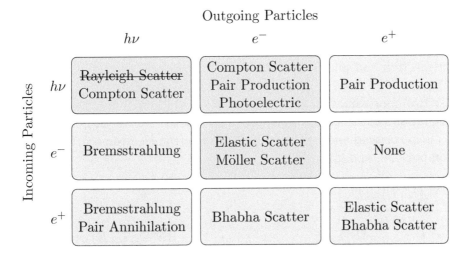

Figure 6.16: The interaction matrix has been updated from Figure 6.9 to indicate the assumptions typically made in commercial external beam radiotherapy dose calculation algorithms. The *greyed-out* matrix elements represent particles and interactions <u>not</u> explicitly included in the algorithm.

1. pair production produces two electrons instead of one electron and one positron

2. photons can produce electrons, but electrons do not produce photons (e.g. bremsstrahlung)

3. bremsstrahlung photon energy is *locally* deposited

4. electron interactions are divided into *soft* and *hard* collisions

5. soft electron collisions are modelled using the *continuous slowing down* approximation (CSDA)

The first three approximations above can be used to update the interaction matrix depicted in Figure 6.16, in order to simplify the coupled photon/electron/positron Boltzmann transport equations as follows:

$$\hat{\Omega}_\gamma \cdot \vec{\nabla} \varphi^\gamma(\vec{r}, \hat{\Omega}_\gamma, E_\gamma) + \Sigma_t^\gamma(\vec{r}, E_\gamma)\, \varphi^\gamma(\vec{r}, \hat{\Omega}_\gamma, E_\gamma)$$
$$= s^{\gamma\gamma}(\vec{r}, \hat{\Omega}_\gamma, E_\gamma) + s^\gamma(\vec{r}, \hat{\Omega}_\gamma, E_\gamma)$$

$$\hat{\Omega}_e \cdot \vec{\nabla} \varphi^{e^-}(\vec{r}, \hat{\Omega}_e, E_e) + \Sigma_t^{e^-}(\vec{r}, E_e)\, \varphi^{e^-}(\vec{r}, \hat{\Omega}_e, E_e)$$
$$= s^{e^- e^-}(\vec{r}, \hat{\Omega}_e, E_e) + s^{\gamma e^-}(\vec{r}, \hat{\Omega}_e, E_e) + s^{e^-}(\vec{r}, \hat{\Omega}_e, E_e)$$

Notice that the total number of Boltzmann transport equations has been reduced from three to <u>two</u> equations. Furthermore, in these two remaining equations, the total number of scattering sources has also been reduced, which in turn limits the extent of coupling between the equations. For example, in the photon equation, only the photon-photon scattering source remains, while in the electron equation, the electron-electron and photon-electron scattering sources are included.

The last two approximations above affect the form of the electron Boltzmann transport equation. The equation is modified as follows. An electron threshold energy, ΔE_{th}, is introduced and used to divide the electron-electron scattering source into two components, and given by

$$\int_E^\infty dE' \, \Sigma_s(E', E) \, \varphi(E')$$

$$= \int_E^{E+\Delta E_{th}} dE' \, \Sigma_s(E', E) \, \varphi(E') + \int_{E+\Delta E_{th}}^\infty dE' \, \Sigma_s(E', E) \, \varphi(E') \quad (6.223)$$

where the first integral on the right hand side accounts for soft collisions, while the second integral corresponds to hard collisions.

The soft collision integral can be simplified using a continuous slowing down approximation (CSDA) approach as follows:

$$\int_E^{E+\Delta E_{th}} dE' \, \Sigma_s(E', E) \, \varphi(E') = \Sigma_\Delta(E) \, \varphi(E) + \frac{\partial}{\partial E} \left[L_\Delta(E) \, \varphi(E) \right] \quad (6.224)$$

where $\Sigma_\Delta(E)$ is the total cross section for soft collisions, as given by

$$\Sigma_\Delta(E) = \int_0^{\Delta E_{th}} d(\Delta E) \, \Sigma_s(E, \Delta E) \quad (6.225)$$

and $L_\Delta(E)$ is the *restricted* stopping power, as given by

$$L_\Delta(E) = \int_0^{\Delta E_{th}} d(\Delta E) \, \Sigma_s(E, \Delta E) \, (\Delta E) \quad (6.226)$$

which represents the average energy lost in soft collisions per unit path length.

The form of the soft collision integral given in Equations 6.224 to 6.226 is related to a technique also used in condensed history algorithms for charged particle transport in Monte Carlo methods (see Chapter 5). In the technique, each hard collision is *explicitly* simulated, and between hard collisions, a charged particle *continuously* loses energy at a rate given by the restricted stopping power.

Therefore, based on the above approximation, the electron linear Boltzmann transport equation can now be written as

$$\left[\hat{\Omega}_e \cdot \vec{\nabla} + \Sigma_t^{e^-}(\vec{r}, E_e) \right] \varphi^{e^-}(\vec{r}, \hat{\Omega}_e, E_e) - \frac{\partial}{\partial E_e} L(E_e) \, \varphi^{e^-}(\vec{r}, \hat{\Omega}_e, E_e)$$

$$= s^{e^- e^-}(\vec{r}, \hat{\Omega}_e, E_e) + s^{\gamma e^-}(\vec{r}, \hat{\Omega}_e, E_e) + s^{e^-}(\vec{r}, \hat{\Omega}_e, E_e)$$

Series Expansion Approximations

As shown earlier, the particle fluence density distribution and differential cross section were expanded in series using spherical harmonics and Legendre polynomials. These expansions are exact if summed to infinity. However, in practice, the scattering order, L, is constrained with a typical value of seven, $0 < l < 7$. Therefore, the numbers of spherical harmonics or Legendre moments kept in the particle fluence density distribution and differential cross section are given by

$$\varphi(\vec{r}, \hat{\Omega}, E) \approx \sum_{l=0}^{7} \sum_{m=-l}^{l} \varphi_l^m(\vec{r}, E) \, Y_l^m(\hat{\Omega}) \tag{6.227}$$

and

$$\Sigma_s(\vec{r}, \hat{\Omega}' \cdot \hat{\Omega}, E', E) \approx \sum_{l=0}^{7} \Sigma_{s,l}(\vec{r}, E', E) \, P_l(\cos \theta_s) \tag{6.228}$$

Discretization Approximations

The overall accuracy of the dose algorithm using practical deterministic solutions of the linear Boltzmann transport equation is ultimately dependent upon the approximations made for the phase space discretization. These approximations as applied to the energy, angular, and spatial variables can be summarized as follows:

Energy Discretization

1. multi-group method used for particle fluence density distributions and total and/or differential cross sections

2. finite element method (linear, discontinuous version) used for energy derivative in electron transport equation

3. cross section library uses 25 photon energy groups and 46 electron energy groups

4. transport cut-off energy used for photons (1 [keV]) and electrons (200 [keV])

Angular Discretization

1. discrete ordinates method used for particle fluence density distributions and differential scattering cross sections

2. Gauss-Chebyshev quadrature used in integration over polar angle

3. Gauss-Legendre quadrature used in integration over azimuthal angle

4. quadrature order ranges between N=4 (32 discrete angles) and N=16 (512 discrete angles) depending on particle energy

Spatial Discretization

1. finite element method (linear, discontinuous) used to divide problem domain into variable-sized elements

2. tetrahedral shape functions used as the finite elements

3. adaptive mesh refinement (AMR) used to select high-gradient and material boundary areas with finer elements and low-gradient areas with coarser elements

4. each finite element consists of a single material density

6.4.5 Solution for External Beam Radiotherapy

In the following, an outline of a solution strategy is provided to solve the coupled photon-electron Boltzmann transport equation, based on all the material presented so far, for a standard, external beam radiotherapy dose calculation.

The final set of linear Boltzmann transport equations,[9] including the breakdown of the photon fluence distribution into *uncollided* and *collided* components, used in the solution can be summarized as

$$\left[\hat{\Omega}_\gamma \cdot \vec{\nabla} + \Sigma_t^\gamma(\vec{r}, E_\gamma) \right] \varphi_{unc}^\gamma(\vec{r}, \hat{\Omega}_\gamma, E_\gamma) = s^\gamma(\vec{r}, \hat{\Omega}_\gamma, E_\gamma) \tag{6.229}$$

$$\left[\hat{\Omega}_\gamma \cdot \vec{\nabla} + \Sigma_t^\gamma(\vec{r}, E_\gamma) \right] \varphi_{col}^\gamma(\vec{r}, \hat{\Omega}_\gamma, E_\gamma) = s_{col}^{\gamma\gamma}(\vec{r}, \hat{\Omega}_\gamma, E_\gamma) + s_{unc}^{\gamma\gamma}(\vec{r}, \hat{\Omega}_\gamma, E_\gamma) \tag{6.230}$$

$$\left[\hat{\Omega}_e \cdot \vec{\nabla} + \Sigma_t^{e^-}(\vec{r}, E_e) \right] \varphi^{e^-}(\vec{r}, \hat{\Omega}_e, E_e) - \frac{\partial}{\partial E_e} L(E_e) \varphi^{e^-}(\vec{r}, \hat{\Omega}_e, E_e)$$
$$= s^{e^- e^-}(\vec{r}, \hat{\Omega}_e, E_e) + s_{col}^{\gamma e^-}(\vec{r}, \hat{\Omega}_e, E_e) + s_{unc}^{\gamma e^-}(\vec{r}, \hat{\Omega}_e, E_e) + s^{e^-}(\vec{r}, \hat{\Omega}_e, E_e) \tag{6.231}$$

The solution to the uncollided component in Equation 6.229 can easily be solved (see Section 6.3.9) using simple analytic ray-tracing methods to give

$$\varphi_{unc}^\gamma(\vec{r}, \hat{\Omega}_\gamma, E_\gamma) = s^\gamma(\hat{\Omega}_\gamma, E_\gamma) \, \delta^{(2)} \left(\hat{\Omega} - \frac{\vec{r} - \vec{r}_0}{|\vec{r} - \vec{r}_0|} \right) \frac{\exp[-\tau(\vec{r}, \vec{r}_0)]}{|\vec{r} - \vec{r}_0|^2} \tag{6.232}$$

---

[9]The continuous formulation (instead of the discrete version) is shown here for clarity purposes.

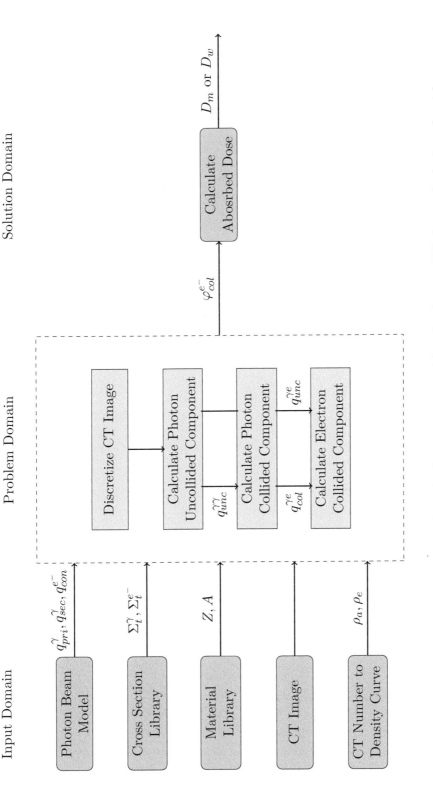

Figure 6.17: Flowchart showing the calculation steps used by the Acuros XB dose calculation algorithm.

Ingredients

The main ingredients required as input to the deterministic Boltzmann transport dose algorithm include:

1. photon beam source model, which includes the primary photon, secondary photon, and electron contaminant components

2. cross section library for photon and electron interactions in various materials

3. CT image of patient or phantom in DICOM format

4. CT number to physical density library for various materials

Calculation Steps

The main calculation steps required to solve the deterministic linear Boltzmann transport equations (see Figure 6.17) can be summarized as follows:

1. discretization of CT image and initialization of beam source model and dose computation grid

2. formulation of discretized version of coupled photon-electron Boltzmann transport equations

3. calculate uncollided component of photon fluence distribution

4. calculate collided component of photon fluence distribution using the uncollided component as a source

5. calculate electron fluence distribution using uncollided and collided components of photon fluence distribution as sources

6. calculate dose (to medium or water) distribution using electron fluence distribution

Limitations

Based on the approximations discussed above, the simplified formulation of the coupled photon-electron linear Boltzmann transport equation is valid for incident photon beam spectra, ranging from 4 MV to 20 MV. This energy range is sufficient for the purposes of clinical, external beam radiotherapy dose calculations in patients and phantoms. In terms of atomic number, dose calculations of materials that range from air to soft tissue to bone, are suitably accurate. Higher atomic number materials, such as those found in metallic implants, limit the accuracy of the dose calculation, due to the enhanced importance of pair production and bremsstrahlung events.

The limitations of the simplified deterministic Boltzmann transport dose algorithm also depend significantly on the level of phase space discretization. Spatial discretization errors typically manifest as over-or under-shooting the local dose solution in a finite element, due to the requirement that a *linear* solution in each element be enforced while maintaining particle balance within the element. Common to discrete ordinates methods that discretize the angular variable are so called ray effects, which are non-physical angular oscillations in the solution. Lastly, energy discretization errors generally appear as solution biases over a large region of the computational grid.

Computational Efficiency

The discrete version of the linear Boltzmann transport equation *explicitly* shows the number of phase space variables needed to solve the equation algebraically. The total number of elements in the transport and scattering matrices can be determined by multiplying together the number of energy groups, angular quadrature angles, and tetrahedral finite elements in the dose computation grid. Based on the numbers provided earlier, these matrices will contain at least $4 \times 100 \times 100 \times 100 \times 32 \times 25 = 3,200,000,000$ elements in the best and worst case scenarios, respectively! If each element is represented as a double-precision floating point number, the memory requirements for each 3D matrix operator would be $204,800,000,000$ bits or $25,600,000,000$ bytes of memory! However, by employing the algebraic solution techniques (energy iteration, source iteration, synthetic acceleration) described earlier, these matrix operations can be simplified in terms of memory consumption and calculated much more quickly.

To give a clinical perspective, dose calculations using deterministic methods of solution for the linear Boltzmann transport equation(s) of complex intensity modulated radiation therapy plans of head and neck, lung, or prostate cancers, typically take on the order of a few minutes. These times are comparable to, if not better than traditional convolution-based dose algorithms, and also provide better accuracy.

6.4.6 Comparison with Convolution and Monte Carlo Results

Several publications have evaluated the accuracy of the Acuros XB dose algorithm against Monte Carlo (MC) simulation for external beam radiotherapy. Vassiliev *et al.* (Vassiliev et al. 2010) investigated the AXB algorithm against a Monte Carlo algorithm (based on EGSnrc) for both 6 and 18 MV photon beams. A slab phantom consisting of layers of soft tissue, bone, and lung materials; and an anthropomorphic breast phantom were investigated. Good agreement between AXB and MC was found: 2% or 1 mm and 2% or 2 mm distance to agreement, for voxel locations with dose values greater than 10% of maximum prescription dose, for the slab and breast phantom, respectively. Bush *et al.* (Bush et al. 2011) tested the accuracy between AXB, MC and AAA dose algorithms, and measurements for 6 and 18 MV photon beams, but in homogeneous and inhomogeneous slab phantoms only. They found that AXB and measurement in a homogeneous phantom were within 1.9% in the inner field region for all energies and field sizes. Similarly, in the case of the inhomogeneous phantom, AXB agreed with MC to within 2% and 3% in high and low density lung, respectively. Han *et al.* (Das 2015) compared AXB with AAA dose calculations for the RPC lung phantom and film or TLD measurements for both static-IMRT and VMAT planning techniques. AXB agreed within 0.5% to 4.5% compared to the TLD measurements, and achieved a gamma value of 98% for 3% or 3mm.

The bottom line is that dose algorithms based on deterministic solutions of the linear Boltzmann transport equation are on par with Monte Carlo solutions in terms of accuracy, and more importantly, can achieve these solutions significantly faster, under the conditions required by external beam radiotherapy.

6.4.7 Summary

The Boltzmann transport equation represents the governing equation for radiation transport. Within the context of external beam radiotherapy problems, the equation can be solved both deterministically and statistically, using grid-based or Monte Carlo-based methods, respectively, for the 3D dose distribution within a patient or phantom volume. Upon first glance, the linear Boltzmann transport equation in its various formulations project fear into the heart of the *average* medical physicist and trainees. As commercial treatment planning systems transition from convolution-based to more accurate and precise dose algorithms, based on the linear Boltzmann transport equation (e.g. Acuros XB, Varian Medical Systems, Palo Alto, CA), then it becomes imperative for the clinical physicist not to view these algorithms simply as a "black box". The goal of the foregoing chapter was to break down the barriers associated with understanding the deterministic methods used for solving the linear Boltzmann transport equation. The chapter was intentionally organized to gradually *build-up* the material, much the same way as learning a new language. First, the "alphabet" (variables) was introduced, followed by the "words" (functions), through to the formation of "sentences" (Boltzmann transport equation), and finally, to the construction of "paragraphs" (practical methods of solution). The author hopes that the material presented here eases the burden towards understanding the application and teaching of the linear Boltzmann transport equation as used in external beam radiotherapy.

BIBLIOGRAPHY

Adams, M. and E. Larsen (2002). Fast iterative methods for discrete-ordinates particle transport calculations. *Progress in Nuclear Energy 40*(1), 3–159.

Arfken, G., H. Weber, and F. Harris (2012). *Mathematical Methods for Physicists*. Elsevier, Oxford, UK.

Boltzmann, L. (1872). Weitere studien uber das warmegleichgewicht unter gas-molekulen (German). *Sitzungsberichte Akad. Wiss. 66*, 275–370.

Boman, E., J. Tervo, and M. Vauhkonen (2005). Modelling the transport of ionizing radiation using the finite element method. *Physics in Medicine and Biology 50*, 265–280.

Bouchard, H. and A. Bielajew (2015). Lorentz force correction to the Boltzmann radiation transport equation and its implications for Monte Carlo algorithms. *Physics in Medicine and Biology 60*, 4963–4971.

Bush, K., I. Gagne, S. Zavgorodni, W. Ansbacher, and W. Beckham (2011). Dosimetric validation of Acuros XB with Monte Carlo methods for photon dose calculations. *Medical Physics 38*(4), 2208–2221.

Chandrasekhar, S. (1960). *Radiative Transfer*. Dover, New York, NY.

Das, T. (2015). Mathematical model for the solution of the Boltzmann transport equation for photons. *Asian J. Appl. Sci. Eng. 4*(1), 61–67.

Duderstadt, J. and W. Martin (1979). *Transport Theory*. Wiley, New York, NY.

Fan, W., C. Drumm, S. Pautz, and C. Turner (2013). Modeling electron transport in the presence of electric and magnetic fields. Technical Report SAND2013-8053, Sandia National Laboratory, Albuquerque, NM.

Gifford, K., J. Horton, T. Wareing, G. Failla, and F. Mourtada (2006). Comparison of a finite-element multi-group discrete-ordinates code with Monte Carlo for radiotherapy calculations. *Physics in Medicine and Biology 51*(9), 2253–2265.

Hensel, H., R. Iza-Teran, and N. Siedow (2006). Deterministic model for dose calculation in photon radiotherapy. *Physics in Medicine and Biology 51*(3), 675–693.

ICRU (2011). Fundamental quantities and units for ionizing radiation. Report 85, International Commission on Radiation Units and Measurements, Bethesda, MD.

Lewis, E. and W. Miller (1984). *Computational Methods of Neutron Transport*. Wiley, New York, NY.

Martin, W. and J. Duderstadt (1977). Finite element solutions of the neutron transport equation with applications to stron heterogeneities. *Nuclear Science and Engineering 62*, 371–390.

Martin, W., C. Yehnert, L. Lorence, and J. Duderstadt (1981). Phase-space finite element methods applied to the first-order form of the transport equation. *Annals of Nuclear Energy 8*, 633–646.

McDermott, P. (2016). *Tutorials in Radiotherapy Physics.* CRC Press, Boca Raton, FL.

Nahum, A., J. Seuntjens, and F. Attix (2017). Macroscopic aspects of the transport of radiation through matter. In P. Andreo and D. Burns (Eds.), *Fundamentals of Ionizing Radiation Dosimetry*, Chapter 6, pp. 279–313. New York, NY: Wiley.

NCRP (1991). Conceptual basis for calculations of absorbed dose distributions. Report 108, National Council on Radiation Protection and Measurements, Bethesda, MD.

Papiez, L. and J. Battista (1994). Radiance and particle fluence. *Physics in Medicine and Biology 39*, 1053–1062.

Podgorsak, E. (2010). *Radiation Physics for Medical Physicists.* Springer, New York, NY.

Roesch, W. (1968). *Mathematical Theory of Radiation Fields*, pp. 229–273. Academic Press, New York, NY.

Roesch, W. and F. Attix (1968). *Basic Concepts of Dosimetry*, pp. 1–41. Academic Press, New York, NY.

Rossi, H. and W. Roesch (1962). Field equations in dosimetry. *Radiation Research 16*(6), 783–795.

St Aubin, J., A. Keyvanloo, and B. Fallone (2016). Discontinuous finite element space-angle treatment of the first order linear Boltzmann transport equation with magnetic fields: Application to MRI-guided radiotherapy. *Medical Physics 43*(1), 195–2.

St Aubin, J., A. Keyvanloo, O. Vassiliev, and B. Fallone (2015). A deterministic solution of the first order linear Boltzmann transport equation in the presence of external magnetic fields. *Medical Physics 42*(2), 780–793.

Vassiliev, O. (2016). *Monte Carlo Methods for Radiation Transport.* Springer, New York, NY.

Vassiliev, O., T. Wareing, I. Davis, J. McGhee, D. Barnett, J. Horton, K. Gifford, G. Failla, U. Titt, and F. Mourtada (2008). Feasibility of a multi-group deterministic solution method for three-dimensional radiotherapy dose calculations. *International Journal of Radiation Oncology, Biology and Physics 72*(1), 220–227.

Vassiliev, O., T. Wareing, J. McGhee, G. Failla, M. Salehpour, and F. Mourtada (2010). Validation of a new grid-based Boltzmann equation solver for dose calculation in radiotherapy with photon beams. *Physics in Medicine and Biology 55*(3), 581–598.

Wareing, T., J. McGhee, J. Morel, and S. Pautz (2001). Discontinuous finite element S_N methods on three-dimensional unstructured grids. *Nuclear Science and Engineering 138*(2), 256–268.

Williams, M., D. Ilas, E. Sajo, D. Jones, and K. Watkins (2003). Deterministic photon transport calculations in general geometry for external beam radiation therapy. *Medical Physics 30*(12), 3183–3195.

Zienkiewicz, O., R. Taylor, and J. Zhu (2005). *The Finite Element Method: Its Basis and Fundamentals* (6th ed.). Elsevier, Burlington, MA.

CHAPTER 7

En Route to 4D Dose Computations

Jerry J. Battista

London Health Sciences Centre and University of Western Ontario

CONTENTS

7.1 SYNOPSIS OF 3D DOSE COMPUTATION ALGORITHMS

I N the opening chapter of this book, we briefly reviewed the history of dose computation algorithms. We described the scientific rationale that guides the optimization of a treatment plan including the synergy between radiation biology and radiation physics. We emphasized the clinical need for maintaining good dose accuracy, while remaining cognizant of overall uncertainty in radiotherapy procedures with practical constraints. Chapters 2 and 3 encapsulated the theory of radiological physics and provided an evolutionary view of dose computation algorithms. In Chapters 4 through 6, details of convolution-superposition, stochastic, and deterministic solutions to the Boltzmann radiation transport equations were presented. During this development, we focused on ten principles that captured key points and unique features for algorithm developments. For convenience, these principles are now paraphrased as "10 Commandments" in Table 7.1. The first two com-

Table 7.1: "Ten Commandments" of dose computation

N	Chapter	Principle
1	1	Accurate dosimetry plays a critical role in treatment planning, delivery verification, and clinical outcome assessments
2	1	Treatment planning concerns the optimization of dose *gradients*
3	2	X-ray energy is deposited exclusively by charged particles
4	2	Absorbed dose $\approx KERMA_c$ only in regions of charged particle equilibrium
5	3,4	The versatility and accuracy of an algorithm improve with greater dimensionality of scatter integration
6	3,4	Algorithm performance under conditions of lateral non-CPE is a strong indicator of clinical reliability
7	3,4	Dose computation is simplified when energy *transferred* from primary photons is separated from energy *propagated* by emerging secondary particles
8	5	Monte Carlo simulations mimic radiation experiments by rolling dice to sample realistic probabilities of particle interactions
9	6	The Boltzmann equations enforce accounting and auditing of particle fluences as they propagate through phase space
10	7	Dose distributions on a computer screen may not match interstitial dose distributions

mandments deal with the clinical importance of accurate dosimetry and treatment planning strategy. The third and fourth commandments reflect basic features of x-radiation absorption. The fifth, sixth, and seventh commandments draw attention to having a 3D model of distinct scattered radiation to assure maximum accuracy and versatility of an algorithm. Pencil beam algorithms are only applied in 2D with partial regard for the third dimension; they therefore have a restricted range of applicability and accuracy, especially in heterogeneous tissue. The sixth commandment draws attention to the longer range of megavoltage charged particles that can spoil equilibrium conditions. Algorithms that intrinsically assume localized absorption of charged particle energy (or equivalently CPE) can mislead the dose optimization and prescription; they are not recommended for application to low-density tissue exposed to small higher energy megavoltage fields. These considerations paved the way to Chapter 4. This chapter explained the relationship between convolution and superposition methods implemented in 2D or 3D. They represent the most widely used methods in commercial software packages of this era. Three-dimensional implementations using the collapsed-cone approximation generally strike an excellent compromise between dose accuracy and reasonable turnaround time for clinical treatment planning. Interstitial atomic number variations are considered in the primary calculation of TERMA but ignored in simple density-scaling of the scatter kernels. Monte Carlo methods (eighth commandment) consider these atomic number effects for both primary and secondary radiation, including radiation passage through hardware components of the accelerator head before reaching the patient (Lobo and Popescu 2010). The method has been widely used for commissioning and quality assurance of weaker algorithms. Another feature of Monte Carlo algorithms is the ability to model the effects of magnetic fields on charged particles when MRI is used concurrently with radiation exposure. More recently, fast Monte

Carlo codes have been specialized for treatment planning while taking advantage of parallel processing. Deterministic Boltzmann algorithms (ninth commandment) are a recent addition and solve similar challenging problems with competitive accuracy and computational efficiency. With access to fast cost-effective workstations, Monte Carlo and Boltzmann methods hold promise not only for treatment planning, but also for consolidating dose optimization, reconstruction (tenth commandment) and adaptive radiotherapy procedures with a unified and consistent infrastructure. The tenth commandment will be elaborated in Section 7.3.

Table 7.2 shows computation times for clinical situations including head and neck, lung, and prostate cancer treatments. No attempt was made to normalize the performance data for different irradiation techniques, run-time conditions, software and hardware configurations, and dose accuracy; the data simply reflect typical execution times achievable in a contemporary environment. The contrast from only a few decades ago is the diversity and maturity of Monte Carlo codes used for dosimetric quality assurance or interactive treatment planning. For quality assurance, the EGSnrc code provides a detailed simulation of the radiation passing through collimators (i.e. BEAM code) and the patient (i.e. DOSXYZ code) (Popescu et al. 2015). At the Vancouver Cancer Centre, the Condor cluster is hosted at a remote location and shared; computation times can vary with network congestion. Execution times for treatment planning using specialized software and local workstations have dropped to less than a minute per plan. A special issue of the journal *Medical Dosimetry* provides an alternative summary of commercial software presented from a dosimetrist's perspective (Saw 2018).

7.2 TIME TO CONSIDER THE FOURTH DIMENSION

Following the introduction of CT scans in the mid-1970's, radiation therapy planning evolved smoothly from a 2D to 3D mode. However, patient anatomy and treatment fields are dynamic entities requiring a 4-dimensional (4D) perspective with consideration of temporal variations occurring in the patient or radiation beams over periods of seconds, minutes, days, or weeks. The extension to a 4-D dose model has the ultimate goal of producing *in silico* dose distributions on computer screens that more accurately portray the dose distribution actually delivered in the patient *in vivo*.

7.2.1 Voxels in Motion

Voxels may be considered as small test tubes containing biological cells that are exposed to radiation in a spatio-temporal domain. Consider the case of voxels that move in a non-uniform radiation field. Voxels need to be traceable to nominal locations (x,y,z) in a home coordinate system anchored in the patient. Dose that accumulates in a moving voxel after a radiation exposure of duration, T, is sim-

Table 7.2: Contemporary dose calculation algorithms*

Target Site	Algorithm	System	Technique	Dose Matrix (voxels)	Voxel (mm)	Time	Remarks
Head and Neck	SVD	Pinnacle	2 Arcs; 182 Beams	100 × 100 × 100	2.5	30 s	8-core Intel 2.9 GHz Xeon
	CCC	Pinnacle	2 Arcs; 182 Beams	100 × 100 × 100	2.5	4.5 m	8-core Intel 2.9 GHz Xeon
	CCC	RayStation	2 Arcs; 182 Beams	100 × 100 × 100	2.5	18.2 s	1 server with a GPU
	AAA	Eclipse	2 Arcs; 182 Beams	100 × 100 × 100	2.5	35 s	13 frame agent server, 30 cores
	Acuros	Eclipse	2 Arcs; 182 Beams	100 × 100 × 100	2.5	1.9 m	1 frame agent, 2X 8-core, 2.9 GHz
	EGSnrc	Stand-Alone	2 Full Arcs	240 × 240 × 159	2.5	39 m	16 x 8-core Intel Xeon 2.0 GHz (2 threads/core)
	PhiMC-DPM	Royal Marsden	IMRT; 9 Beams	256 × 256 × 116	1.6	8.2 - 18.8 s	0.9-0.6% precision; Xeon V3
Lung	SVD	Pinnacle	2 Arcs; 108 Beams	100 × 100 × 100	2.5	21 s	8-core Intel 2.9 GHz Xeon
	CCC	Pinnacle	2 Arcs; 108 Beams	100 × 100 × 100	2.5	2.6 m	8-core Intel 2.9 GHz Xeon
	CCC	RayStation	2 Arcs; 108 Beams	100 × 100 × 100	2.5	13 s	1 server with GPU
	AAA	Eclipse	2 Arcs; 108 Beams	100 × 100 × 100	2.5	20 s	13 frame agents, 30 cores were used
	Acuros	Eclipse	2 Arcs; 108 Beams	100 × 100 × 100	2.5	47 s	1 frame agent, 2X 8-core, 2.9 GHz
	EGSnrc	Stand-Alone	2 Partial Arcs	259 × 259 × 180	2.5	60 m	16 x 8-core Intel Xeon 2.0 GHz (2 threads/core)
Prostate	SVD	Pinnacle	2 Arcs; 182 Beams	147 × 118 × 86	3	35 s	8-core Intel 2.9 GHz Xeon
	CCC	Pinnacle	2 Arcs; 182 Beams	147 × 118 × 86	3	5.1 m 3 s	8-core Intel 2.9 GHz Xeon
	CCC	RayStation	2 Arcs; 182 Beams	147 × 118 × 86	3	19 s	1 server with GPU
	AAA	Eclipse	2 Arcs; 182 Beams	147 × 118 × 86	3	26 s	13 frame agents, 30 cores were used
	Acuros	Eclipse	2 Arcs; 182 Beams	147 × 118 × 86	3	1.33 m	1 frame agent, 2X 8-core, 2.9 GHz
	GPUMCD	Monaco	IMRT; 7Beams	344 × 286 × 172	2	2.7 m	2% precision, B-Field ON
	GPUMCD	Monaco	IMRT; 7Beams	324 × 265 × 152	2	1 m	2% precision, B-Field OFF
	EGSnrc	Stand-Alone	Full Arc	240 × 240 × 175	2.5	60 m	16 x 8-core Intel Xeon 2.0 GHz (2 threads/core)
	PhiMC-DPM	Royal Marsden	IMRT; 9 Beams	256 × 256 × 234	2	10.8-26.7 s	1.1-0.7%; Xeon V3

[a] Data courtesy of Odette Cancer Centre (GPUMCD), London Regional Cancer Program (SVD, CCC, AAA, Acuros), Vancouver Cancer Centre (EGSnrc) (Popescu et al. 2015), and published work (PhiMC-DPM) (Ziegenhein et al. 2015). Algorithms are used for treatment planning, except for EGSnrc used for quality assurance. Typical computation times.

ply the integral of dose rate along its trajectory. A simple analogy might assist with physical interpretation. Imagine gathering water droplets during a rain storm, starting with an empty cup. The amount of water accumulated after an integration period (T) is predictable if the cup trajectory and the rate of downpour ($\dot{D}$) are known at every location and moment in time. The accumulated dose is given by the following for a voxel, indexed by its home coordinates (x,y,z):

$$D(x, y, z) = \int_0^T \dot{D}[(x(t), y(t), z(t), t] \, \mathrm{d}t \tag{7.1}$$

We can also re-interpret this expression as a *line integral* over the trajectory, [x(t),y(t),z(t)], traversed in small path length steps (dl). The dwell time at each location is related to the velocity (v) of the voxel, i.e. dt = d$l/v(t)$. The positional and velocity information can be extracted from 4D imaging sequences and deformable image registration techniques (Section 7.2.1.1). The instantaneous *dose rate* experienced by a tissue voxel in motion is calculated with spatio-temporal considerations as follows:

$$\dot{D}(x(t), y(t), z(t), t) = \frac{\partial D}{\partial x}\frac{\mathrm{d}x}{\mathrm{d}t} + \frac{\partial D}{\partial y}\frac{\mathrm{d}y}{\mathrm{d}t} + \frac{\partial D}{\partial z}\frac{\mathrm{d}z}{\mathrm{d}t} + \frac{\partial D}{\partial t} \tag{7.2}$$

This accounts for all dose rate perturbations occurring when the voxel wanders off to nearby positions and if the external dose rate changes. The partial (∂D) derivatives with respect to position ($\partial x, \partial y, \partial z$) represent the dose gradients evaluated at (x,y,z,t) (Foster et al. 2013). The derivatives of position with respect to time describe the velocity components of a moving voxel. The last partial derivative with respect to time denotes the externally-driven dose rate. It changes with the treatment machine dose rate and dynamic beam collimation. If a voxel remains stationary or moves in a *uniform* radiation field, voxels will only experience this local dose rate.

Note that we assumed that voxel displacements did not affect the fundamental dose rate distribution. However, the shuffling of voxels could indirectly affect the interstitial dose rate distribution. For example, a low-density voxel moving into a beam path changes the fluence to downstream voxels lying in its shadow. This is called a *dose deformation effect* it can be accounted for by re-computing the dose at every moment in time using a series of time-stamped CT images from a 4D scan. The effect is generally small if radiological path lengths to voxels of interest are robust. For example, during respiration, the geometric depth to an interior point in lung changes as the chest cavity expands or contracts. However, the radiological depth to a voxel may not change significantly because of compensating changes in the density of intervening voxels. On the other hand, a systematic weight loss of a patient over the course of therapy should be taken into account by dose re-computation with the altered anatomy.

7.2.1.1 Deformable Image Registration

Deformable image registration (DIR) is becoming a critical tool for tracking movement of voxels occurring during delivery of single dose fraction or across dose fractions (Oliveira and Tavares 2014; Saenz et al. 2016; Nie et al. 2016; Vickress et al. 2016; Singhrao et al. 2014; Yeo et al. 2013; Kirby et al. 2013; Nie et al. 2013; Brock 2010). The goal of any deformable registration algorithm is to establish a geographic correspondence between individual voxels observed in two different image sets. In our context, an image is understood to be a 3D or 4D matrix of voxels. Many algorithms rely on identification of homologous isolated landmarks or contour sets, resulting in a holistic best-fit deformation vector field (DVF) that describes how *all* voxels are interconnected across image pairs. The reference image is normally chosen at initial treatment planning (e.g. CT simulation). However, if the patient is re-planned during a course of therapy, registration to the most recent image set can be used for dose summation of subsequent fractions. A set of DVFs between images used for initial or subsequent treatment plans establishes a bucket-brigade link for proper accumulation of dose to voxels across the entire treatment course.

Consider a static dose distribution for a conformal prostate treatment, as shown in Figure 7.1. We can focus on the spatial dose gradient terms within the bracketed term of Equation 7.1. Tissue displacements were mapped across a pair of CT images taken on different treatment days using deformable image registration. The green vectors in Figure 7.1C range up to ≈ 5 mm in length. The displacements of voxels were due mainly to a change in rectal gas volume seen as a dark void. If a tissue voxel is located near point P and moves along a steep dose gradient, the dose changes significantly. To the contrary, the same excursion near point Q would result in minimal dose changes.

When used repeatedly for dose accumulation, the accuracy of the DVF becomes important (Kierkels et al. 2018; Kim et al. 2017; Kirby et al. 2016); inaccurate mapping leads to uncertainty in position and dose of tissue voxels and affects dose-volume parameters. The accuracy of the deformation field can be tested by checking the mapping of reliable landmarks or contours. Contour-driven algorithms include potential for inter-observer contouring variability. Algorithms that rely on biomechanical modelling require knowledge of tissue material properties that are not readily measured in individual patients. Note also that DIR algorithms do not inherently account for the dose deformation effect. An example of DIR mapping uncertainty is shown in Figure 7.2, demonstrating intra-fraction effects during respiration (Vickress et al. 2017). The migration of a single landmark voxel is shown by the arrow across a pair of 4D CT images recorded at end-inspiration (A) and end-expiration (B). The two distributions were recomputed for each anatomy to account for the dose deformation effect but are only slightly different, as expected if radiological paths are invariant (Section 7.2.1). The error circle represents a sphere of uncertainty of fuzziness in the placement of the arrow tip due to limitations of the

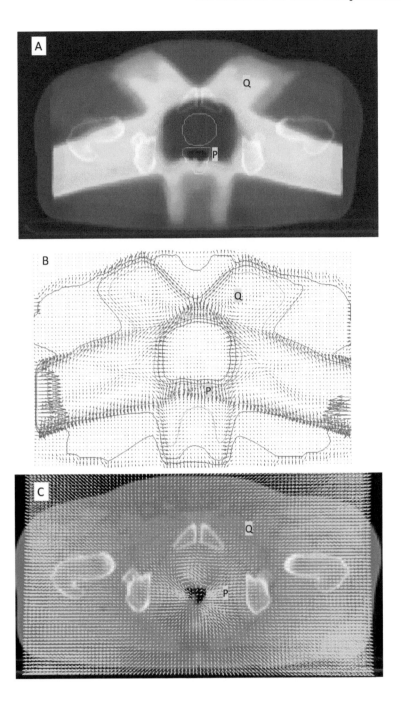

Figure 7.1: A. Dose distribution for a typical conformal prostate treatment.
B. Dose gradient vector map. C. Deformation vector field showing voxel displacements. Data and graphics courtesy of Kyle Foster (Foster et al. 2013).

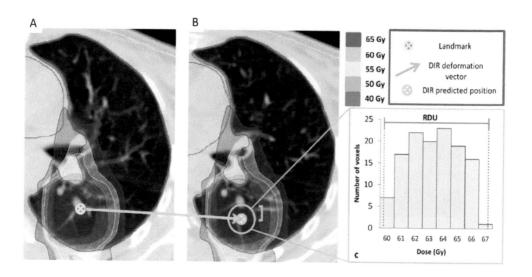

Figure 7.2: A. Dose distribution and landmark for a lung treatment at end-inspiration. B. Dose distribution and the migrated landmark at end-expiration, shown by the arrow tip. The error sphere (blue) represents uncertainty in the arrow tip placement due to DIR limitations. C. Histogram of dose values within the sphere of uncertainty. RDU = Range of dose uncertainty. Reproduced with permission (Vickress et al. 2017).

DIR algorithm (MIM Maestro, MIM Software Inc, Cleveland, OH, USA). A typical sphere has a radius in the range of 2 to 5 mm. The histogram represents dose values present within that sphere, for a prescribed dose of $D_{95} = 60$ Gy. This study evaluated DIR uncertainty for a total of 300 landmark points in 10 representative 4D CT lung studies. To recap, the DIR algorithm introduces dose uncertainty, beyond the inherent accuracy limits of the dose calculation algorithm.

7.2.1.2 Interplay Effects

Three-dimensional treatment planning intrinsically assumes that the patient anatomy is static and that all radiation fields are applied instantaneously. This is certainly not realistic for multiple IMRT fields delivered on the order of several minutes. For sliding-window IMRT, VMAT, tomotherapy, or robotic delivery (e.g. Cyberknife), beams are applied while collimator elements, gantry, and treatment couch are in motion. *Dose distributions are not deposited instantly!* Generally, the beam delivery subsystems are not synchronized with motion of tissue voxels driven by independent physiological processes. Simultaneous out-of-phase movements will momentarily overexpose or underexpose some voxels. The dose to a voxel increases if it moves into a high dose region but decreases if a collimator momentarily occludes it; voxels can be at the wrong place at the wrong time! Considering the permutations and combinations of moving parts and tissue elements, the dose rate anomalies at individual voxels become a complex function of space and time (i.e. Equation 7.2). (Netherton et al. 2018; Fernandez 2016; Engelsman et al. 2001; Jiang et al. 2003; Bortfeld et al. 2004). These effects can be minimized through 4D treatment planning involving synchronization techniques (breath-holding, beam-gating, real-time tracking) (Keall et al. 2018; Nguyen et al. 2017; Caillet et al. 2017).

The most intuitive effect of any motion is a blurring of the dose near a region of high dose gradient such as a beam edge and this can be modelled by a convolution kernel (Craig et al. 2003). To avoid loss of tumour control, the width of blurring can be accommodated by setting appropriate margins for the planning target volume (PTV). The second effect is a rippling or oscillatory dose pattern seen in tomotherapy delivery. This interference phenomenon can be understood by considering tissue and machine components moving with different frequencies, amplitudes, and phases. This incoherent motion will cause moiré or beating patterns in dose. The third effect is due to general asynchrony of collimator leaf and target movements; it is highly specific to the treatment technique and target site. All these effects are not normally displayed in treatment plans but could potentially be included in 4D treatment planning system, using 4D images of the patient and machine log files that document every programmed step of the treatment delivery.

To demonstrate interplay effects, consider an example from fan-beam helical tomotherapy (Kim et al. 2009). Dose patterns were predicted by computer simulation and verified by film dosimetry within a thorax phantom with a moving lung insert. In the following examples, the range of target motion was 3.0 cm, and the period was intentionally slowed to 8 seconds to elicit the potential effects. Figure 7.3 shows results for a helical exposure intended to protect an off-axis critical organ within the yellow band (right of the black dotted line). This example was chosen to highlight the potential deleterious effect occurring during *single-fraction* treatment. The dose anomalies can be avoided through smart selection of rotational pitch, gantry rotation speed, and number of rotations. Furthermore, localized dose anomalies will average out over multi-fraction treatments. If the dose discrepancies remain unacceptable, then breath-holding or gated collimation can be applied to prevent interference patterns of dose (Kim et al. 2005, 2010).

7.2.2 Interfractional Changes in Anatomy

Over the full course of treatment, the patient can undergo progressive time-dependent changes. These are caused by early growth or subsequent shrinkage of the tumour volume and lymph nodes as well as shrinkage of nearby normal organs such as the parotid glands in head and neck radiotherapy. Differential filling of the bladder and bowel causes tissue displacements as was illustrated in Figure 7.1. In addition, the patient may undergo a general loss in body weight. An analysis of

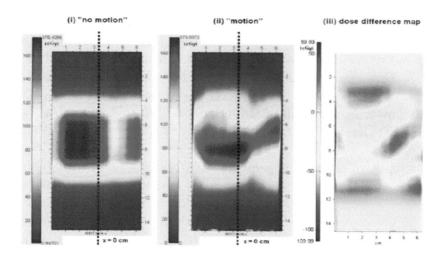

Figure 7.3: IMRT dose distribution delivered to dosimetric film via helical tomotherapy with (ii) and without (i) film motion. The dose difference map is also shown (iii). A conformal delivery (i) was intended to shield a cord-like structure in the yellow gap to the right of the rotation axis (black dotted line). Simulated respiratory motion caused this dose to be shifted into the critical structure (ii). Reproduced with permission (Kim et al. 2009).

electronic portal images (McDermott et al. 2006) revealed that 57% of lung cases and 37% of head and neck cases exhibited significant changes throughout radiation treatment. In Figure 7.4, changes in lung are evident from beginning to end of radiation treatment (Hugo et al. 2017). All images correspond to the end-inhalation phase of the respiratory cycle. Similar images are available for research purposes from the Cancer Imaging Archive (http://www.cancerimagingarchive.net/). Volume changes in the gross target volume (GTV) during a full course of tomotherapy were tracked quantitatively using frequent megavoltage CT (MVCT) scans (Woodford et al. 2007). The temporal profiles of GTV changes were found to be specific to each patient. Figure 7.5 shows a subset of patients for which the tumour regression followed a biphasic time course, with moderate change initially followed by an abrupt decrease to a secondary plateau. In some cases, changes were significant enough to warrant a treatment re-planning action.

7.2.3 Intrafractional Changes in Anatomy

During radiation exposure, respiration can cause cyclic displacement of internal organs and tumours. Swallowing can also cause intermittent shifting of laryngeal tumours (e.g. glottis). X-ray fluoroscopy and ultrasonography were the first imaging techniques to fully demonstrate internal organ movements in two dimensions (2D). Progress has since been made using 4D CT or dynamic MRI techniques (Choudhury et al. 2017). Motion mitigation strategies include breath-holding, beam-gating, or

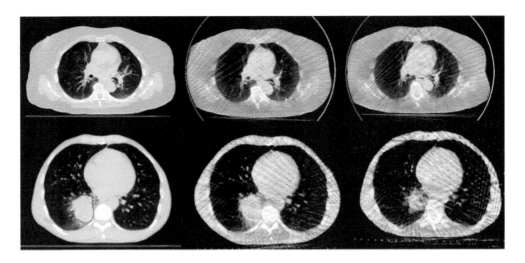

Figure 7.4: All CT images obtained at the end-inhalation phase of respiration. Images on the left are obtained by 4D CT scanning. The upper left image highlights an implanted fiducial marker. Interfractional changes in lung from beginning (middle panel) to end (right panel) of a radiation treatment course, reconstructed from cone beam CT (CBCT) data sets. Reproduced with permission (Hugo et al. 2017).

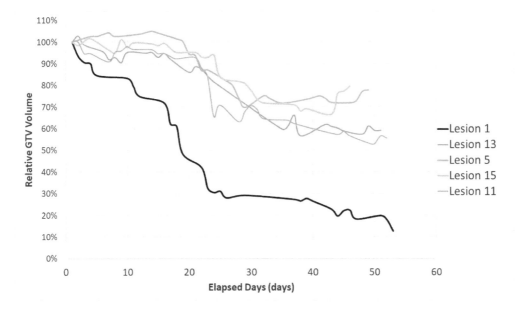

Figure 7.5: Interfractional reductions in gross tumour volume (GTV) of different patients during a course of lung tomotherapy. Data courtesy of Slav Yartsev, London Regional Cancer Program (Woodford et al. 2007).

live target tracking (Keall et al. 2018; Nguyen et al. 2017; Caillet et al. 2017). Target motion can be frozen in time by coached breath-holding or stroboscopic techniques with treatment fields flashed on at precise moments of cyclic target motion.

7.2.3.1 Image Guidance with MRI

Gantry-mounted CT systems greatly improved the precision of target positioning, and enabled dose reconstruction procedures (Section 7.3.1). However, on-board CT imaging cannot produce 4D data sets in real-time due to the slow gantry rotation, although phase-gated image reconstruction is feasible for off-line reconstruction (Hugo et al. 2017; O'Brien et al. 2016). The advent of MRI-guided radiotherapy (Keall et al. 2014; Lagendijk et al. 2014; Fallone 2014) provides real-time MR images, while enhancing soft tissue discrimination. There are a few complications to this approach, however. MR images are subject to geometric distortion and require correction when applied to high-precision radiotherapy. In addition, MRI pixel values do not reflect tissue electron density. Image segmentation of tissue regions can be used, however, to assign standard tissue compositions.

The union of an MR imaging system and linear accelerator also adds a new *twist* to dose computation algorithms (pun intended). The main magnetic field (B_0) produces a strong Lorentz force that bends the trajectory of charged particles set-in-motion by x-rays in tissue (Malkov and Rogers 2016). This distorts dose distributions and can affect target dose coverage or normal tissue sparing. Monte Carlo or Boltzmann methods can incorporate the effects of magnetic deflections (Oborn et al. 2017; Bouchard and Bielajew 2015; St. Aubin et al. 2015). The magnitude of the dose perturbations depends on the relative orientations of the x-ray beams and magnetic field, as well as the field strength. The *MRIdian* commercial system marketed by ViewRay (Oakwood Village, Ohio) uses a lower magnetic field strength (0.35 Tesla) to minimize this issue but still uses a Monte Carlo algorithm. The Alberta system also uses a lower field strength (0.6 Tesla) and further optimizes the B-field orientation. Figure 7.6 shows a five-field conformal lung treatment plan, using 6 MV x-rays with a concurrently applied field of 1.5 Tesla (Kirkby et al. 2010), as used in the Elekta Unity system developed in the Netherlands. The figures show dose difference maps with the magnetic field switched on or off. When the magnetic field direction is oriented perpendicular to beam axes (Figures a-d), dose differences approach 30% at tissue interfaces. With the magnetic field re-oriented parallel to the beam axes instead (e-f), dose perturbations are dampened and confined mainly within the beam paths.

7.2.4 Dose Verification in 4D

New dosimetric approaches are needed to measure time-integrated dose distributions for dynamic procedures (Colvill et al. 2018). With a pragmatic approach, existing dosimeters are moved on a programmable platform during data acquisition. A rep-

Transverse B-Field

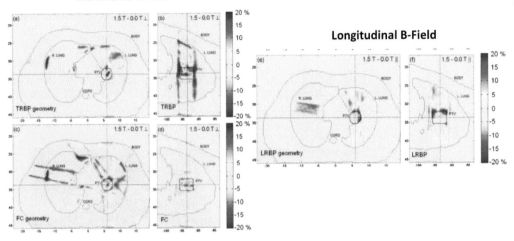

Figure 7.6: Effect of a strong magnetic field (1.5 Tesla) on a 5-field conformal 6 MV x-ray treatment, shown as dose difference (%) between the magnetic switched *on* or *off*. Transverse field options (a-d); longitudinal field options (e-f). Reproduced with permission (Kirkby et al. 2010).

resentative sample of dosimetric instruments is shown in Figure 7.7. These rigid systems do not generally have a 3-dimensional array of dosimeters and are therefore limited to point or planar assessments for front-line verification of complex dynamic procedures. Tissue-equivalent gel dosimeters continue to be developed for more comprehensive and direct 4-dimensional dose mapping (Schreiner 2015), with a potential to mimic tissue and dose deformation effects (Yeo et al. 2013).

Experimental tissue-like phantoms have been developed for verification of treatment procedures (Niebuhr et al. 2016). Virtual lung phantoms can also be used to model 4D dose deposition (Segars et al. 2018; Kyriakou and McKenzie 2011). In the experimental approach for lung, dose sensors placed within a lung-like surrounding material are moved under computer control (Steidl et al. 2012). More anthropomorphic models use compressible lung-like foams that simulate time-dependent variations in density (Cherpak et al. 2011; Kashani et al. 2007). Future phantoms will likely have components produced by 3D printing techniques (Anderson 2017; Liao et al. 2017; Madamesila et al. 2016; Mayer et al. 2015).

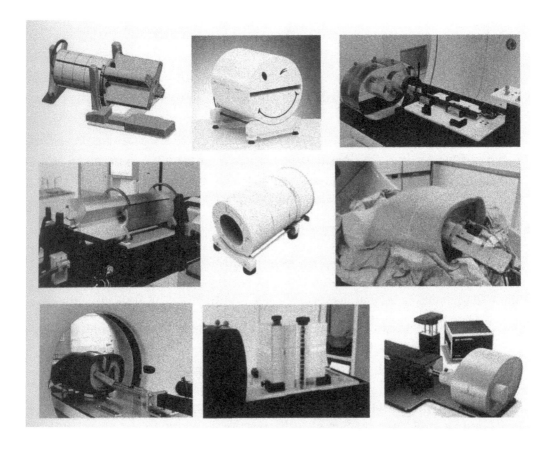

Figure 7.7: 4D dosimetry systems that have moving elements during dose data acquisition. Some thoracic phantoms contain lung-like materials with a time-varying density. Reproduced with permission (Colvill et al. 2018).

7.3 DOSE-ADAPTIVE RADIOTHERAPY (DART)

7.3.1   3D Dose Reconstruction

Principle #10: "The dose distribution of a treatment plan on a computer screen may not match the interstitial dose distribution"

While major geometric changes in anatomy can trigger re-planning of treatment during a course of therapy, an analysis of dosimetric impact provides much better information for making an objective decision. The decision cannot be taken lightly because it has major impact on radiation oncology resources, both in terms of personnel and equipment usage. Moreover, if the residual number of dose fractions is small, the intervention becomes progressively less effective for the added labour and inconvenience to the patient. The concurrent availability of on-board CT systems and electronic portal imaging devices (EPIDs) makes it possible to reconstruct 3D *in vivo* dose distributions *du jour* in individual patients for quality assurance (Branchini et al. 2017). This topic has been comprehensively reviewed elsewhere (van Elmpt et al. 2008, 2009; Mijnheer 2018). Alternatively, CT images and machine log files can be used to replay a virtual treatment execution and predict the resultant 3D dose distribution (Katsuta et al. 2017; Popescu et al. 2015). In this section, we provide an example of EPID dosimetry and stress the important role of the dose computation algorithm. EPIDs were developed originally for verification of patient positioning in the beam's eye view. However, by combining beam projection data with on-line CT imaging, 3D dose reconstruction becomes feasible. The basic steps of dose reconstruction are shown in Figure 7.8.

The EPID pixel data must be corrected for a number of instrumental issues (McCurdy 2013). Removal of significant scatter originating in the patient (Jaffray et al. 1994) is essential to isolate the pure *primary* exit fluence transmitted through the patient. This fluence is then backprojected through the CT matrix of attenuation coefficients. Megavoltage fan-beam CT scanners on tomotherapy units provide the best electron density data; CBCT scanners, like EPIDs, require removal of patient-specific scatter contributions to the image projections (MacFarlane et al. 2018; Usui et al. 2017; Rong et al. 2010; Hatton et al. 2009). The resulting pure primary incident fluence is then used to re-compute the dose distribution using forward projection *via* a dose algorithm of the user's choice. This 3D distribution is finally checked for convergence to a planned dose distribution using a 3D–γ or DVH assay.

At the Netherlands Cancer Institute (NKI), 5,000 treatments per year are verified in this way using EPID images and automatic dose infraction alerts calling for follow-up in 30% of cases (Mijnheer 2018). At the MAASTRO clinic also in the Netherlands, all patient cases are also subjected to EPID verification dosimetry to detect machine-related, plan-related, or patient-related discrepancies on-line (Persoon et al. 2013). A sample process for VMAT verification is shown in Figure 7.9.

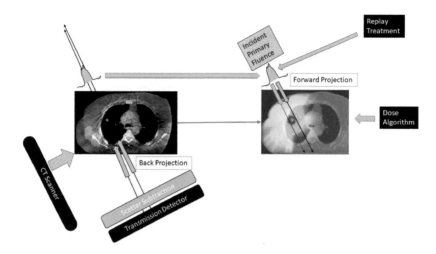

Figure 7.8: Basic procedure for 3D dose reconstruction using on-line CT and EPID images or treatment machine log files. EPID voxel signals are corrected to yield transmitted *primary* fluence and then backprojected through the CT data set. This primary incident fluence is then forward-projected to yield the dose distribution per field, then summed over all fields. Alternatively, this can be achieved by replaying the treatment machine program, as recorded in log files.

An example of dose discrepancies between planned dose distributions and EPID-reconstructed dose distributions based on CBCT scans is illustrated in Figure 7.10. A quad set of tolerances is specified for values of γ_{mean}, $\gamma_{1\%}$ or 99%-tile cut-off, γ pass rate, and difference in isocentric dose ($\Delta Disoc$). The input γ threshold parameters are normally set at 3%/3 mm. The γ values are evaluated in the usual manner (as in Section 1.3.1) but a prefix + or - is added to indicate overdosage (yellow-red) or underdosage (green). This provides an alarm system at the treatment console to flag unacceptable dose deviations that require further investigation and possibly treatment re-planning. Violations are displayed for the operator in a direct way using a traffic light icon. The activated red light indicates several dose infractions (Mijnheer et al. 2017).

It is important to remain cognizant of current limitations in reconstructing 3D dose distributions: (1) EPID pixel signals must be calibrated for the treatment beam energy and post-processed to subtract the exit fluence due to in-patient scattering, (2) dose accuracy is limited by inaccuracy of electron densities extracted from cone-beam CT data (MacFarlane et al. 2018), (3) dose accuracy is subject to limitations of the dose calculation algorithm *per se*. The use of the same algorithm for both treatment planning and dose reconstruction ensures *consistency* but it does not necessarily reflect the true dose distribution, and (4) there are pitfalls in relying *solely* on a gamma evaluation for dose comparisons detecting clinically-

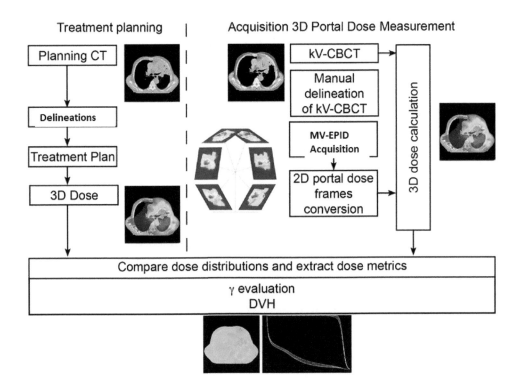

Figure 7.9: Workflow for 3D dose verification of VMAT with an EPID. The treatment plan serves as the reference condition (left) and a dose reconstruction is shown on the right. Gamma and DVH analyses are used to judge the match. Reproduced with permission (Persoon et al. 2013).

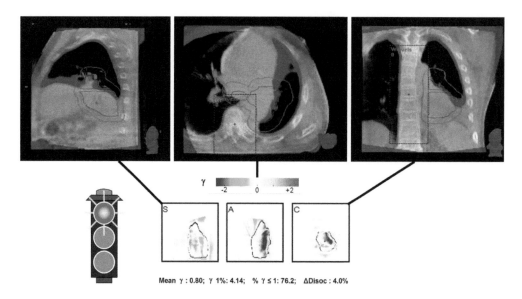

Figure 7.10: Dose reconstruction from EPID projections for hypo-fractionated VMAT treatment of lung in sagittal (S), axial (A), and coronal(C) sections. CBCT scans *du jour* (green) and planning CT scans (purple) are overlaid. Gamma value maps (3%/3mm threshold) within the 50% isodose volume (black line), are displayed below each image, comparing the delivered to planned dose distributions. The red light was tripped by unacceptable values listed below the γ maps. Data courtesy of Ben Mijnheer of the Netherlands Cancer Institute and adapted with permission (Mijnheer 2018). Use of the traffic light icon was permitted under Creative Commons licensing (https://www.allaboutcircuits.com).

relevant dose-volume violations (DVH, TCP, NTCP) (Schreiner et al. 2013; Nelms et al. 2013). Whichever assay is adopted, end-to-end testing with a range of input parameters and treatment failure modes should verify the sensitivity and specificity for detecting clinically-relevant deviations (Steers and Fraass 2016; Caivano et al. 2014; Cozzolino et al. 2014).

7.3.2   Dose Accumulation Over Time

The dose distribution computed during treatment planning is only a forecast of the dose distribution expected to be delivered *in vivo* when all dose fractions are accumulated. Cumulative dose distribution can modelled by summing a series of 3D reconstructed dose distributions. Using the original treatment plan as a reference template, deformable image registration techniques (Section 7.2.1.1) can be applied to this situation (Wong et al. 2018; Samant et al. 2018; Shessel et al. 2018; Marshall et al. 2017; Kim et al. 2017; Avanzo et al. 2017; Oh and Kim 2017; Nassef et al. 2016; Park et al. 2016; Yoon et al. 2016; Weistrand and Svensson 2015; Veiga et al.

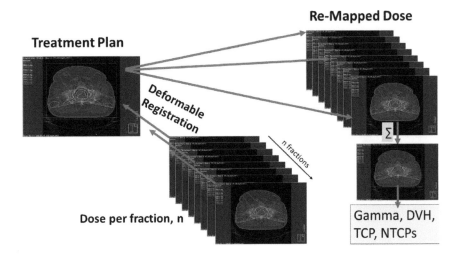

Figure 7.11: Steps involved in summation of dose distributions acquired at each treatment fraction. For illustrative purposes, *only the central slice* of the 3D dose distribution is illustrated.

2015; Rigaud et al. 2015; Yang et al. 2011). The process is illustrated in Figure 7.11 for a single axial dose slice of the 3D dose distribution. The dose distribution per fraction (center) is remapped to the treatment plan (left) before addition in the accumulating dose distribution (right). Ideally, this could be done per delivered dose fraction to test convergence to the original treatment plan and clinical objectives. (NTCP). Lack of convergence could require re-planning of treatment over the residual number of fractions.

An example of a dose propagation program implemented in Pinnacle software scripts (Philips Healthcare, Radiation Oncology Systems) is sketched in Figure 7.12 (Battista et al. 2013). The purpose of this simulation was to compare a range of dose-adaptive strategies. A prostate treatment with 5-field IMRT with 18 MV x-rays was modelled and daily on-line CT images were used for dose reconstruction. Prostate displacement was mainly caused by variations in bladder and rectal filling. Different scenarios were modelled for 13 patients with a variable frequency in CT imaging, and geometric or dosimetric adaptation. PTV volumes were established using anisotropic 7/10 mm (anterior/posterior) margins or an isotropic 5 mm margin. Figure 7.12 shows the program loops for each patient (outer loop) and dose fraction (inner loop). Dose reconstruction was accomplished by recomputing the dose distributions *de novo* on the CT image set of the day, assuming perfect execution by the treatment machine hardware. The fractional dose distribution was remapped to the treatment planning space using surface-based deformable registration prior to each dose summation. DVH results for the larger PTV margin are presented in Figure 7.13. Without image guidance (row A), daily offsets in patient

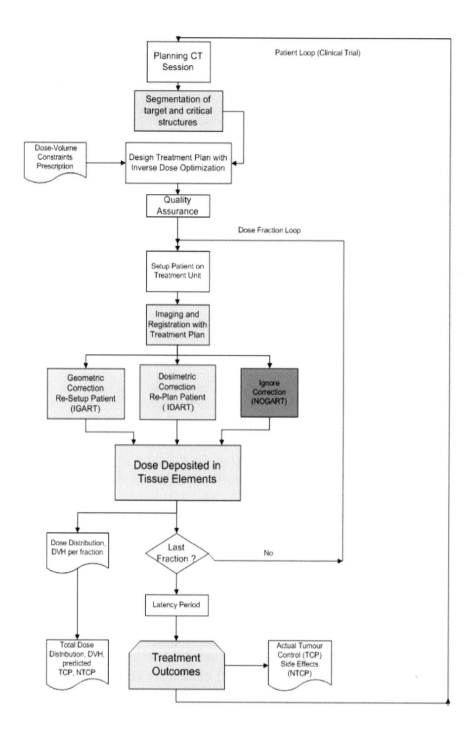

Figure 7.12: Flowchart for simulation of a multi-fraction dose accumulation, with (IGART) and without (NOGART) CT image guidance, and treatment re-planning (IDART) (Battista et al. 2013).

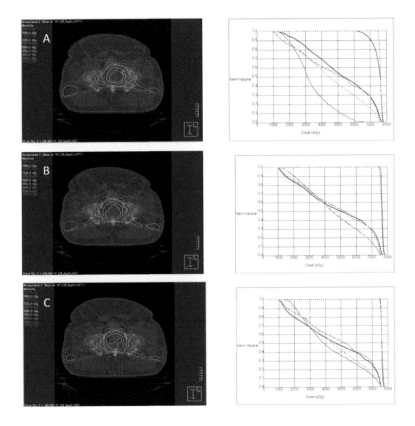

Figure 7.13: Comparison of dose distributions and corresponding DVH curves for three different adaptation scenarios A. No image guidance. B. Daily re-positioning. C. Daily re-planning. Non-isotropic margins of 10/7 mm (anterior/posterior) were applied for GTV-to-CTV. DVH curves are dashed for the original treatment plan and solid for delivered treatment. The DVH curves are shown in red for GTV, blue for bladder, and green for rectum. Data and graphics courtesy of Carol Johnson, London Regional Cancer Program (Battista et al. 2013).

position caused unacceptable blurring of the dose within the gross target volume (GTV), as seen in the DVH curve (red). A geographic miss is also obvious in the dose distribution, while collateral sparing of the rectum is also observed (solid green curve). In other words, uncorrected anterior field placements miss both the tumour and rectum! With more precise positioning (row B) or re-planning (row C), the DVH curve for the rectum shifts back towards the planned curve. Similar patterns were observed for an isotropic 5 mm margin, but results showed a more acute need for treatment re-planning. Predictions of complication-free tumour control were also modelled as the product of TCP and (1-NTCP) for rectal tissue. Results are presented in Figure 7.14 for a range of scenarios, including variable PTV margin sizes, patient re-positioning, and treatment re-planning schedules. The treatment plan is robust, provided that some form of CT image guidance is applied.

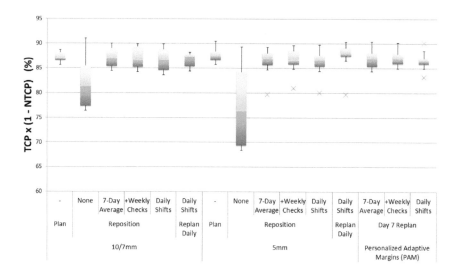

Figure 7.14: Predicted probability of prostate tumour control *without* rectal complications for different adaptation scenarios with large non-isotropic margins of 10/7 mm (anterior/posterior), tighter isotropic margin (5 mm), or individual adaptive margins for the CTV. Box plot represents the 25 to 75%-tile cut-off limits; error bars are set at 1.5 × quartiles; X marker in the figure denotes outlier cases. Data and graphics courtesy of Carol Johnson, London Regional Cancer Program (Battista et al. 2013).

7.4 BIOLOGICALLY-ADAPTIVE RADIOTHERAPY (BART)

> "...despite all this technological and clinical innovation, experience shows that in many cases it is still not possible to control local tumour growth. There's a simple explanation for this anomaly, summed up neatly by the following dictum (credited to the Canadian medical physicist Harold Johns): **"If you can't see it, you can't hit it, and if you can't hit it, you can't cure it!"**. Put another way: clinicians still know very little about the target volume. To address this shortcoming, radiation oncology as a discipline needs to reinvent itself once more and pursue an ambitious development roadmap that will ultimately enable radiation oncologists and physicists to characterize the tumour in terms of the 3M's - morphology, movement and molecular profiling - before, during and after a course of treatment."

> Dr. Wolfgang Schlegel *circa* 2010

The concept of bio-targeting the hallmarks of cancer cells was articulated decades ago (Pavlova and Thompson 2016; Ling et al. 2000) but its practical implementation has remained elusive. Cancer is a microscopic disease while treatments still rely on macroscopic imaging techniques to establish the gross target volume.

The typical density of human cells is 1 million cells per mm^3. For a voxel seen in a medical image, several million cells are therefore sampled and their signals averaged in the image data. Perhaps only an invisible clonogenic subset of these cells is ultimately responsible for treatment failure, but which subset? Hypoxic cells have long been suspected of limiting treatment efficacy, for example (Thorwarth and Alber 2010; Thorwarth et al. 2007). Peripheral, distant, circulating, communicating, or dormant cells can limit long term survival, even after an aggressive dose of radiotherapy.

7.4.1 Evolving Functional and Molecular Imaging

Image guidance in the broadest sense includes *6-D's* of applications (Greco and Ling 2008; Grau et al. 2008; Wang et al. 2016): (1) detection, diagnosis and staging of disease, (2) delineation of target volumes and organs at risk, (3) determining attributes of tumours and their niche environment, (4) design of dose distributions during treatment planning, (5) dose delivery quality assurance and (6) deciphering response to treatment during or after a course of therapy. Some advances have certainly been made in functional (Thorwarth 2015) and molecular (Neveu et al. 2016; Yankeelov et al. 2014) imaging. Hybrid imaging systems such as PET-CT (Beyer et al. 2000) or PET-MRI scanners (Catana et al. 2013; Bailey et al. 2015) are complementary and simplify the fusion of anatomy, physiology, and molecular findings. These advances will ultimately provide the canvas for detailed dose painting at the voxel level (Bentzen 2011). Adaptation of treatment to tumour response will require an imaging technique that can distinguish cells that are alive, dormant, dying, or destined to die (Bentzen et al. 2010).

Functional imaging includes mapping techniques that probe the functional capacity of an organ. For example, CT scans, radioisotope scans, and MRI can map lung function in terms of perfusion and ventilation. Treatment planning can include considerations of these features to spare as much lung as possible (i.e. conformal avoidance of functional regions). Similarly, dynamic contrast-enhanced CT or MRI can track the flow of an injected contrast agent to obtain several vascular parameters such as flow rate, blood volume, and oxygen transport with major impact on early diagnosis of cardiovascular disease including stroke. This technique also has important implications for oncology as it can provide insights into oxygenation patterns and the synergy of anti-angiogenic therapy and radiation therapy.

Molecular imaging techniques "monitor and record the spatio-temporal distribution of molecular or cellular processes for biochemical, biologic, diagnostic, or therapeutic applications" (Thakur and Lentle 2006). There has been progress in identifying specific markers of tumour burden, metabolism, hypoxia (Neveu et al. 2016) and proliferative capacity. A summary of molecular markers and imaging modes is provided in Table 7.3. The development of image contrast and therapeu-

Table 7.3: Functional and molecular imaging probes (Yankeelov et al. 2014)

Marker for	PET	MRI or MRS
Glucose Metabolism	^{18}F -DeoxyGlucose (FDG)	
Amino Acid Metabolism	^{11}C - Choline (Cho)	
	^{18}F - Choline	
	^{11}C - Methionine	
	^{18}F -Tyrosine	
Cell Proliferation	^{18}F - Thymidine (FLT)	
Cell Apoptosis	^{99m}Tc - Annexin V	
Cellularity		Diffusion
Hemodynamics/Perfusion	^{15}Oxygen	Dynamic Contrast* or Diffusion
		*also possible with Perfusion CT
Angiogenesis	^{64}Cu -DOTA-VEGF	
Hypoxia	^{18}F - Misonidazole	BOLD
	^{64}Cu - ATSM	
Breast Cancer	^{18}F-ES (Estrogen)	
Breast Cancer	^{18}F-PR (Progesterone)	
Breast Cancer	^{68}Ga HER2	
Prostate Cancer	^{18}F or ^{68}Ga PSMA	
Prostate Cancer		(Cho+Amine+Creatine):Citrate Ratio
Glioma		NAA:Cr and Cho:Cr Ratio

tic agents requires progress in tumour biology, nanotechnology, and theranostics - imaging that guides therapeutic intervention (Li et al. 2016; Botchway et al. 2015; Penet et al. 2012). This will complement genomic expression profiles (Hamburg and Collins 2010; Lambin et al. 2017) that will triage patients into highly personalized treatment with a stronger expectation of effectiveness. The potential for custom therapy spans across many tumour sites, including brain (Fink et al. 2015), head and neck (Grégoire et al. 2015), and prostate (Pathmanathan et al. 2016).

7.5 EVOLVING COMPUTER TECHNOLOGY

In the past, limitations in computing power restricted the algorithms that could be practically applied to radiotherapy. Approximate algorithms were necessary to make treatment planning interactive with a reasonable turnaround time for clinical application. Clinically, a timely start of treatments has obvious advantages for rapidly-proliferating tumours and it also provides psychosocial benefit to patients waiting in queue. Future developments such as frequent on-line dose reconstruction (Spreeuw et al. 2016) will escalate computational demand even further. Persistent advances in computer technology will have a positive impact not only on treatment planning but also on multi-dimensional imaging, dose optimization, and adaptive radiotherapy with a greater need for automation. The size, power consumption, and cost of processors have plummeted precipitously and uninterrupted. The speed of program execution and graphical displays fortunately continues to increase exponentially, prodded by mobile phones, interactive gaming, and the movie industry that continue to explore virtual reality. Central processing units (CPUs) can be programmed for parallel computing using many-integrated-core (MIC) techniques with standard computer languages (e.g. C++); this facilitates portability and code sharing across computer platforms. Graphics processing units (GPUs) can also be custom programmed for scientific applications with specialized languages such as Nvidia's CUDA (Compute Unified Device Architecture). These developments will have profound implications for future treatment planning workstations (Jia and Jiang 2018).

In principle, the theoretical maximum speed (S_{max}) performance in scientific number crunching can be estimated for a computer system, as follows:

$$S_{max} \ (TFLOP/s) = 0.001 \times n_{srv} \times n_{proc} \times n_{core} \times n_{th} \times FLOP_c \times C \qquad (7.3)$$

where n_{srv} is the number of servers, n_{proc} is the number of processors per server, each with n_{core} cores running n_{th} threads concurrently. Each core can complete $FLOP_c$ operations *per clock cycle* where C is the fundamental clock frequency (GHz) of the system. For example, a single GPU (i.e. Nvidia Titan Xp) with 12 million transistors runs 3,840 cores at a frequency of 1.4 GHz, each completing 2 FLOPs per clock cycle (single thread). This processor achieves a peak calculation speed of 10.7 TeraFLOP/s (TFLOP/s), and costs less than $1,200 (USD). Various parallel architectures on central and graphics processors are yielding comparable peak performances while maintaining affordability (Rodriguez and Brualla 2018; Pratx and Xing 2011; Jia et al. 2011; Jia and Jiang 2018).

Figure 7.15 summarizes the performance of contemporary CPU and GPU systems. The measurement of computational speed is a science onto itself; standardized tests reflect performance under ideal conditions. The achievable performance not only depends on raw computational rate (TFLOP/s) but also on limits imposed by memory size, access speed, and software management of multiple processes. As the number of co-processors is increased, scaling can become sub-linear, stifled by job control overhead and data traffic congestion (Rodriguez and Brualla 2018; Jia and Jiang 2018; Ziegenhein et al. 2015).

A review of Monte Carlo codes classified over 20 programs according to their purpose for either treatment planning or dose verification (Brualla et al. 2017). This topic was covered in Chapter 5 (Table 5.4). Seven systems were evaluated using CPU processors with aggregate computational speeds up to 0.64 TFLOP/s, on the low end of the range depicted in Figure 7.15. Benchmark tests were conducted using simple photon and electron fields incident on a homogeneous water phantom with voxel dimension 3 mm. Run times varied greatly [10 to 10,000 seconds], depending on the MC code (customized *versus* general purpose) and sophistication of the radiation source modelling. Since the time of the review, additional programs

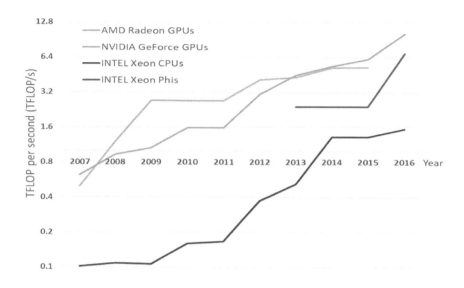

Figure 7.15: Computational power of GPUs and CPUs in terms of *peak* TFLOP per second (TFLOP/s) for single-precision arithmetic. Data source: Karl Rupp (https://www.karlrupp.net/wp-content/uploads/2013/06/gFLOPs-sp.png), with permission under creative commons license, Attribution 4.0 International.

have been released (Lechner et al. 2017; Lamb et al. 2017). Using MIC technology (Rodriguez and Brualla 2018), significant acceleration was achieved using a multi-core CPU (dual 12-core Xeon E5-2670V3). The authors recently upgraded *PRIMO* (www.primoproject.net) which is a self-contained system based originally on the PENELOPE code but now supporting the fast DPM code. Fast Monte Carlo (MC) codes specialized for clinical treatment planning, such as XVMC used in iPlan (BrainLab, Feldkirchen, Germany) and Monaco (Elekta Instruments, Stockholm, Sweden) systems typically execute an overall task size of 10 TFLOPS for acceptable precision of a few percent in dose (Brualla et al. 2017). Using a moderate computational rate of 1 TFLOP/s (from Figure 7.15), Monte-Carlo-based treatment planning should become possible within a time frame of a few seconds per plan!

For example, a multi-core CPU-based implementation can now produce a clinical 9-field IMRT treatment plans in about 10 seconds (Ziegenhein et al. 2015). The dose precision was 1% of the maximum dose for a small voxel size ($\approx 2mm$). The resultant dose distribution is shown in Figure 7.16. Cloud-based computing resources will render treatment plans within time frames that are rapidly converging to the *1 second mark* (Ziegenhein et al. 2017; Jahnke et al. 2012). Pratx and Xing (Pratx and Xing 2011) described GPU technology applied to general medical physics and list over a dozen applications. The PENELOPE code was ported to achieve a

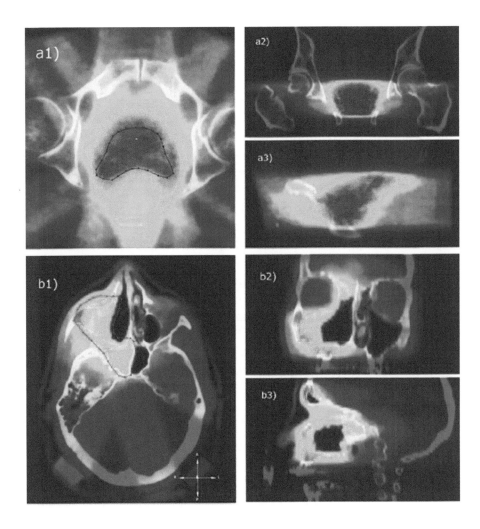

Figure 7.16: IMRT dose distributions for a prostate case (top panels) and head and neck case (bottom panels). These were calculated in about 10 seconds for 1% precision using PhiMC Monte Carlo code on a multi-core CPU (XeonV3). Reproduced under a creative commons license (https://creativecommons.org/licenses/by/3.0/) (Ziegenhein et al. 2015).

speed gain of ≈ 30. Similarly, the GPUMCD code was implemented on a GPU (Nvidia GTX480), with speed gains also on the order of 1000 over general-purpose EGSnrc/DOSXYZ code running on a CPU (Intel i7 860) (Hissoiny et al. 2011). Execution times for GPUMCD were on the order of one second. The preference for CPU or GPU acceleration will continue to be debated, with due considerations for platform-independent programming and scalability.

7.6 CONCLUDING REMARKS

Dose algorithms are paramount to successful clinical radiation therapy of cancer. Dose accuracy will continue to improve, opening opportunities for *interactive* treatment re-optimization *at the treatment console*. When meshed with objective assays of clinical outcomes assessed under more tightly-controlled dosimetry conditions, progress in modelling of *in-vivo* human radiobiology of tumours and normal tissue is sure to follow (El Naqa 2018; Jaffray et al. 2010).

It is projected that by the year 2030, the number of new cancer cases around the world will approach 23 million per year (Atun et al. 2015). If radiotherapy resources can be provided on a global scale, approximately 2.5 million patients could benefit *annually* from loco-regional tumour control, and 950,000 would experience prolonged survival. With longer life spans, an extra capacity of 10-25% will be required to account for complex re-treatments. These cases will require cautious treatment planning to avoid overexposure of previously-treated normal tissues reaching tolerance dose limits. In conclusion, all these cancer patients will be counting on trustworthy dose computation algorithms as described in this book.

Editor's sign-off to punctuate the end of book writing,
originally cut from a styrofoam block by Michael Sharpe.

BIBLIOGRAPHY

Anderson, P. (2017). Clinical Applications of 3D Printing. *Spine 42*(7S), S30–S31.

Atun, R., D. Jaffray, M. Barton, F. Bray, M. Baumann, B. Vikram, T. Hanna, F. Knaul, Y. Lievens, T. Lui, M. Milosevic, B. O'Sullivan, D. Rodin, E. Rosenblatt, J. Van Dyk, M. Yap, E. Zubizarreta, and M. Gospodarowicz (2015). Expanding global access to radiotherapy. *The Lancet Oncology 16*(10), 1153–1186.

Avanzo, M., S. Barbiero, M. Trovo, J. Bissonnette, R. Jena, J. Stancanello, G. Pirrone, F. Matrone, E. Minatel, C. Cappelletto, C. Furlan, D. Jaffray, and G. Sartor (2017). Voxel-by-voxel correlation between radiologically radiation-induced lung injury and dose after image-guided, intensity modulated radiotherapy for lung tumours. *Physica Medica 42*, 150–156.

Bailey, D., B. Pichler, B. Gückel, H. Barthel, A. Beer, J. Bremerich, J. Czernin, A. Drzezga, C. Franzius, V. Goh, M. Hartenbach, H. Iida, A. Kjaer, C. la Fougère, C. Ladefoged, I. Law, K. Nikolaou, H. Quick, O. Sabri, J. Schäfer, M. Schäfers, H. Wehrl, and T. Beyer (2015). Combined PET/MRI: Multi-modality multiparametric imaging is here: summary report of the 4th International Workshop on PET/MR Imaging; February 23 2015, Tübingen, Germany. *Molecular Imaging and Biology 17*(5), 595–608.

Battista, J., C. Johnson, D. Turnbull, J. Kempe, K. Bzdusek, J. Van Dyk, and G. Bauman (2013). Dosimetric and radiobiological consequences of computed tomography-guided adaptive strategies for intensity modulated radiation therapy of the prostate. *International Journal of Radiation Oncology, Biology, and Physics 87*(5), 874–880.

Bentzen, S. (2011). Molecular-imaging-based dose painting: a novel paradigm for radiation therapy prescription. *Seminars in Radiation Oncology 21*(2), 101–110.

Bentzen, S., L. Constine, J. Deasy, A. Eisbruch, A. Jackson, L. Marks, R. Ten Haken, and E. Yorke (2010). Quantitative analyses of normal tissue effects in the clinic (QUANTEC): an introduction to the scientific issues. *International Journal of Radiation Oncology, Biology, and Physics 76*(Supplement 3), 3–9.

Beyer, T., D. Townsend, T. Brun, and P. Kinahan (2000). A combined PET/CT scanner for clinical oncology. *The Journal of Nuclear Medicine 41*(8), 1369–1379.

Bortfeld, T., S. Jiang, and E. Rietzel (2004). Effects of Motion on the Total Dose Distribution. *Seminars in Radiation Oncology 14*(1), 41–51.

Botchway, S., J. Coulter, and F. Currell (2015). Imaging intracellular and systemic in vivo gold nanoparticles to enhance radiotherapy. *British Journal of Radiology 88*(1054), 1–13.

Bouchard, H. and A. Bielajew (2015). Lorentz force correction to the Boltzmann radiation transport equation and its implications for Monte Carlo algorithms. *Physics in Medicine and Biology 60*(13), 4963–4971.

Branchini, M., C. Fiorino, I. Dell'Oca, M. Belli, L. Perna, N. Di Muzio, R. Calandrino, and S. Broggi (2017). Validation of a method for dose-of-the-day calculation in head-and-neck tomotherapy by using planning CT to MVCT deformable image registration. *Physica Medica 39*, 73–79.

Brock, K. (2010). Results of a multi-institution deformable registration accuracy study (MIDRAS). *International Journal of Radiation Oncology, Biology, and Physics 76*(2), 583–596.

Brualla, L., M. Rodriguez, and A. Lallena (2017). Monte Carlo systems used for treatment planning and dose verification. *Strahlentherapie und Onkologie 193*(4), 243–259.

Caillet, V., J. Booth, and P. Keall (2017). IGRT and motion management during lung SBRT delivery. *Physica Medica 44*, 113–122.

Caivano, R., G. Califano, A. Fiorentino, M. Cozzolino, C. Oliviero, P. Pedicini, S. Clemente, C. Chiumento, and V. Fusco (2014). Clinically relevant quality assurance for intensity modulated radiotherapy plans: gamma maps and DVH-based evaluation. *Cancer Investigation 32*(3), 85–91.

Catana, C., A. Guimaraes, and B. Rosen (2013). PET and MR imaging: the odd couple or a match made in heaven? *Journal of Nuclear Medicine 54*(5), 815–824.

Cherpak, A., M. Serban, J. Seuntjens, and J. Cygler (2011). 4D dose-position verification in radiation therapy using the RADPOS system in a deformable lung phantom. *Medical Physics 38*(1), 179–187.

Choudhury, A., G. Budgell, R. MacKay, S. Falk, C. Faivre-Finn, M. Dubec, M. van Herk, and A. McWilliam (2017). The future of image-guided radiotherapy. *Clinical Oncology 29*(10), 662–666.

Colvill, E., J. Booth, and P. Keall (2018). 4D Dosimetry. In B. Mijnheer (Ed.), *Clinical 3D Dosimetry in Modern Radiation Therapy.*, Chapter 11, pp. 281–299. Boca Raton, Florida: CRC Press, Taylor & Francis Group.

Cozzolino, M., C. Oliviero, G. Califano, S. Clemente, P. Pedicini, R. Caivano, C. Chiumento, A. Fiorentino, and V. Fusco (2014). Clinically relevant quality assurance (QA) for prostate RapidArc plans: gamma maps and DVH-based evaluation. *Physica Medica 30*(4), 462–472.

Craig, T., J. Battista, and J. Van Dyk (2003). Limitations of a convolution method for modeling geometric uncertainties in radiation therapy. I. The effect of shift invariance. *Medical Physics 30*(8), 2001–2011.

El Naqa, I. (2018). *A Guide to Outcome Modeling in Radiotherapy and Oncology: Listening to the Data.* Boca Raton, Florida: CRC Press, Taylor & Francis Group.

Engelsman, M., E. Damen, K. De Jaeger, K. van Ingen, and B. Mijnheer (2001). The effect of breathing and set-up errors on the cumulative dose to a lung tumor. *Radiotherapy and Oncology 60*(1), 95–105.

Fallone, B. (2014). The rotating biplanar LINAC magnetic resonance imaging (MRI) system. In *Seminars in Radiation Oncology*, Volume 24, pp. 200–202. Elsevier.

Fernandez, D. (2016). Interplay Effects in Highly Modulated Stereotactic Body Radiation Therapy Lung Cases Treated with Volumetric Modulated Arc Therapy. Master's thesis, Louisiana State University.

Fink, J., M. Muzi, M. Peck, and K. Krohn (2015). Multimodality brain tumor imaging: MR imaging, PET, and PET/MR imaging. *Journal of Nuclear Medicine 56*(10), 1554–1561.

Foster, K., J. Kempe, E. Osei, J. Battista, and R. Barnett (2013). Assessing the Severity of Interfraction Organ Motion during Image Guided Radiotherapy. In *Proceedings of Annual Meeting of Canadian Organization of Medical Physicists (COMP)*, Montreal, Quebec.

Grau, C., L. Muren, M. Høyer, J. Lindegaard, and J. Overgaard (2008). Image-guided adaptive radiotherapy - Integration of biology and technology to improve clinical outcome. *Acta Oncologica 47*(7), 1182–1185.

Greco, C. and C. Ling (2008). Broadening the scope of image-guided radiotherapy (IGRT). *Acta Oncologica 47*(7), 1193–1200.

Grégoire, V., J. Langendijk, and S. Nuyts (2015). Advances in radiotherapy for head and neck cancer. *Journal of Clinical Oncology 33*(29), 3277–3284.

Hamburg, M. and F. Collins (2010). The path to personalized medicine. *New England Journal of Medicine 363*(4), 301–304.

Hatton, J., B. McCurdy, and P. Greer (2009). Cone beam computerized tomography: the effect of calibration of the Hounsfield unit number to electron density on dose calculation accuracy for adaptive radiation therapy. *Physics in Medicine and Biology 54*(15), N329.

Hissoiny, S., B. Ozell, H. Bouchard, and P. Després (2011). GPUMCD: A new GPU-oriented Monte Carlo dose calculation platform. *Medical Physics 38*(2), 754–764.

Hugo, G., E. Weiss, W. Sleeman, S. Balik, P. Keall, J. Lu, and J. Williamson (2017). A longitudinal four dimensional computed tomography and cone beam computed tomography dataset for image guided radiation therapy research in lung cancer. *Medical Physics 44*(2), 762–771.

Jaffray, D., J. Battista, A. Fenster, and P. Munro (1994). X-ray scatter in mega-voltage transmission radiography: physical characteristics and influence on image quality. *Medical Physics 21*(1), 45–60.

Jaffray, D., P. Lindsay, K. Brock, J. Deasy, and W. Tomé (2010). Accurate accumulation of dose for improved understanding of radiation effects in normal tissue. *International Journal of Radiation Oncology, Biology, and Physics 76*(3), S135–S139.

Jahnke, L., J. Fleckenstein, F. Wenz, and J. Hesser (2012). GMC: a GPU implementation of a Monte Carlo dose calculation based on Geant4. *Physics in Medicine and Biology 57*(5), 1217–1229.

Jia, X., X. Gu, Y. Graves, M. Folkerts, and S. Jiang (2011). GPU-based fast Monte Carlo simulation for radiotherapy dose calculation. *Physics in Medicine and Biology 56*(22), 7017–7031.

Jia, X. and S. Jiang (2018). *Graphics Processing Unit-Based High Performance Computing in Radiation Therapy*. Boca Raton, Florida: CRC Press, Taylor & Francis Group.

Jiang, S., C. Pope, K. Al Jarrah, J. Kung, T. Bortfeld, and G. Chen (2003). An experimental investigation on intra-fractional organ motion effects in lung IMRT treatments. *Physics in Medicine and Biology 48*(12), 1773–1784.

Kashani, R., J. Balter, K. Lam, and D. Litzenberg (2007). A deformable phantom for dynamic modeling in radiation therapy. *Medical Physics 34*(1), 199–201.

Katsuta, Y., N. Kadoya, Y. Fujita, E. Shimizu, K. Matsunaga, K. Sawada, H. Matsushita, K. Majima, and K. Jingu (2017). Patient-Specific quality assurance using Monte Carlo dose calculation and Elekta log files for prostate volumetric-modulated arc therapy. *Technology in Cancer Research and Treatment 16*(6), 1220–1225.

Keall, P., M. Barton, and S. Crozier (2014). The Australian magnetic resonance imaging LINAC program. In *Seminars in Radiation Oncology*, Volume 24, pp. 203–206. Elsevier.

Keall, P., D. Nguyen, R. O'Brien, P. Zhang, L. Happersett, J. Bertholet, and P. Poulsen (2018). Review of real-time 3-dimensional image guided radiation therapy on standard-equipped cancer radiation therapy systems: Are we at the tipping point for the era of real-time radiation therapy? *International Journal of Radiation Oncology, Biology, and Physics 102*, in press.

Kierkels, R., L. den Otter, E. Korevaar, J. Langendijk, A. van der Schaaf, A. Knopf, and N. Sijtsema (2018). An automated, quantitative, and case-specific evaluation of deformable image registration in computed tomography images. *Physics in Medicine and Biology 63*(4), 45026.

Kim, B., J. Chen, T. Kron, and J. Battista (2009). Motion-induced dose artifacts in helical tomotherapy. *Physics in Medicine and Biology 54*(19), 5707–5734.

Kim, B., J. Chen, T. Kron, and J. Battista (2010). Feasibility study of multi-pass respiratory-gated helical tomotherapy of a moving target via binary MLC closure. *Physics in Medicine and Biology 55*(22), 6673–6694.

Kim, B., T. Kron, J. Battista, and J. Van Dyk (2005). Investigation of dose homogeneity for loose helical tomotherapy delivery in the context of breath-hold radiation therapy. *Physics in Medicine and Biology 50*(10), 2387–2404.

Kim, H., J. Chen, J. Phillips, J. Pukala, S. Yom, and N. Kirby (2017). Validating dose uncertainty estimates produced by AUTODIRECT: An automated program to evaluate deformable image registration accuracy. *Technology in Cancer Research and Treatment 16*(6), 885–892.

Kim, J., J. Kim, and J. Park (2017). Uncertainties of cumulative dose assessment for prostate IMRT. *Translational Cancer Research 6*(2), S357–S359.

Kirby, N., J. Chen, H. Kim, O. Morin, K. Nie, and J. Pouliot (2016). An automated deformable image registration evaluation of confidence tool. *Physics in Medicine and Biology 61*(8), N203.

Kirby, N., C. Chuang, U. Ueda, and J. Pouliot (2013). The need for application-based adaptation of deformable image registration. *Medical Physics 40*(1), 011702.

Kirkby, C., B. Murray, S. Rathee, and B. Fallone (2010). Lung dosimetry in a linac-MRI radiotherapy unit with a longitudinal magnetic field. *Medical Physics 37*(9), 4722–4732.

Kyriakou, E. and D. McKenzie (2011). Dynamic modeling of lung tumor motion during respiration. *Physics in Medicine and Biology 56*(10), 2999–3013.

Lagendijk, J., B. Raaymakers, C. Van Den Berg, M. Moerland, M. Philippens, and M. Van Vulpen (2014). MR guidance in radiotherapy. *Physics in Medicine and Biology 59*(21), R349–R369.

Lamb, J., J. Ginn, D. O'Connell, N. Agazaryan, M. Cao, D. Thomas, Y. Yang, M. Lazea, P. Lee, and D. Low (2017). Dosimetric validation of a magnetic resonance image gated radiotherapy system using a motion phantom and radiochromic film. *Journal of Applied Clinical Medical Physics 18*(3), 163–169.

Lambin, P., R. Leijenaar, T. Deist, J. Peerlings, E. de Jong, J. van Timmeren, S. Sanduleanu, R. Larue, A. Even, and A. Jochems (2017). Radiomics: the bridge between medical imaging and personalized medicine. *Nature Reviews: Clinical Oncology 14*(12), 749.

Lechner, W., H. Fuch, and D. Georg (2017). PO-0805: Commissioning of the new Monte Carlo algorithm SciMoCa for a VersaHD LINAC. *Radiotherapy and Oncology 123*, S429–S430.

Li, X., X. Zhang, X. Li, J. Chang, X. Li, X. Zhang, X. Li, and J. Chang (2016). Multimodality imaging in nanomedicine and nanotheranostics. *Cancer Biology and Medicine 13*(3), 339–348.

Liao, Y., L. Wang, X. Xu, H. Chen, J. Chen, G. Zhang, H. Lei, R. Wang, S. Zhang, and X. Gu (2017). An anthropomorphic abdominal phantom for deformable image registration accuracy validation in adaptive radiation therapy. *Medical Physics 44*(6), 2369–2378.

Ling, C., J. Humm, S. Larson, H. Amols, Z. Fuks, S. Leibel, and J. Koutcher (2000). Towards Multidimensional Radiotherapy (Md-CRT): Biological Imaging and Biological Conformality. *International Journal of Radiation Oncology, Biology, and Physics 47*(3), 551–560.

Lobo, J. and I. Popescu (2010). Two new DOSXYZnrc sources for 4D Monte Carlo simulations of continuously variable beam configurations, with applications to RapidArc, VMAT, TomoTherapy and CyberKnife. *Physics in Medicine and Biology 55*(16), 4431.

MacFarlane, M., D. Wong, D. Hoover, E. Wong, C. Johnson, J. Battista, and J. Chen (2018). Patient-specific calibration of cone beam computed tomography data sets for radiotherapy dose calculations and treatment plan assessment. *Journal of Applied Clinical Medical Physics 19*(2), 249–257.

Madamesila, J., P. McGeachy, J. Barajas, and R. Khan (2016). Characterizing 3D printing in the fabrication of variable density phantoms for quality assurance of radiotherapy. *Physica Medica 32*(1), 242–247.

Malkov, V. and D. Rogers (2016). Charged particle transport in magnetic fields in EGSnrc. *Medical Physics 43*(7), 4447–4458.

Marshall, A., V. Kong, B. Chan, J. Moseley, A. Sun, P. Lindsay, and J. Bissonnette (2017). Comparing delivered and planned radiation therapy doses using deformable image registration and dose accumulation for locally advanced non-small cell lung cancer. *International Journal of Radiation Oncology, Biology, and Physics 99*(2), E696–E697.

Mayer, R., P. Liacouras, A. Thomas, M. Kang, L. Lin, and C. Simone (2015). 3D printer-generated thorax phantom with mobile tumor for radiation dosimetry. *Review of Scientific Instruments 86*(7), 074301.

McCurdy, B. (2013). Dosimetry in radiotherapy using a-Si EPIDs: Systems, methods, and applications focusing on 3D patient dose estimation. *Journal of Physics: Conference Series 444*(1), 012002.

McDermott, L., M. Wendling, J. Sonke, M. van Herk, and B. Mijnheer (2006). Anatomy changes in radiotherapy detected using portal imaging. *Radiotherapy and Oncology 79*(2), 211–217.

Mijnheer, B. (2018). Patient-Specific Quality Assurance: In-Vivo 3D Dose Verification. In *Clinical 3D Dosimetry in Modern Radiation Therapy*, Chapter 18, pp. 457–485. Boca Raton, Florida: CRC press, Taylor & Francis Group.

Mijnheer, B., R. Rozendaal, I. Olaciregui Ruiz, P. González, R. Van Oers, and A. Mans (2017). New developments in EPID-based 3D dosimetry in the Netherlands Cancer Institute. *Journal of Physics: Conference Series 847*(1), 012033.

Nassef, M., A. Simon, G. Cazoulat, A. Duménil, C. Blay, C. Lafond, O. Acosta, J. Balosso, P. Haigron, and R. de Crevoisier (2016). Quantification of dose uncertainties in cumulated dose estimation compared to planned dose in prostate IMRT. *Radiotherapy and Oncology 119*(1), 129–136.

Nelms, B., M. Chan, G. Jarry, M. Lemire, J. Lowden, C. Hampton, and V. Feygelman (2013). Evaluating IMRT and VMAT dose accuracy: practical examples of failure to detect systematic errors when applying a commonly used metric and action levels. *Medical Physics 40*(11), 111722.

Netherton, T., Y. Li, P. Nitsch, S. Shaitelman, P. Balter, S. Gao, A. Klopp, M. Muruganandham, and L. Court (2018). Interplay effect on a 6 MV Flattening-Filter-Free linear accelerator with high dose rate and fast multi-leaf collimator motion treating breast and lung phantoms. *Medical Physics (in press)*.

Neveu, M., G. De Preter, V. Marchand, A. Bol, J. Brender, K. Saito, S. Kishimoto, P. Porporato, P. Sonveaux, V. Grégoire, O. Feron, B. Jordan, M. Krishna, and B. Gallez (2016). Multimodality imaging identifies distinct metabolic profiles in-vitro and in-vivo. *Neoplasia 18*(12), 742–752.

Nguyen, D., R. O'Brien, J. Kim, C. Huang, L. Wilton, P. Greer, K. Legge, J. Booth, P. Poulsen, and J. Martin (2017). The first clinical implementation of a real-time six degree of freedom target tracking system during radiation therapy based on Kilovoltage Intrafraction Monitoring (KIM). *Radiotherapy and Oncology 123*(1), 37–42.

Nie, K., C. Chuang, N. Kirby, S. Braunstein, and J. Pouliot (2013). Site-specific deformable imaging registration algorithm selection using patient-based simulated deformations. *Medical Physics 40*(4), 040911.

Nie, K., J. Pouliot, E. Smith, and C. Chuang (2016). Performance variations among clinically available deformable image registration tools in adaptive radiotherapy: how should we evaluate and interpret the result? *Journal of Applied Clinical Medical Physics 17*(2), 328–340.

Niebuhr, N., W. Johnen, T. Güldaglar, A. Runz, G. Echner, P. Mann, C. Möhler, A. Pfaffenberger, O. Jäkel, and S. Greilich (2016). Technical Note: Radiological properties of tissue surrogates used in a multimodality deformable pelvic phantom for MR-guided radiotherapy. *Medical Physics 43*(2), 908–916.

Oborn, B., M. Gargett, T. Causer, S. Alnaghy, N. Hardcastle, P. Metcalfe, and P. Keall (2017). Experimental verification of dose enhancement effects in a lung phantom from inline magnetic fields. *Radiotherapy and Oncology 125*(3), 433–438.

O'Brien, R., B. Cooper, C. Shieh, U. Stankovic, P. Keall, and J. Sonke (2016). The first implementation of respiratory-triggered 4D-CBCT on a linear accelerator. *Physics in Medicine and Biology 61*(9), 3488.

Oh, S. and S. Kim (2017). Deformable image registration in radiation therapy. *Radiation Oncology Journal 35*(2), 101.

Oliveira, F. and J. Tavares (2014). Medical image registration: a review. *Computer Methods in Biomechanics and Biomedical Engineering 17*(2), 73–93.

Park, S., T. McNutt, W. Plishker, H. Quon, J. Wong, R. Shekhar, and J. Lee (2016). SCUDA: A software platform for cumulative dose assessment. *Medical Physics 43*(10), 5339–5346.

Pathmanathan, A., E. Alexander, R. Huddart, and A. Tree (2016). The delineation of intraprostatic boost regions for radiotherapy using multimodality imaging. *Future Oncology 12*(21), 2495–2511.

Pavlova, N. and C. Thompson (2016). The emerging hallmarks of cancer metabolism. *Cell Metabolism 23*(1), 27–47.

Penet, M., Z. Chen, S. Kakkad, M. Pomper, and Z. Bhujwalla (2012). Theranostic imaging of cancer. *European Journal of Radiology 81*(1), S124–S126.

Persoon, L., A. Egelmeer, M. Öllers, S. Nijsten, E. Troost, and F. Verhaegen (2013). First clinical results of adaptive radiotherapy based on 3D portal dosimetry for lung cancer patients with atelectasis treated with volumetric-modulated arc therapy (VMAT). *Acta Oncologica 52*(7), 1484–1489.

Popescu, I., P. Atwal, J. Lobo, J. Lucido, and B. McCurdy (2015). Patient-specific QA using 4D Monte Carlo phase space predictions and EPID dosimetry. *Journal of Physics: Conference Series 573*(1), 012004.

Pratx, G. and L. Xing (2011). GPU computing in medical physics: A review. *Medical Physics 38*(5), 2685–2697.

Rigaud, B., A. Simon, J. Castelli, M. Gobeli, J. Ospina Arango, G. Cazoulat, O. Henry, P. Haigron, and R. De Crevoisier (2015). Evaluation of deformable image registration methods for dose monitoring in head and neck radiotherapy. *BioMed Research International ID26268*, 1–16.

Rodriguez, M. and L. Brualla (2018). Many-integrated core (MIC) technology for accelerating Monte Carlo simulation of radiation transport: A study based on the code DPM. *Computer Physics Communications 225*, 28–35.

Rong, Y., J. Smilowitz, D. Tewatia, W. Tomé, and B. Paliwal (2010). Dose calculation on kV cone beam CT images: an investigation of the HU-density conversion stability and dose accuracy using the site-specific calibration. *Medical Dosimetry 35*(3), 195–207.

Saenz, D., H. Kim, J. Chen, S. Stathakis, and N. Kirby (2016). The level of detail required in a deformable phantom to accurately perform quality assurance of deformable image registration. *Physics in Medicine and Biology 61*(17), 6269.

Samant, S., S. Lee, and S. Samant (2018). GPU-Based Unimodal Deformable Image Registration in Radiation Therapy. In X. Jia and S. Jiang (Eds.), *Graphics Processing Unit-Based High Performance Computing in Radiation Therapy*, Chapter 9, pp. 147–166. Boca Raton, Florida: CRC Press, Taylor & Francis Group.

Saw, C. (2018). Special Issue: 3D Treatment Planning Systems. *Medical Dosimetry 43*(2), 103–206.

Schreiner, L. (2015). True 3D chemical dosimetry (gels, plastics): Development and clinical role. *Journal of Physics: Conference Series 573*(012003), 1–11.

Schreiner, L., O. Holmes, and G. Salomons (2013). Analysis and evaluation of planned and delivered dose distributions: Practical concerns with γ- and χ- evaluations. *Journal of Physics: Conference Series 444*(012016), 1–9.

Segars, W., B. Tsui, J. Cai, F. Yin, G. Fung, and E. Samei (2018). Application of the 4-D XCAT Phantoms in Biomedical Imaging and Beyond. *IEEE Transactions on Medical Imaging 37*(3), 680–692.

Shessel, A., V. Kong, B. Chan, J. Moseley, A. Sun, and J. Bissonnette (2018). Deformable image registration and dose accumulation for locally advanced non-small cell lung cancer, comparing delivered and planned radiotherapy doses. *Journal of Medical Imaging and Radiation Sciences 49*(1), S15.

Singhrao, K., N. Kirby, and J. Pouliot (2014). A three-dimensional head and neck phantom for validation of multi-modality deformable image registration for adaptive radiotherapy. *Medical Physics 41*(12), 121709.

Spreeuw, H., R. Rozendaal, I. Olaciregui Ruiz, P. González, A. Mans, B. Mijnheer, and M. van Herk (2016). Online 3D EPID based dose verification: Proof of concept. *Medical Physics 43*(7), 3969–3974.

St. Aubin, J., A. Keyvanloo, O. Vassiliev, and B. Fallone (2015). A deterministic solution of the first order linear Boltzmann transport equation in the presence of external magnetic fields. *Medical Physics 42*(2), 780–793.

Steers, J. and B. Fraass (2016). IMRT QA: Selecting gamma criteria based on error detection sensitivity. *Medical Physics 43*(4), 1982–1994.

Steidl, P., D. Richter, C. Schuy, E. Schubert, T. Haberer, M. Durante, and C. Bert (2012). A breathing thorax phantom with independently programmable 6D tumour motion for dosimetric measurements in radiation therapy. *Physics in Medicine and Biology 57*(8), 2235.

Thakur, M. and B. Lentle (2006). Report of a summit on molecular imaging. *American Journal of Roentgenology 186*(2), 297–299.

Thorwarth, D. (2015). Functional imaging for radiotherapy treatment planning: Current status and future directions - A review. *British Journal of Radiology 88*(ID20150056), 1–9.

Thorwarth, D. and M. Alber (2010). Implementation of hypoxia imaging into treatment planning and delivery. *Radiotherapy and Oncology 97*(2), 172–175.

Thorwarth, D., S. Eschmann, F. Paulsen, and M. Alber (2007). Hypoxia dose painting by numbers: a planning study. *International Journal of Radiation Oncology, Biology, and Physics 68*(1), 291–300.

Usui, K., K. Ogawa, and K. Sasai (2017). Analysis of dose calculation accuracy in cone beam computed tomography with various amounts of scattered photon contamination. *International Journal of Medical Physics, Clinical Engineering and Radiation Oncology 6*(03), 233–251.

van Elmpt, W., L. McDermott, S. Nijsten, M. Wendling, P. Lambin, and B. Mijnheer (2008). A literature review of electronic portal imaging for radiotherapy dosimetry. *Radiotherapy and Oncology 88*(3), 289–309.

van Elmpt, W., S. Nijsten, S. Petit, B. Mijnheer, P. Lambin, and A. Dekker (2009). 3D in-vivo dosimetry using megavoltage cone-beam CT and EPID dosimetry. *International Journal of Radiation Oncology, Biology, and Physics 73*(5), 1580–1587.

Veiga, C., A. Lourenço, S. Mouinuddin, M. van Herk, M. Modat, S. Ourselin, G. Royle, J. Mcclelland, and G. Royle (2015). Toward adaptive radiotherapy for head and neck patients : Uncertainties in dose warping due to the choice of deformable registration algorithm. *Medical Physics 42*(2), 760–769.

Vickress, J., J. Battista, R. Barnett, J. Morgan, and S. Yartsev (2016). Automatic landmark generation for deformable image registration evaluation for 4D CT images of lung. *Physics in Medicine and Biology 61*(20), 7236–7245.

Vickress, J., J. Battista, R. Barnett, and S. Yartsev (2017). Representing the dosimetric impact of deformable image registration errors. *Physics in Medicine and Biology 62*, N391–N403.

Wang, S., X. Chen, Y. Li, and Y. Zhang (2016). Application of Multimodality Imaging Fusion Technology in Diagnosis and Treatment of Malignant Tumors under the Precision Medicine Plan. *Chinese Medical Journal 129*(24), 2991.

Weistrand, O. and S. Svensson (2015). The ANACONDA algorithm for deformable image registration in radiotherapy. *Medical Physics 42*(1), 40–53.

Wong, O., A. McNiven, B. Chan, J. Moseley, J. Lee, L. Le, C. Ren, J. Waldron, J.and Bissonnette, and M. Giuliani (2018). Evaluation of differences between estimated delivered dose and planned dose in nasopharynx patients using deformable image registration and dose accumulation. *Journal of Medical Imaging and Radiation Sciences 49*(1), S2–S3.

Woodford, C., S. Yartsev, A. Dar, G. Bauman, and J. Van Dyk (2007). Adaptive Radiotherapy Planning on Decreasing Gross Tumor Volumes as Seen on Megavoltage Computed Tomography Images. *International Journal of Radiation Oncology, Biology, and Physics 69*(4), 1316–1322.

Yang, D., S. Brame, I. El Naqa, A. Aditya, Y. Wu, S. Murty Goddu, S. Mutic, J. Deasy, and D. Low (2011). DIRART: A software suite for deformable image registration and adaptive radiotherapy research. *Medical Physics 38*(1), 67–77.

Yankeelov, T., R. Abramson, and C. Quarles (2014). Quantitative multi-modality imaging in cancer research and therapy. *Nature Reviews Clinical Oncology 11*(11), 670.

Yeo, U., J. Supple, M. Taylor, R. Smith, T. Kron, and R. Franich (2013). Performance of 12 DIR algorithms in low contrast regions for mass and density conserving deformation. *Medical Physics 40*(10), 101701–101701.

Yeo, U., M. Taylor, J. Supple, R. Smith, T. Kron, and R. Franich (2013). Deformable gel dosimetry I: application to external beam radiotherapy and brachytherapy. *Journal of Physics: Conference Series 444*(1).

Yoon, J., J. Jung, J. Kim, B. Yi, and I. Yeo (2016). Four dimensional dose reconstruction through in vivo phase matching of cine images of electronic portal imaging device. *Medical Physics 43*(7), 4420–4430.

Ziegenhein, P., I. Kozin, C. Kamerling, and U. Oelfke (2017). Towards real-time photon Monte Carlo dose calculation in the cloud. *Physics in Medicine Biology 62*(11), 4375–4389.

Ziegenhein, P., S. Pirner, C. Ph Kamerling, and U. Oelfke (2015). Fast CPU-based Monte Carlo simulation for radiotherapy dose calculation. *Physics in Medicine and Biology 60*, 6097–6111.

Index

T - #0476 - 071024 - C448 - 254/178/20 - PB - 9780367780517 - Gloss Lamination